The Science of Flavonoids

The Science of Flavonoids

Edited by

Erich Grotewold

The Ohio State University
Columbus, Ohio, USA

Erich Grotewold
Department of Plant Cellular and Molecular Biology
The Ohio State University
Columbus, Ohio 43210
USA
grotewold.1@osu.edu

The structure is based on a homology model of the *Arabidopsis* F3′H enzyme developed by Sanjeewa Rupasinghe and Dr. Mary Schuler (provided by Dr. Brenda Winkel).

Library of Congress Control Number: 2007939565

ISBN-13: 978-0-387-74550-3   e-ISBN-13: 978-0-387-28822-2

Printed on acid-free paper

9 8 7 6 5 4 3 2 1

springer.com

PREFACE

There is no doubt that among the large number of natural products of plant origin, debatably called secondary metabolites because their importance to the eco-physiology of the organisms that accumulate them was not initially recognized, flavonoids play a central role. These compounds and their derived pigments have contributed to shaping our knowledge of modern genetics, providing colorful tools to investigate a number of central plant problems, including the biology of transposons, the regulation of gene expression, gene silencing, and the organization of metabolic pathways. The legacy left by several outstanding chemists who have devoted their lives to the understanding of the chemistry of flavonoids is being carried by a growing number of scientists who take interdisciplinary approaches to continue to advance our knowledge of the pathway and develop new means to manipulate the synthesis of these compounds, which have significant potential in providing solutions to plant and animal illnesses alike.

The interdisciplinary nature of the research currently being carried out in the area of flavonoids is part of the spirit that this book has tried to capture. Chemistry, biochemistry, genetics, and cellular and molecular biology are all parts of the toolbox that the investigator has at hand in addressing fundamental biological questions regarding the biosynthesis, storage, regulation, evolution, and biological activities of flavonoids. These tools are combined in each of the nine chapters that form this book to address what I have perceived to be some of the most significant challenges currently being pursued in the area of the biology of flavonoids. If specific topics have been left out, such as, for example, the metabolic engineering of flavonoids, it is only because in my opinion the number of reviews in this subject exceeds the quantity of novel relevant primary research publications.

Chapter 1 provides a novel look at flavonoids from the perspective of stereochemistry. Chapter 2 provides an overview of the state of the art in flavonoid isolation and characterization. Historic and up-to-date perspectives on the biosynthesis of flavonoids are provided in Chapter 3. Chapter 4 integrates the studies in several plants to provide models on how the multiple branches of flavonoid biosynthesis might be regulated. Chapter 5 explores the poorly understood mechanisms that underlie the trafficking of flavonoids within cells. A review of the contributions that flavonoids and derived pigments have and continue to provide to geneticists and molecular biologists is provided in Chapter 6. Chapter 7 illustrates models that may help to explain the evolution of flavonoids and the corresponding regulatory and biosynthetic genes. Chapter 8 delves into the expanding field of the role that flavonoids play in health, and Chapter 9 provides a review on the role of flavonoids as plant-signaling molecules.

I want to finish by thanking the authors who contributed to this book and for their patience in bearing with the multiple revisions of their submissions. I also want to acknowledge the several reviewers who provided me with comments on the chapters. Most wholeheartedly I want to thank Sarat Subramaniam for his help with the formatting and editing of the book.

CONTENTS

CONTENTS

CHAPTER 1

THE STEREOCHEMISTRY OF FLAVONOIDS

JANNIE P.J. MARAIS,[a] BETTINA DEAVOURS,[b] RICHARD A. DIXON,[b] AND DANEEL FERREIRA[a,c]

[a]National Center for Natural Products Research, Research Institute of Pharmaceutical Sciences, School of Pharmacy, The University of Mississippi, University, MS 38677 USA; [b]Plant Biology Division, Samuel Roberts Noble Foundation, 2510 Sam Noble Parkway, Ardmore OK 73401 USA; [c]Department of Pharmacognosy, Research Institute of Pharmaceutical Sciences, School of Pharmacy, The University of Mississippi, MS 38677 USA

1. INTRODUCTION

The study of flavonoid chemistry has emerged, like that of most natural products, from the search for new compounds with useful physiological properties. Semisynthetic endeavors of oligoflavonoids are in most instances confined to those substitution patterns exhibited by monomeric natural products that are available in quantities sufficient for preparative purposes. In order to alleviate these restrictions, several programs focusing on synthesis of enantiomeric pure flavonoid monomers have been undertaken. However, synthesis of the desired enantiomer in optically pure forms remains a daunting objective and is limited to only a few types of compounds. Chalcone epoxides, α- and β-hydroxydihydrochalcones, dihydro-flavonols, flavan-3-ols, flavan-3,4-diols, isoflavans, isoflavanones, and pterocarpans thus far have been synthesized in reasonable yields and purity.

2. NOMENCLATURE

The term "flavonoid" is generally used to describe a broad collection of natural products that include a C_6-C_3-C_6 carbon framework, or more specifically a phenylbenzopyran functionality. Depending on the position of the linkage of the aromatic ring to the benzopyrano (chromano) moiety, this group of natural products may be divided into three classes: the flavonoids (2-phenylbenzopyrans) **1**, isoflavonoids (3-benzopyrans) **2**, and the neoflavonoids (4-benzopyrans) **3**. These

1

groups usually share a common chalcone precursor, and therefore are biogenetically and structurally related.

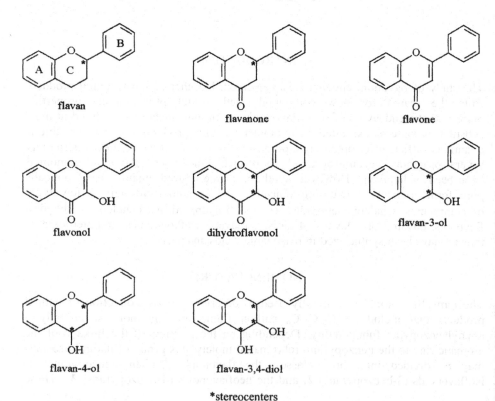

2.1. 2-Phenylbenzopyrans (C$_6$-C$_3$-C$_6$ Backbone)

Based on the degree of oxidation and saturation present in the heterocyclic C-ring, the flavonoids may be divided into the following groups:

flavan

flavanone

flavone

flavonol

dihydroflavonol

flavan-3-ol

flavan-4-ol

flavan-3,4-diol

*stereocenters

2.2. *Isoflavonoids*

The isoflavonoids are a distinctive subclass of the flavonoids. These compounds possess a 3-phenylchroman skeleton that is biogenetically derived by 1,2-aryl migration in a 2-phenylchroman precursor. Despite their limited distribution in the plant kingdom, isoflavonoids are remarkably diverse as far as structural variations are concerned. This arises not only from the number and complexity of substituents on the basic 3-phenylchroman system, but also from the different oxidation levels and presence of additional heterocyclic rings. Isoflavonoids are subdivided into the following groups:

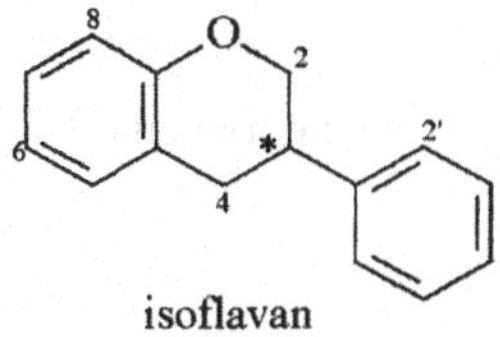

isoflavan

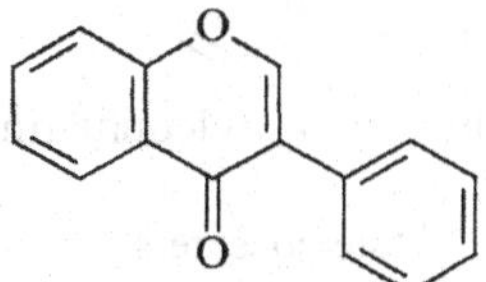

isoflavone

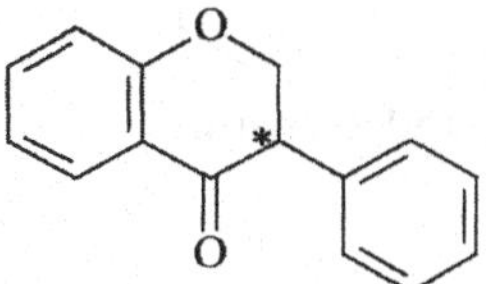

isoflavanone

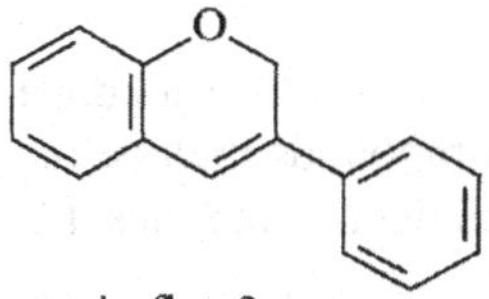

isoflav-3-ene

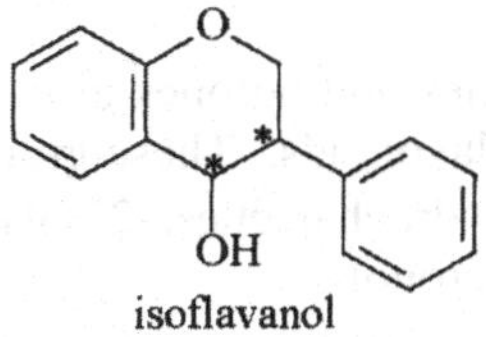

isoflavanol

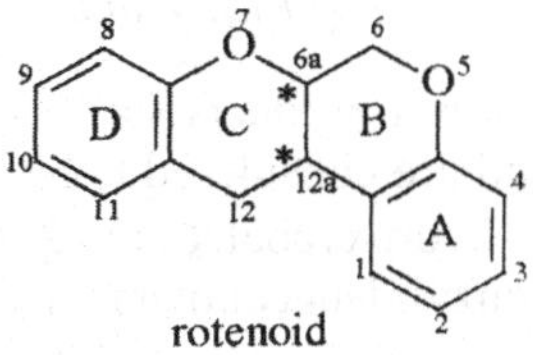

rotenoid

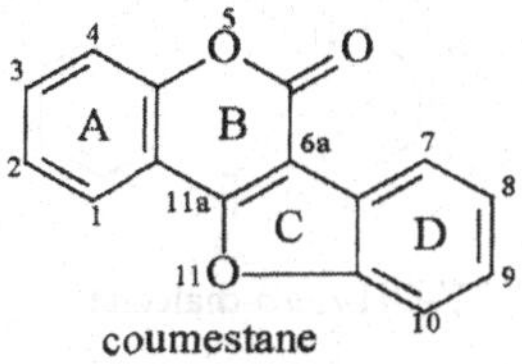

coumestane

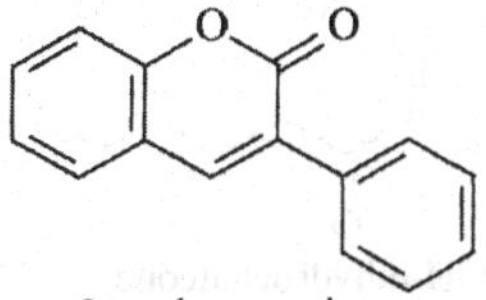

3-arylcoumarin

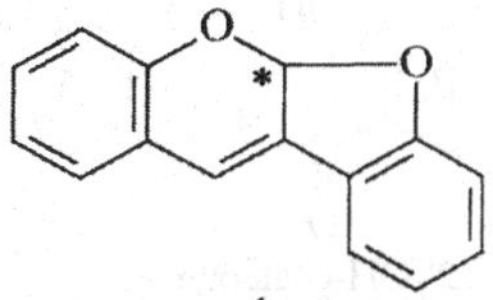

coumaronochromene

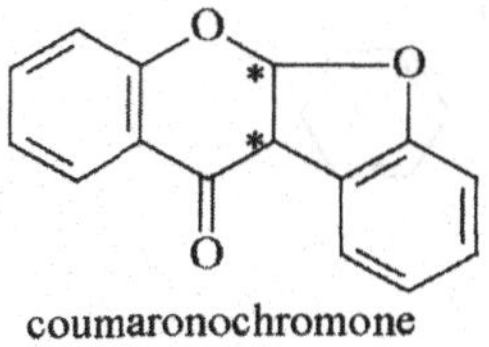

coumaronochromone

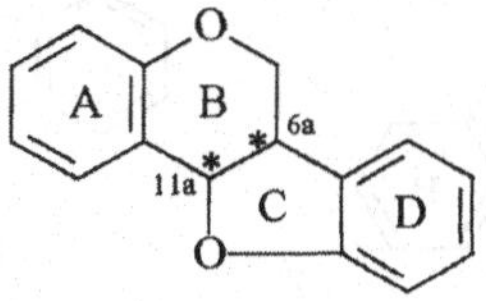

pterocarpan

*stereocenters

2.3. Neoflavonoids

The neoflavonoids are structurally and biogenetically closely related to the flavonoids and the isoflavonoids and comprise the 4-arylcoumarins (4-aryl-2*H*-1-benzopyran-2-ones), 3,4-dihydro-4-arylcoumarins, and neoflavenes.

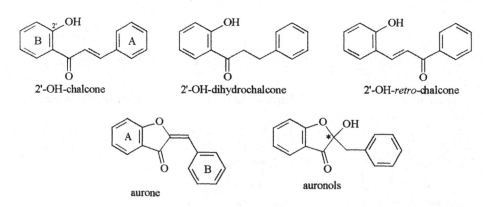

4-arylcoumarin   3,4-dihydro-4-arylcoumarin   neoflavene

*stereocenters

2.4. Minor Flavonoids

Natural products such as chalcones and aurones also contain a C_6-C_3-C_6 backbone and are considered to be minor flavonoids. These groups of compounds include the 2′-hydroxychalcones, 2′-OH-dihydrochalcones, 2′-OH-*retro*-chalcone, aurones (2-benzylidenecoumaranone), and auronols.

2'-OH-chalcone   2'-OH-dihydrochalcone   2'-OH-*retro*-chalcone

aurone   auronols

*stereocenters

3. SYNTHESIS OF FLAVONOIDS

3.1. Chalcones, Dihydrochalcones, and Racemic Flavonoids

Chalcones and dihydrochalcones are considered to be the primary C_6-C_3-C_6 precursors and constitute important intermediates in the synthesis of flavonoids. Chalcones are readily accessible via two well-established routes comprising a base-catalyzed aldol condensation or acid-mediated aldolization of 2-hydroxy-acetophenones **4** and benzaldehydes **5** (Von Konstanecki and Rossbach, 1896; Augustyn et al., 1990a) (Scheme 1.1). The base-catalyzed aldol condensation is usually the preferred route toward chalcone **6** formation, since under acidic conditions cyclization of the ensuing chalcone leads to formation of corresponding racemic flavanones **7** (Claisen and Claparède, 1881). Dihydrochalcones **8** are generally obtained via reduction (H_2/Pd) of the preceding chalcones (Scheme 1.1).

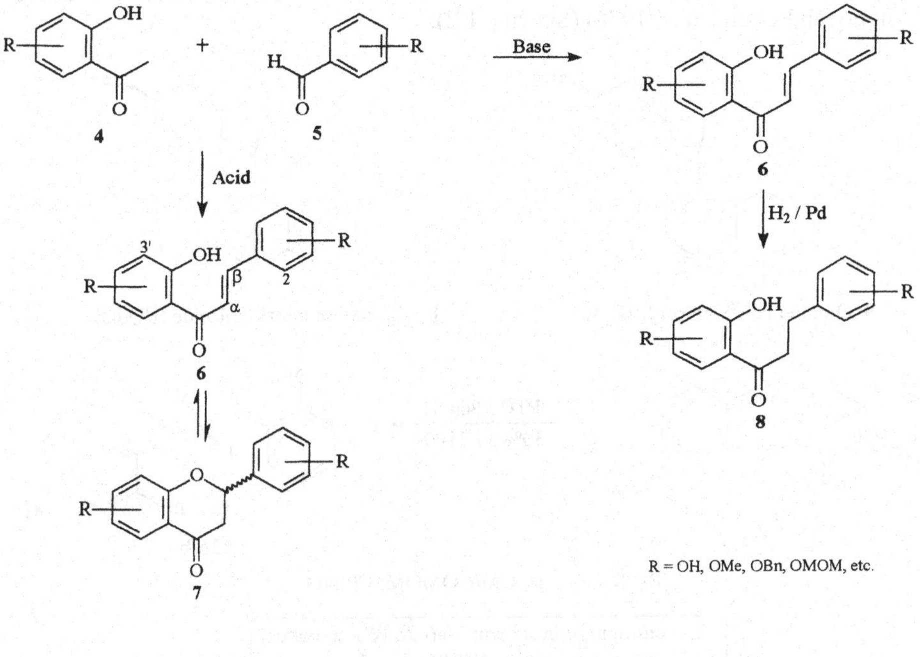

Scheme 1.1 *Acid- and base-catalyzed synthesis of chalcones, racemic flavanones, and dihydrochalcones.*

3.2. Asymmetric Epoxidation of Chalcones

Asymmetric epoxidation of olefinic bonds plays a crucial role in introducing chirality in the synthesis of several classes of optically active natural compounds. Sharpless (Katsuki and Sharpless, 1980; Johnson and Sharpless 1993) and Jacobson (1993) developed viable protocols for the enantioselective epoxidation of allylic alcohols and unfunctionalized olefins. However, attempts regarding the enantio-selective epoxidation of α,β-unsaturated ketones, in particular chalcones, have met with limited success.

Wynberg and Greijdanus (1978) first reported the utilization of quinine benzylchloride **9** (BQC) and quinidine benzylchloride (BQdC) **10** as chiral phase-transfer catalysts (PTC). Since then, the use of PTC has emerged as one of the preferred methods for the asymmetric epoxidation of α,β-unsaturated ketones and led to the first stereoselective synthesis of (-)- and (+)-*trans*-chalcone epoxides **12a/b** [yield: 38–92%; enantiomeric excess (ee): 25–48%] (Helder et al., 1976; Wynberg and Greijdanus, 1978) (Scheme 1.2).

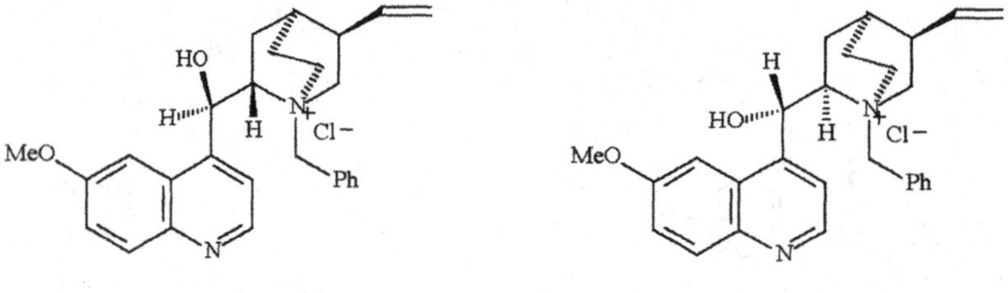

9 Quinine benzylchloride (BQC)   **10** Quinidine benzylchloride (BQdC)

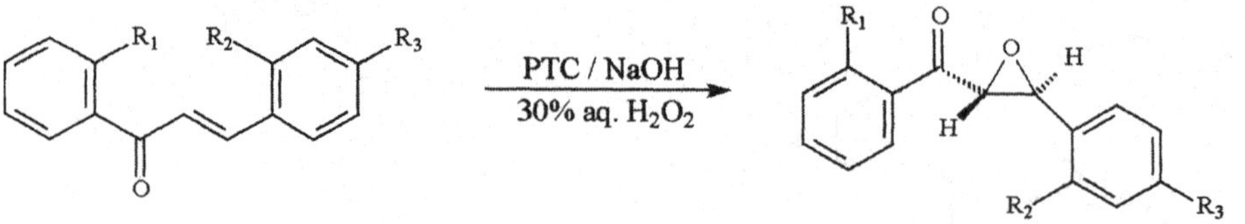

11   **12a/b**

R$_1$, R$_2$, R$_3$ = H, OMe, OMOM, Cl, NO$_2$

| **a** = configuration shown: (-)-(αR, βS)-epoxides |
| **b** = enantiomer: (+)-(αS, βR)-epoxides |

Scheme 1.2 Epoxidation of chalcones 11 with BQC 9 and BQdC 10 as PTC.

Except for the poor ee, this protocol demonstrated the preferential formation of (-)-(αR,βS)-**12a** and (+)-(αS,βR)-**12b** epoxides, with BQC **9** and BQdC **10** used, respectively, as PTC. This resulted in several investigations of alternative catalysts and reaction conditions to enhance the enantioselectivity of the epoxidation of enones (Table 1.1). However, these attempts were limited to nonchalcone enones and a few non- and monooxygenated chalcone substrates, which lacked natural product oxygenation patterns.

Table 1.1 *Asymmetric epoxidation of electron-deficient olefins*

Type of reaction and reaction conditions	*References*
1. Bovine serum albumin catalyzed epoxidation: Bovine serum albumin (BSA) under Weitz–Scheffer conditions and *aq* NaOCl with α- and β-cyclodextrins as catalysts.	Colonna et al., 1985; Colonna and Manfredi, 1986
2. Zinc-mediated asymmetric epoxidation: Metal-based catalytic systems: Epoxidation of α,β-unsaturated ketones with O_2 in the presence of Et_2Zn and (*R*,*R*)-*N*-methylpseudoephedrine.	Enders et al., 1996, 1997
Metal-based polymeric catalytic systems: Polybinaphthyl zinc catalyst for the asymmetric epoxidation of enones in the presence of Bu^tOOH (TBPH).	Yu et al., 1999
3. Lanthanide–BINOL systems: Asymmetric epoxidation of enones using lanthanoid complexes: Several kinds of heterobimetallic chiral catalysts [La- and Yb–BINOL and La- and Yb-3-hydroxymethyl–BINOL complexes] are useful for this procedure, using TBHP and cumene hydroperoxide (CMHP).	Bougauchi et al., 1997; Daikai et al., 1998
Enantioselective epoxidation of α,β-enones by utilizing chiral La(*O-i-*Pr)$_3$-(*S*)-6,6′-dibromo-BINOL and Gd(*O-i-*Pr)$_3$-(*S*)-6,6′-diphenyl–BINOL catalysts and CMHP.	Chen et al., 2001
4. Diethyl tartrate–metal peroxides: Modified Sharpless protocol, with chiral metal alkyl peroxides as nucleophilic oxidants: Using (+)-diethyl tartrate [(+)-DET] as chiral modifier in the presence of Li-TBHP and *n*-BuLi, yielded the (+)-chalcone epoxide. The (–)-chalcone epoxide was obtained simply via replacing *n*-BuLi with *n*-Bu$_2$Mg.	Elston et al., 1997
5. Phase-tranfer catalyst: Enantioselective epoxidation of chalcones utilizing Cinchona alkaloid-derived quaternary ammonium phase-transfer catalysts bearing an *N*-anthracenylmethyl function with sodium hypochlorite as oxidant.	Lygo and Wainwright, 1998, 1999
Use of chiral quaternary cinchonidinium and dihydrocinchonidinium cations for the nucleophilic epoxidation of various α,β-enones, utilizing KOCl in stoichiometric amounts as oxidant, at –40°C.	Lygo and To, 2001; Corey and Zhang, 1999

Table 1.1 (continued)

Type of reaction and reaction conditions	References
Catalytic asymmetric epoxidation of enones promoted by *aq.* H_2O_2 with chiral ammonium salts (cinchonine or quinidine derivatives).	Arai et al., 2002
Epoxidation of enones under mild reaction conditions, using a new chiral quaternary ammonium bromide with dual functions as phase transfer catalyst.	Ooi et al., 2004;
Asymmetric epoxidation with optically active hydroperoxides (cumyl hydroperoxide) and mediated by cinchonine- and cinchonidine-derived phase-transfer catalyst	Adam et al., 2001
Epoxidation of chalcones using the phase-transfer catalyst, chiral monoaza-15-crown-5 lariat ethers, synthesized from D-glucose, galactose, and mannitol, with TBHP as oxidant.	Bakó et al., 1999, 2004
Use of optically active solvents [2-(*N*,*N*-diethylamino)-1-butanol or 2-(*N*,*N*-di-*n*-butylamino)-1-butanol], *n*-Bu$_4$NBr as PTC and alkaline H_2O_2.	Singh and Arora, 1987

6. Epoxidation with chiral dioxiranes:

Involving dimethyldioxirane (DMDO) type of epoxidation utilizing chiral dioxiranes generated *in situ* from potassium peroxomonosulfate (oxone) and asymmetric ketones.	Wang and Shi, 1997; Wang et al., 1997, 1999
2-Substituted-2,4-endo-dimethyl-8-oxabicyclo[3.2.1]octan-3-ones as catalyst for the asymmetric epoxidation of alkenes with oxone.	Klein and Roberts, 2002

7. Polyamino acid-catalyzed epoxidation:

Julia–Colonna asymmetric epoxidation, originally employs a three-phase system comprising alkaline H_2O_2, an organic solvent (hexane or toluene) and an insoluble polymer (poly-L-/-D-alanine or -leucine).	Julia et al., 1980; Colonna et al., 1983;
Asymmetric epoxidation using a nonaqueous two-phase system of urea hydrogen peroxide (UHP) in THF or tert-butyl methyl ether, with immobilized poly-L-/-D-leucine.	Banfi et al., 1984
Julia–Colonna stereoselective epoxidation under nonaqueous conditions using polyamino acid (poly-L-/-D-alanine or β-leucine) on silica (PaaSiCat).	Adger et al., 1997;
β-Peptides as catalyst: poly-β-leucine in Julia–Colonna asymmetric epoxidation.	Bentley et al., 1997
Polyethylene glycol (PEG)-bound poly-L-leucine acts as a THF-soluble catalyst for the Julia–Colonna asymmetric epoxidation of enones.	Geller and Roberts, 1999; Carde et al., 1999 Coffey et al., 2001 Flood et al., 2001

As a feasible alternative to the utilization of enzymes as catalysts in organic reactions, Julia and Colonna (Julia et al., 1980, 1982; Colonna et al., 1983; Banfi et al., 1984) investigated the use of synthetic peptides in the epoxidation of chalcones. Because of the potential use of polyoxygenated chalcone epoxides as chirons in the enantiomeric synthesis of flavonoids and to determine the effect of different levels of oxygenation and substitution patterns on the poly-amino acid-catalyzed epoxidation, this protocol was extended to a series of chalcones exhibiting aromatic oxygenation patterns usually encountered in the naturally occurring flavonoids (Bezuidenhoudt et al., 1987; Augustyn et al., 1990a) (Table 1.2).

Table 1.2 Asymmetric epoxidation of chalcones *20a/b–26a/b* using poly-L- and poly-D-alanine as catalysts

Epoxides	R_1	R_2	R_3	R_4	Alanine	% yield	$[\alpha]^{278}$	% ee
(-)-20a	H	H	H	H	L	65	-50	38
(+)-20b	H	H	H	H	D	57	+75	53
(-)-21a	H	H	H	OMe	L	64	-76	66
(+)-21b	H	H	H	OMe	D	38	+52	46
(-)-22a	OMe	H	H	OMe	L	74	-122	84
(+)-22b	OMe	H	H	OMe	D	26	+77	53
(-)-23a	OMe	H	OMe	OMe	L	46	-79	62
(+)-23b	OMe	H	OMe	OMe	D	34	+31	25
(-)-24a	OMe	OMe	H	OMe	L	*	*	32
(+)-24b	OMe	OMe	H	OMe	D	*	*	20
(-)-25a	OMe	OMe	OMe	OMe	L	*	*	*
(+)-25b	OMe	OMe	OMe	OMe	D	*	*	*
(-)-26a	OMOM	H	H	OMe	L	43	*	70
(+)-26b	OMOM	H	H	OMe	D	36	*	36

*Not reported

The triphasic system comprising poly-L- or poly-D-alanine, alkaline H_2O_2, and organic solvent (CCl_4 or toluene) was utilized during the enantioselective epoxidation of chalcones **13–19**, to afford epoxides **20a/b-26a/b** in moderate yield and ee.

Although the Julia asymmetric epoxidation has proved to be a reliable reaction to afford polyoxygenated chalcone epoxides in good yield and moderate to high ee's, this protocol is not without limitations, since reaction times are often unacceptably long and require continuous addition of oxidant and base. Degradation of the poly-amino acid under such reaction conditions also poses difficulties. Bentley and Roberts found satisfactory solutions to many of these problems by conducting the asymmetric epoxidation in a two-phase non-aqueous system consisting of oxidant, a nonnucleophilic base, immobilized poly-amino acid, and an organic solvent (Itsuno et al., 1990; Lasterra-Sanchez et al., 1996; Bentley et al., 1997). This procedure afforded chiral enone epoxides in high yields and optical purity with a substantial reduction in reaction times and also was extended successfully to chalcone substrates (Nel et al., 1998, 1999a; Van Rensburg et al., 1996, 1997a) (See also Sections 3.3 and 3.4).

3.3. α- and β-Hydroxydihydrochalcones

α- and β-Hydroxydihydrochalcones constitute rare groups of C_6-C_3-C_6 metabolites presumably sharing a close biogenetic relationship with the α-methyldeoxybenzoins and isoflavonoids (Bhakuni et al., 1973; Shukla et al., 1973; Bezuidenhoudt et al., 1981; Beltrami et al., 1982; Ferrari et al., 1983; Thakkar and Cushman, 1995). Wynberg prepared an aromatic deoxy α-hydroxydihydrochalcone via catalytic hydrogenation of the corresponding chalcone (Marsman and Wynberg, 1979). However, by utilizing the versatile epoxidation methodology, Bezuidenhoudt et al. (1987) and Augustyn et al. (1990a, 1990b) extended this protocol to the enantioselective synthesis of a series of α-hydroxydihydrochalcones. Treatment of (-)-**20a-26a** and (+)-chalcone epoxides **20b-26b** with either Pd-BaSO$_4$/H$_2$ or Pd-C/H$_2$ afforded (+)-**27a-33a** and (-)-α-hydroxydihydrochalcones **27b-33b**, respectively, in moderate to high yields and moderate ee's (Table 1.3).

Although several procedures, comprising diverse reagents, such as benzeneselenolate ion, samarium diiodide, aluminium amalgam/ultrasound, and metallic lithium in liquid ammonia, have been used for the regioselective reductive ring opening of α,β-epoxyketones to form the β-hydroxyketone (Molander and Hahn, 1986; Otsubo et al., 1987; Moreno et al., 1993; Engman and Stern, 1994), the most general reagent for these conversions is tributyltin hydride (TBTH)/azobisisobutyronitrile (AIBN) (Hasegawa et al., 1992). This method was applied to a series of chalcone epoxides comprising the methyl ethers of substrates with natural hydroxylation patterns (Nel et al., 1998, 1999a).

Table 1.3 Synthesis of α-hydroxydihydrochalcones **27a/b–33a/b**

20a/b	$R_1 = R_2 = R_3 = R_4 = H$
21a/b	$R_1 = R_2 = R_3 = H$, $R_4 = OMe$
22a/b	$R_2 = R_3 = H$, $R_1 = R_4 = OMe$
23a/b	$R_2 = H$, $R_1 = R_3 = R_4 = OMe$
24a/b	$R_3 = H$, $R_1 = R_2 = R_4 = OMe$
25a/b	$R_1 = R_2 = R_3 = R_4 = OMe$
26a/b	$R_1 = OMOM$, $R_2 = R_3 = H$, $R_4 = OMe$

a = configuration shown
b = enantiomer

27a/b	$R_1 = R_2 = R_3 = R_4 = H$
28a/b	$R_1 = R_2 = R_3 = H$, $R_4 = OMe$
29a/b	$R_2 = R_3 = H$, $R_1 = R_4 = OMe$
30a/b	$R_2 = H$, $R_1 = R_3 = R_4 = OMe$
31a/b	$R_3 = H$, $R_1 = R_2 = R_4 = OMe$
32a/b	$R_1 = R_2 = R_3 = R_4 = OMe$
33a/b	$R_1 = OMOM$, $R_2 = R_3 = H$, $R_4 = OMe$

Substrate (% ee)	Catalyst – H_2	Product	% yield	% ee
(-)-20a (38)	Pd / BaSO$_4$	(+)-27a	92	27
(+)-20b (53)	Pd / BaSO$_4$	(-)-27b	61	54
(-)-21a (66)	Pd / BaSO$_4$	(+)-28a	51	61
(+)-21b (46)	Pd / BaSO$_4$	(-)-28b	72	48
(-)-22a (84)	Pd / BaSO$_4$	(+)-29a	88	76
(+)-22b (53)	Pd / BaSO$_4$	(-)-29b	70	52
(-)-23a (62)	10% Pd / C	(+)-30a	42	61
(+)-23b (25)	10% Pd / C	(-)-30b	40	16
(-)-24a (32)	5% Pd / C	(+)-31a	*	24
(+)-24b (20)	5% Pd / C	(-)-31b	*	19
(-)-25a (*)	10% Pd / C	(+)-32a	*	14
(+)-25b (*)	10% Pd / C	(-)-32b	*	16
(-)-26a (70)	Pd / BaSO$_4$	(+)-33a	50	65
(+)-26b (36)	Pd / BaSO$_4$	(-)-33b	46	32

*Not reported

Since the Julia asymmetric epoxidation of chalcones often gives disappointing stereoselectivity, Nel et al. (1998, 1999a) also used the improved two-phase nonaqueous system with poly-amino acids as asymmetric catalysts, recently developed by Bentley and Roberts (Lasterra-Sanchez et al., 1996; Bentley et al., 1997). Treatment of enones **14-18** with immobilized poly-L-leucine (PLL)/urea-hydrogen peroxide complex (UHP) and 1,8-diazabicyclo[5.4.0]undec-7-ene (DBU) in dry THF, afforded the (-)-(αR,βS)-trans-epoxychalcones **21a-25a** in moderate to high yields (21-80%) and improved optical purity (53-95% ee). The enantiomeric (+)-(αS,βR)-trans-epoxychalcones **21b-25b** were similarly obtained using immobilized poly-D-leucine (PDL) (yield, 19-76%; ee, 50-90%). The chalcone epoxides **21a/b-25a/b** were then treated with TBTH/AIBN in refluxing benzene to afford the (R)- **34a-38a** and (S)-2'-O-methoxymethyl-β-hydroxydihydrochalcones **34b-38b** in excellent yields (70-90%) and without loss of optically purity (Table 1.4).

Table 1.4 *β-Hydroxydihydrochalcone formation*

14 $R_1 = R_2 = R_3 = H$, $R_4 = OMe$
15 $R_2 = R_3 = H$, $R_1 = R_4 = OMe$
16 $R_2 = H$, $R_1 = R_3 = R_4 = OMe$
17 $R_3 = H$, $R_1 = R_2 = R_4 = OMe$
18 $R_1 = R_2 = R_3 = R_4 = OMe$

Urea-hydrogen / DBU
THF / rt.
Poly-L-leucine or
poly-D-leucine

$(\alpha R, \beta S)$- or $(\alpha S, \beta R)$-epoxychalcones
21a/b $R_1 = R_2 = R_3 = H$, $R_4 = OMe$
22a/b $R_2 = R_3 = H$, $R_1 = R_4 = OMe$
23a/b $R_2 = H$, $R_1 = R_3 = R_4 = OMe$
24a/b $R_3 = H$, $R_1 = R_2 = R_4 = OMe$
25a/b $R_1 = R_2 = R_3 = R_4 = OMe$

TBTH / AIBN
benzene / reflux

(R)- or *(S)*-β-hydroxydihydrochalcone
34a/b $R_1 = R_2 = R_3 = H$, $R_4 = OMe$
35a/b $R_2 = R_3 = H$, $R_1 = R_4 = OMe$
36a/b $R_2 = H$, $R_1 = R_3 = R_4 = OMe$
37a/b $R_3 = H$, $R_1 = R_2 = R_4 = OMe$
38a/b $R_1 = R_2 = R_3 = R_4 = OMe$

a = configuration shown
b = enantiomer

Chalcone	Poly amino acid	Chalcone-epoxide	% yield	% ee	β-hydroxy-dihydro-chalcone	% yield	% ee
14	PLL	21a	71	85	34a	73	85
14	PDL	21b	69	81	34b	70	80
15	PLL	22a	80	95	35a	83	91
15	PDL	22b	76	90	35b	90	88
16	PLL	23a	64	88	36a	78	84
16	PDL	23b	61	87	36b	81	85
17	PLL	24a	36	60	37a	79	55
17	PDL	24b	33	61	37b	76	61
18	PLL	25a	21	53	38a	83	48
18	PDL	25b	19	50	38b	78	47

3.4. Dihydroflavonols

Although the Algar-Flynn-Oyamada (AFO) protocol (Geissman and Fukushima, 1948; Dean and Podimuang, 1965) and the Weeler reaction were mainly used for the synthesis of aurones, it was demonstrated that these reactions can be adapted for the formation of racemic dihydroflavonols (Saxena et al., 1985; Patonay et al., 1993; Donnelly and Doran, 1975; Donnelly et al., 1979; Donnelly and Emerson, 1990; Donnelly and Higginbotham, 1990) in moderate to good yields.

Cyclization of 2'-hydroxy-α,3,4,4'-tetramethoxychalcone **39** with sodium acetate in ethanol furnished both 3,3',4',7-*O*-tetramethyl-2,3-*trans*-**40** and 3,3',4',7-*O*-tetramethyl-2,3-*cis*-dihydroflavonols **41** in 22% and 11% yields, respectively (Scheme 1.3). However, this method was not applicable to cycli-zation of α-OH-chalcones (Van der Merwe et al., 1972; Ferreira et al., 1975).

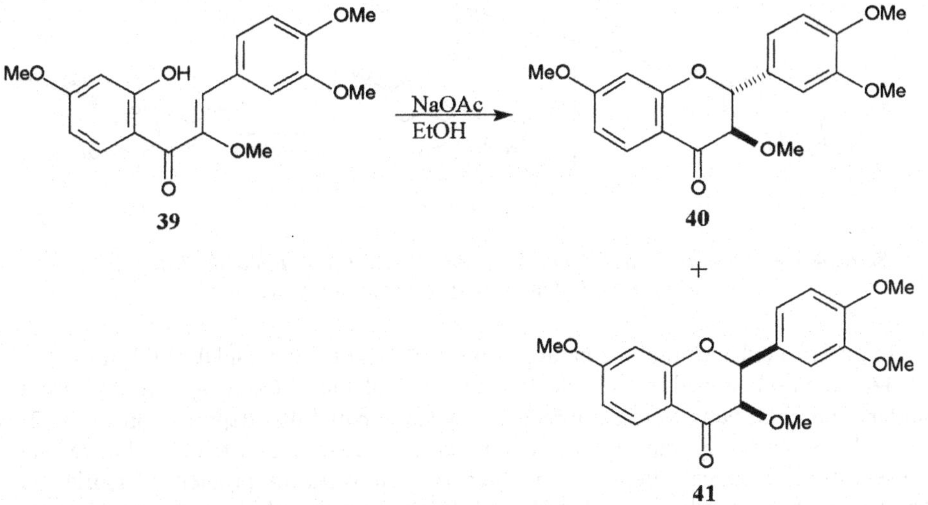

Scheme 1.3 *Chalcone cyclization with NaOAc in EtOH to yield trans- and cis-dihydroflavonols.*

Initial attempts toward acid catalyzed cyclization of the chalcone epoxide to the corresponding (2*R*,3*R*)-2,3-*trans*- **44a** and (2*S*,3*R*)-2,3-*cis*-dihydroflavonols **45a** were hampered by two difficulties, i.e., aryl migration with formation of 4',7-dimethoxyisoflavone **43** and the epimerization/racemization of the thermodynamically less stable (2*S*,3*R*)-2,3-*cis*-4',7-dimethoxydihydroflavonol **45a** to yield (2*S*,3*S*)-2,3-*trans*-dihydroflavonol **44b** (Augustyn et al., 1990a) (Scheme 1.4). The "loss" of optical purity in the **22a** → **44a** conversion indicates competition between protonation of the heterocyclic oxygen and hydrolysis of the 2'-*O*-acetal functionality, hence leading to a considerable degree of S_N1 character for the cyclization step with concomitant racemization at C-β of a presumed carbocationic intermediate **42**, yielding dihydroflavonols **44a** and **45a**. The thermodynamically less stable (2*S*,3*R*)-2,3-*cis*-dihydroflavonol **45a** is rapidly racemized at C-3 to give a mixture of **45a** and **44b** under the prevailing acidic conditions. Formation of the isoflavone **43** is attributed to acid-catalyzed cleavage of the highly reactive oxirane functionality prior to deprotection.

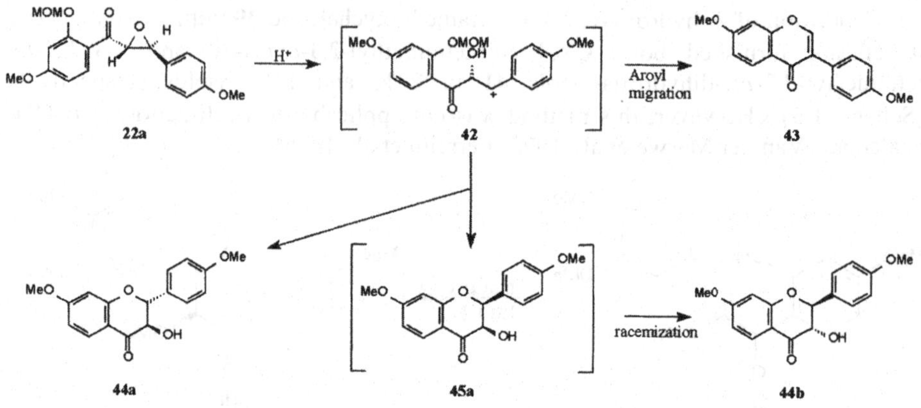

Scheme 1.4 *Attempts toward synthesis of (2R,3R)-2,3-trans-* **44a** *and (2S,3R)-2,3-cis-dihydroflavonols* **45a** *using acid-catalyzed cyclization.*

In order to enhance the S_N2 nature of the ring closure step, and thus the formation of **44a**, methods aimed at the selective removal of the 2'-*O*-methoxymethyl group under mild conditions were explored. It was anticipated that deprotection of the 2'-*O*-methoxymethyl group with concomitant cyclization would enhance the preservation of optical integrity. In order to circumvent the problem of isoflavone formation, Van Rensburg et al. (1996, 1997a) investigated methods aimed at the initial nucleophilic opening of the oxirane functionality, followed by deprotection and cyclization. The excellent nucleophilic and nucleofugic properties of mercaptans (Barrett et al., 1989) prompted evaluation of thiols in the presence of Lewis acids and resulted in the selection of the phenylmethanethiol–tin(IV) chloride (BnSH/SnCl₄) system as the reagent of choice for the oxirane cleavage (Chini et al., 1992). Treatment of the series of chalcone epoxides **21a/b-25a/b** with BnSH/SnCl₄ selectively cleaved the C_β-O bond of the oxirane functionality at –20°C and effectively deprotected the methoxymethyl group at 0°C to give the corresponding α,2'-dihydroxy-β-benzylsulfanyldihydrochalcones **46a/b-50a/b** as diastereomeric mixtures (*syn: anti, ca.* 2.3:1) in 86-93% yield. Treatment of these α-hydroxy-β-benzylsulfanyldihydrochalcones **46a/b-50a/b** with the thiophilic Lewis acid, silver tetrafluoroborate (AgBF₄) in CH₂Cl₂ at 0°C, gave the 2,3-*trans*-dihydroflavonols **44a/b, 51a/b-54a/b** in good yield and albeit in low proportions for the first time also the 2,3-*cis* analogues **45a/b, 55a/b-58a/b** (Table 1.5).

Table 1.5 *Asymmetric synthesis of dihydroflavonols*

$$30\% \text{ H}_2\text{O}_2 \text{ / } 6\text{M NaOH}$$
poly-*L*- or poly-*D*-alanine
CCl$_4$ / rt. / 36-96 h

14 R$_1$ = R$_2$ = R$_3$ = H, R$_4$ = OMe
15 R$_1$ = R$_4$ = OMe, R$_2$ = R$_3$ = H
16 R$_1$ = R$_3$ = R$_4$ = OMe, R$_2$ = H
17 R$_1$ = R$_2$ = R$_4$ = OMe, R$_3$ = H
18 R$_1$ = R$_2$ = R$_3$ = R$_4$ = OMe

21a/b R$_1$ = R$_2$ = R$_3$ = H, R$_4$ = OMe
22a/b R$_1$ = R$_4$ = OMe, R$_2$ = R$_3$ = H
23a/b R$_1$ = R$_3$ = R$_4$ = OMe, R$_2$ = H
24a/b R$_1$ = R$_2$ = R$_4$ = OMe, R$_3$ = H
25a/b R$_1$ = R$_2$ = R$_3$ = R$_4$ = OMe

BnSH / SnCl$_4$
CH$_2$Cl$_2$ / -20 - 0°C

51a/b R$_1$ = R$_2$ = R$_3$ = H, R$_4$ = OMe
44a/b R$_1$ = R$_4$ = OMe, R$_2$ = R$_3$ = H
52a/b R$_1$ = R$_3$ = R$_4$ = OMe, R$_2$ = H
53a/b R$_1$ = R$_2$ = R$_4$ = OMe, R$_3$ = H
54a/b R$_1$ = R$_2$ = R$_3$ = R$_4$ = OMe

AgBF$_4$ / CH$_2$Cl$_2$ / 0°C

46a/b R$_1$ = R$_2$ = R$_3$ = H, R$_4$ = OMe
47a/b R$_1$ = R$_4$ = OMe, R$_2$ = R$_3$ = H
48a/b R$_1$ = R$_3$ = R$_4$ = OMe, R$_2$ = H
49a/b R$_1$ = R$_2$ = R$_4$ = OMe, R$_3$ = H
50a/b R$_1$ = R$_2$ = R$_3$ = R$_4$ = OMe

55a/b R$_1$ = R$_2$ = R$_3$ = H, R$_4$ = OMe
45a/b R$_1$ = R$_4$ = OMe, R$_2$ = R$_3$ = H
56a/b R$_1$ = R$_3$ = R$_4$ = OMe, R$_2$ = H
57a/b R$_1$ = R$_2$ = R$_4$ = OMe, R$_3$ = H
58a/b R$_1$ = R$_2$ = R$_3$ = R$_4$ = OMe

a = configuration shown
b = enantiomer

Epoxide	% yield	% ee	Dihydro-chalcone	% yield	Dihydro-flavonol	% yield	% ee	*trans:cis*
21a	99	84	**46a**	86	**51a / 55a**	86	83	93 : 7
21b	98	69	**46b**	90	**51b / 55b**	83	69	94 : 6
22a	98	86	**47a**	93	**44a / 45a**	71	84	79 : 21
22b	98	74	**47b**	90	**44b / 45b**	72	75	83 : 17
23a	99	67	**48a**	89	**52a / 56a**	81	68	85 : 15
23b	98	58	**48b**	91	**52b / 56b**	79	58	86 : 14
24a	97	70	**49a**	89	**53a / 57a**	65	69	78 : 22
24b	97	53	**49b**	89	**53b / 57b**	64	53	84 : 16
25a	79	49	**50a**	91	**54a / 58a**	61	47	82 : 18
25b	76	49	**50b**	88	**54b / 58b**	63	44	80 : 20

A highly enantioselective synthetic method (99%, ee) was reported by Jew et al. (2000) for optically pure (2*R*,3*R*)-dihydroflavonols, by using catalytic asymmetric dihydroxylation and an intramolecular Mitsunobu reaction as key steps (Scheme 1.5).

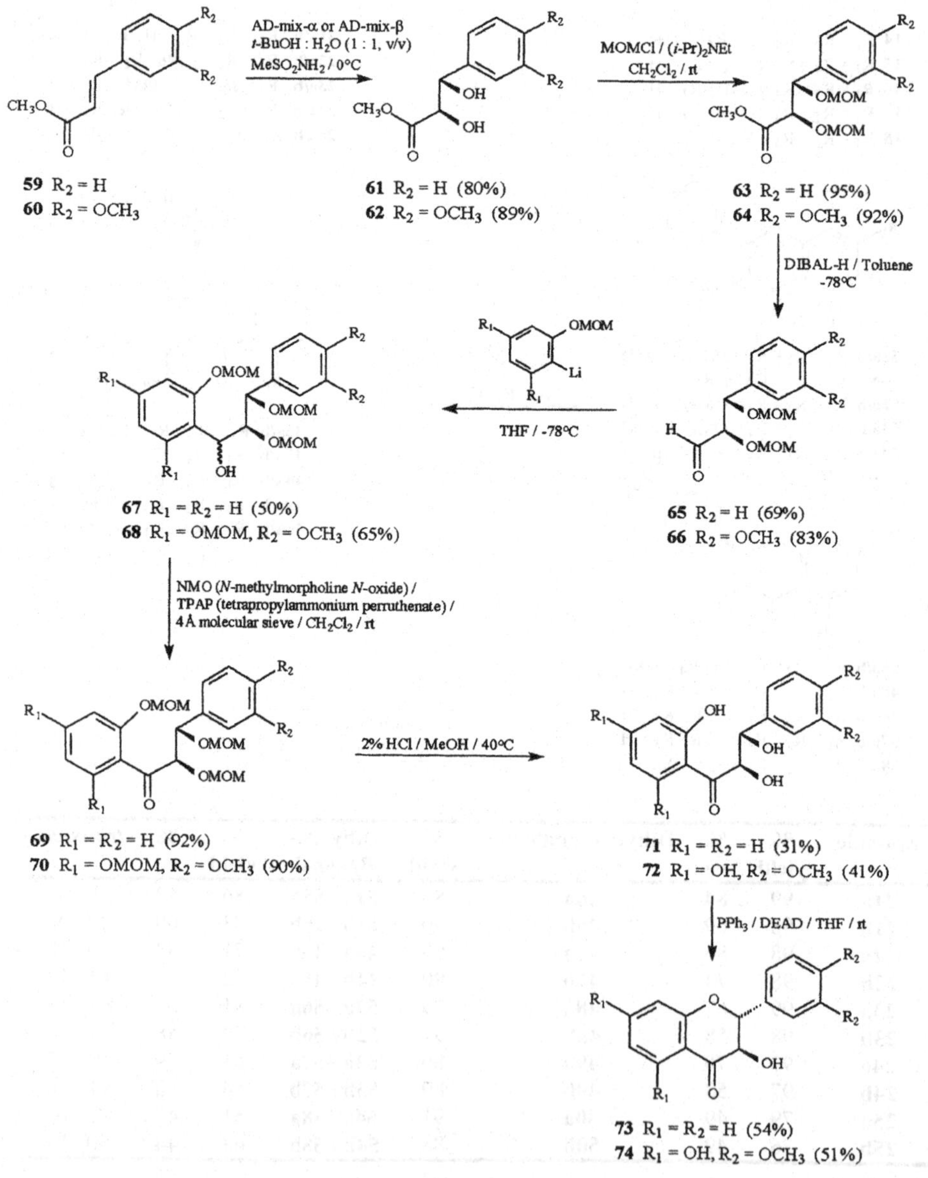

AD-mix-α or AD-mix-β
t-BuOH : H2O (1 : 1, v/v)
MeSO2NH2 / 0°C
MOMCl / (i-Pr)2NEt
CH2Cl2 / rt
DIBAL-H / Toluene
-78°C
THF / -78°C
NMO (N-methylmorpholine N-oxide) /
TPAP (tetrapropylammonium perruthenate) /
4 Å molecular sieve / CH2Cl2 / rt
2% HCl / MeOH / 40°C
PPh3 / DEAD / THF / rt
59 R2 = H
60 R2 = OCH3
61 R2 = H (80%)
62 R2 = OCH3 (89%)
63 R2 = H (95%)
64 R2 = OCH3 (92%)
65 R2 = H (69%)
66 R2 = OCH3 (83%)
67 R1 = R2 = H (50%)
68 R1 = OMOM, R2 = OCH3 (65%)
69 R1 = R2 = H (92%)
70 R1 = OMOM, R2 = OCH3 (90%)
71 R1 = R2 = H (31%)
72 R1 = OH, R2 = OCH3 (41%)
73 R1 = R2 = H (54%)
74 R1 = OH, R2 = OCH3 (51%)

***Scheme 1.5** Synthesis of dihydroflavonol 73 and 3′,4′-di-O-methyltaxifolin 74.*

Sharpless asymmetric dihydroxylation of **59** and **60** with AD-mix-α gave the 2*R*,3*S*-diols **61** and **62** in excellent yields (80% and 89%, respectively) and ee (99%). This was followed by protection of the C-2 and C-3 hydroxyl groups with MOMCl and reduction with diisobutylaluminium hydride to give the corresponding aldehydes **65** and **66**. Addition of aryllitium to aldehydes **65** and **66** afforded the secondary alcohols **67** and **68**. Oxidation of **67** and **68** produced the corresponding ketones **69** and **70**, which were deprotected under acidic conditions to give the pentahydroxyketones **71** and **72**. An intramolecular Mitsunobu (Mitsunobu, 1981) reaction afforded dihydroflavonol **73** and 3',4'-di-*O*-methyltaxifolin **74**, respectively. The absolute configuration of the newly formed stereogenic center C-2 of **73** and **74** were assigned as 2*R*, consistent with the S_N2-mechanism of the Mitsunobu reaction.

3.5. Flavan-3-ols and Flavan-3,4-diols

Flavan-3-ols, e.g., (+)-catechin and (-)-epicatechin, represent the largest class of naturally occurring C_6-C_3-C_6 monomeric flavonoids. Flavan-3-ols also have received considerable interest over the last few years because of their importance as the constituent units of proanthocyanidins (Porter, 1988, 1994; Ferreira and Bekker, 1996; Ferreira and Li, 2000; Ferreira and Slade, 2002; Ferreira et al., 2005). Progress in the study of these complex phenolics is often hampered by the limited availability of naturally occurring flavan-3-ol nucleophiles with 2,3-*trans,* and especially 2,3-*cis,* configuration. One of the most common ways for the synthesis of flavan-3-ols and the closely related flavan-3,4-diol analogues involves the reductive transformation of dihydroflavonols. Reduction of the dihydroflavonols **75a/b** with sodium borohydride in methanol affords the 2,3-*trans*-3,4-*trans*-flavan-3,4-diols **76a/b**, while reduction in an aprotic solvent like dioxane yielded the C_4-epimers **77a/b** exclusively (Scheme 1.6) (Takahashi et al., 1984; Onda et al., 1989). Such reversal in the direction of the hydride attack could probably be explained in terms of the presence of hydrogen bonding in aprotic solvents.

Catechin **80** represents the only flavan-3-ol synthesized from the corresponding dihydroflavonol (Weinges, 1958; Freudenberg and Weinges, 1958). Consecutive treatment of 2,3-*trans*-3-*O*-acetyldihydroquercetin tetra-*O*-benzyl ether **78** with LiAlH$_4$ and H$_2$/Pd gave the free phenolic flavan-3-ol **79** in optically pure form (Scheme 1.7). ^{13}C-Labeled (±)-catechin recently was synthesized by utilizing osmium-catalyzed dihydroxylation of a flav-3-ene intermediate as a key step to yield the 2,3-*trans*-3,4-*cis*-isomer with high diastereoselectivity. The first attempt included ten steps, starting from K[13]CN (Nay et al., 2000). A slightly different but improved approach was later developed by the same group (Arnaudinaud et al., 2001a, 2001b) for the formation of ^{13}C-labeled (-)-procyanidin B-3. Improved yields were reported and the number of steps to the pivotal intermediate flav-3-ene was reduced. A disadvantage using these protocols is that enantiomeric mixtures are formed that require more refined and usally more expensive separation methods.

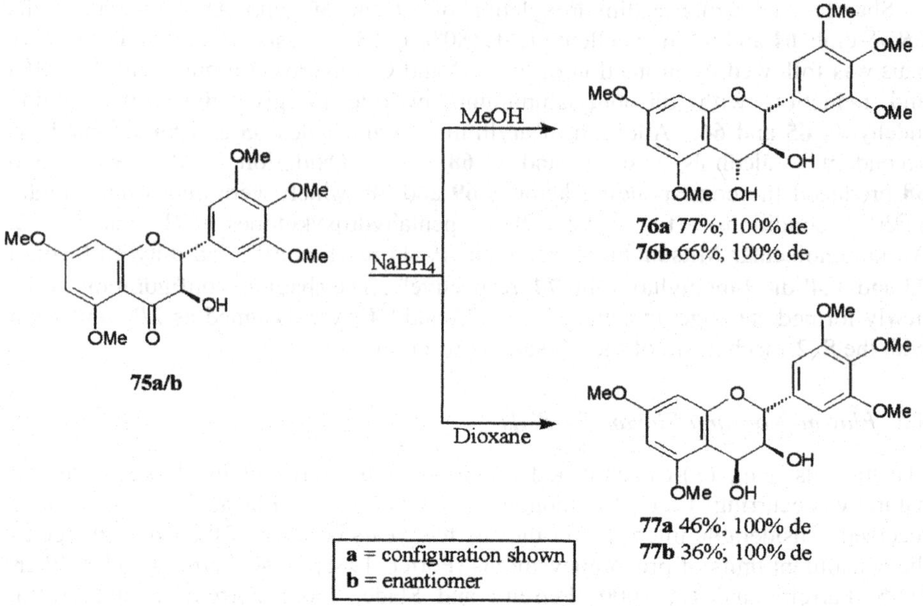

Scheme 1.6 *Reduction of dihydroflavonols with NaBH$_4$ to afford flavan-3,4-diols*

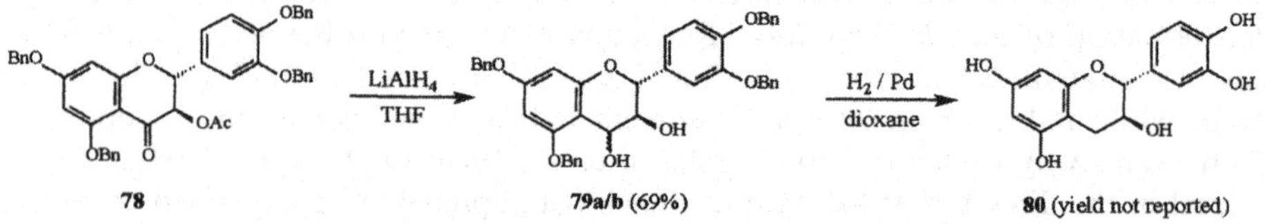

Scheme 1.7 *Reduction of 2,3-trans-3-O-acetyldihydroquercetin tetra-O-benzylether 78 to yield catechin 80.*

(+)-[^{13}C]-Catechin **84a** and (-)-[^{13}C]-epicatechin **87** were isolated in high ee, respectively, by the formation of their tartaric acid derivatives (Nay et al., 2001). The resolution process included the esterification of the 3-OH group of **81a/b** with L-dibenzoyltartaric acid monomethyl ester to give a mixture of diastereomers **82** and **85** (92%) (Scheme 1.8). The (+)-catechin derivative **82** was crystallized in hexane/dichloromethane (3:1) (diastereometic excess [de] > 99%), while the (-)-*ent*-catechin derivative **85** remained in solution. The diastereomeric pure (de = 99%) (-)-*ent*-catechin derivative **86** also was isolated by crystallization after hydrolysis (MeOH/H$_2$O/KOH) of **85**, following esterification with D-tartaric acid. (+)-[^{13}C]-catechin **84a** was isolated in a high yield and ee (99%) after hydrolysis and reduction/deprotection steps. Epimerization at C-2 of (-)-[^{13}C]-*ent*-catechin **84b**, using 1% (w/v) *aq*. Na$_3$PO$_4$, led to an equilibrium mixture of (-)-**84b** and (-)-[^{13}C]-epicatechin **87** in an approximate 3:1 ratio after 20 hr at 25°C (ee >99%).

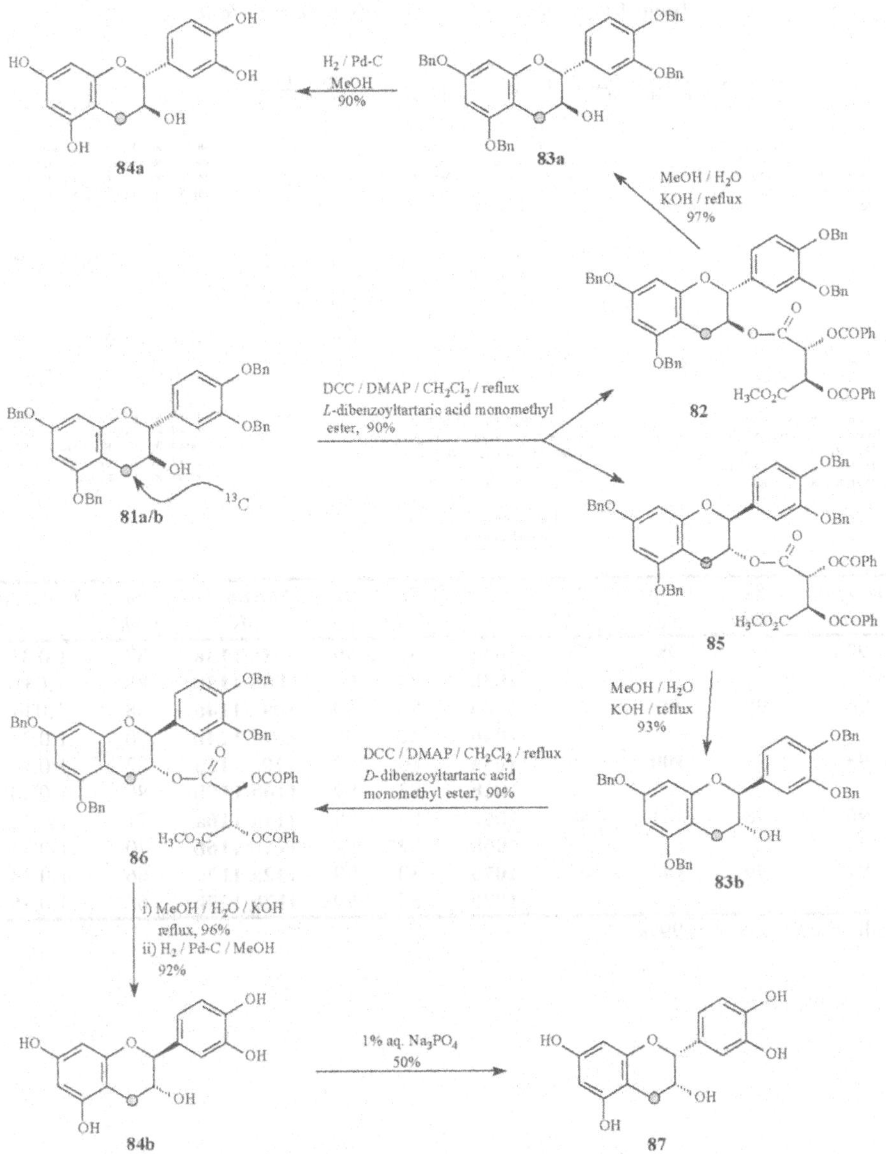

Scheme 1.8 *Synthesis via resolutions of (+)-[^{13}C]-catechin* **84a** *and (-)-[^{13}C]-epicatechin* **87***.*

In order to address the issue of stereocontrol at C-2 and C-3 of the flavan-3-ol molecular framework, Van Rensburg et al. (1997b, 1997c) designed a concise protocol based on the transformation of *retro*-chalcones into 1,3-diaryl-propenes (Table 1.6). These compounds are then subjected to asymmetric dihydroxylation to give polyoxygenated diarylpropan-1,2-diols, which are used as chirons for essentially enantiopure flavan-3-ols. This protocol included a base-catalyzed condensation of the appropriately oxygenated acetophenones and benzaldehydes to

Table 1.6 *Synthesis of flavan-3-ols **108a/b-117a/b***

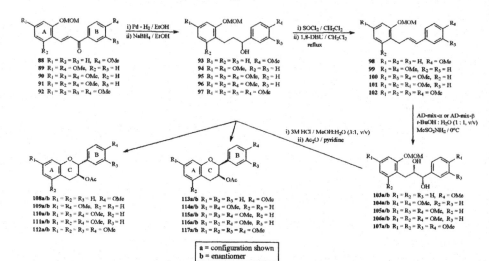

88 $R_1 = R_2 = R_3 = H$, $R_4 = OMe$
89 $R_1 = R_4 = OMe$, $R_2 = R_3 = H$
90 $R_1 = R_3 = R_4 = OMe$, $R_2 = H$
91 $R_1 = R_2 = R_4 = OMe$, $R_3 = H$
92 $R_1 = R_2 = R_3 = R_4 = OMe$

93 $R_1 = R_2 = R_3 = H$, $R_4 = OMe$
94 $R_1 = R_4 = OMe$, $R_2 = R_3 = H$
95 $R_1 = R_3 = R_4 = OMe$, $R_2 = H$
96 $R_1 = R_2 = R_4 = OMe$, $R_3 = H$
97 $R_1 = R_2 = R_3 = R_4 = OMe$

98 $R_1 = R_2 = R_3 = H$, $R_4 = OMe$
99 $R_1 = R_4 = OMe$, $R_2 = R_3 = H$
100 $R_1 = R_3 = R_4 = OMe$, $R_2 = H$
101 $R_1 = R_2 = R_4 = OMe$, $R_3 = H$
102 $R_1 = R_2 = R_3 = R_4 = OMe$

108a/b $R_1 = R_2 = R_3 = H$, $R_4 = OMe$
109a/b $R_1 = R_4 = OMe$, $R_2 = R_3 = H$
110a/b $R_1 = R_3 = R_4 = OMe$, $R_2 = H$
111a/b $R_1 = R_2 = R_4 = OMe$, $R_3 = H$
112a/b $R_1 = R_2 = R_3 = R_4 = OMe$

113a/b $R_1 = R_2 = R_3 = H$, $R_4 = OMe$
114a/b $R_1 = R_4 = OMe$, $R_2 = R_3 = H$
115a/b $R_1 = R_3 = R_4 = OMe$, $R_2 = H$
116a/b $R_1 = R_2 = R_4 = OMe$, $R_3 = H$
117a/b $R_1 = R_2 = R_3 = R_4 = OMe$

103a/b $R_1 = R_2 = R_3 = H$, $R_4 = OMe$
104a/b $R_1 = R_4 = OMe$, $R_2 = R_3 = H$
105a/b $R_1 = R_3 = R_4 = OMe$, $R_2 = H$
106a/b $R_1 = R_2 = R_4 = OMe$, $R_3 = H$
107a/b $R_1 = R_2 = R_3 = R_4 = OMe$

a = configuration shown
b = enantiomer

Propan-1-ols	% yield	Prop-enes	% yield	1,2-diols	% yield	% ee	Flavan-3-ols	% yield	Trans:cis
93	99	98	73	103a	82	99	108a/113a	87	1:0.33
				103b	84	99	108b/113b	88	1:031
94	98	99	74	104a	86	99	109a/114a	88	1:036
				104b	82	99	109b/114b	90	1:0.33
95	99	100	70	105a	85	99	110a/115a	82	1:0.32
				105b	83	99	110b/115b	80	1:0.30
96	98	101	68	106a	80	99	111a/116a	71	1:0.32
				106b	83	99	111b/116b	70	1:0.33
97	99	102	66	107a	80	99	112a/117a	66	1:0.34
				107b	87	99	112b/117b	65	1;0.35

In all cases, the ee was 99%.

afford the (E)-*retro*-chalcones **88–92** $(J_{\alpha,\beta}$ 15.8–16.0 Hz). Consecutive reduction (Pd-H$_2$ and NaBH$_4$), followed by elimination {SOCl$_2$ and 1,8-diazabicyclo[5.4.0]undec-7-ene (1,8-DBU)} of the ensuing alcohols **93–97** afforded the (E)-1,3-diarylpropenes (deoxodihydrochalcones) **98–102** $(J_{1,2}$ 16 Hz) in resonable overall yield (65–73%). Owing to the excellent results obtained (Sharpless et al., 1977, 1992; Kwong et al., 1990; Jeong et al., 1992; Amberg et al., 1993; Gobel and Sharpless, 1993; Wang et al., 1993; Kolb et al., 1994a, 1994b; Norrby et al., 1994) during asymmetric dihydroxylation (AD reaction) of olefins with AD-mix-α or AD-mix-β, these stereoselective catalysts were utilized for the introduction of chirality at C-2 and C-3 of the flavan-3-ol framework. Thus, treatment of the protected (E)-propenes **98–102** at 0°C with AD-mix-α in the two phase system 'BuOH: H$_2$O (1:1) afforded the (+)-(1*S*,2*S*)-*syn*-diols **103a–107a** $(J_{1,2}$ 5.8 -6.5 Hz) in high yields (80–86%) and optical purity (99% ee). The (-)-(1*R*,2*R*)-*syn*-diols **103b–107b** were similarly obtained by using AD-mix-β (yield: 82–87%, 99% ee). Application of the Lewis acid-catalyzed phenylmethanethiol ring-opening and cyclization of chalcone epoxides in the synthesis of dihydroflavonols (see Section 3.4) (Van Rensburg et al., 1996, 1997a) to cyclization of the diols, however, resulted in slow (24 hr) and low percentage conversion (10–20%) into flavan-3-ols.

In order to transform the diols more effectively into the corresponding flavan-3-ols, methods aimed at the selective removal of the 2'-*O*-methoxymethyl group and subsequent ring closure under mild acidic conditions were explored. Simultaneous deprotection and cyclization of diols **103a/b–107a/b** in the presence of 3M HCl in MeOH, followed by acetylation, yielded the 2,3-*trans*- (yield: 48-68%) **108a/b–112a/b** and for the first time 2,3-*cis*-flavan-3-ols (yield:17–22%) **113a/b–117a/b** in excellent enantiomeric excess (>99%). Assignment of the absolute configuration of the resulting flavan-3-ol derivatives **108a/b–117a/b** by ^{1}H-NMR and CD data confirmed the configuration of the diols as derived from the Sharpless model.

The potential of this protocol in the chemistry of the oligomeric proanthocyanidins is evident, especially in view of its aptitude to the synthesis of free phenolic analogues. The latter analogues are as conveniently accessible by simply using more labile protecting groups instead of *O*-methyl ethers. This was illustrated by Nel et al. (1999b) by synthesis of the 4',7'-dihydroxyflavan-3-ol diastereomers to confirm (2*R*,3*S*)-guibourtinidol as a new natural product. Owing to the acid lability of methoxymethyl derivative, the MOM functionality was used as a protecting group. This method was extended to the synthesis of the full range of flavan-3-ols, comprising different oxygenated phenolic substitutions as found in nature (Nel et al., 1999c).

3.6. Isoflavonoids

Synthetic routes to optically pure pterocarpans, exhibiting the aromatic oxygenation patterns of naturally occurring isoflavonoids, are limited by the lack of readily accessible starting materials. These restrictions and the challenge to form the tetracyclic ring system with stereocontrol led to the development of various synthetic approaches. Synthetic endeavors toward pterocarpans comprise Heck arylation (Ishiguro et al., 1982; Narkhede et al., 1990), the reduction and cyclization of the corresponding 2'-hydroxyisoflavanones (Krishna Prasad et al., 1986), cycloaddition reactions of 2H-chromenes with 2-alkoxy-1,4-benzoquinones (Engler et al., 1990; Subburaj et al., 1997), and 1,3-Michael–Claisen annulation (Ozaki et al., 1988, 1989). Only two methods, i.e., asymmetric dihydroxylation of an isoflav-3-ene (Pinard et al., 1998) and subsequent "hydrogenative cyclization" or 1,4-benzoquinone cyclo-addition reactions utilizing chiral Ti(IV) complexes (Engler et al., 1991, 1999), permitted enantioselective access to this class of compounds.

3.6.1. Isoflavans

Given the fact that the configuration at C-3 would dictate the configuration at C-2 or C-4 in the 3-phenylchroman framework, a series of isoflavans were synthesized, which would then afford stereoselective access to other classes of chiral isoflavonoids (Versteeg et al., 1995, 1998, 1999). The protocol involved the stereoselective α-benzylation of phenylacetic acid derivatives, subsequent reductive removal of the chiral auxiliary, and cyclization into the isoflavans (Scheme 1.9). Owing to the efficiency of the asymmetric alkylation reactions of chiral imide enolates, (4S,5R)-(+)- and (4R,5S)-(-)-1,5-dimethyl-4-phenyl-2-imidazolidinones **118a** and **118b** were used as chiral auxiliaries in the benzylation reactions (Close, 1950; Roder et al., 1984; Evans et al., 1987; Cardillo et al., 1988; Drewes et al., 1993). The basicity of the imidazolidinones was decreased by utilizing the trimethylsilyl ethers **119a** and **119b** in the acylation step using the phenylacetyl chlorides **120-122**. The ensuing N-acyl imidazolidinones **123a/b-125a/b** were then alkylated with the appropriate 2-O-methoxymethylbenzyl bromides **126** and **127** in good to excellent yields with only one diastereomer isolated (de > 99%). Removal of the chiral auxiliary was effected by reductive deamination using LiAlH$_4$ in THF for imides **128a/b-130a/b** and a saturated solution of LiBH$_4$ in ether for analogues **131a/b-133a/b** to give the 2,3-diarylpropan-1-ols **134a/b-139a/b** (Cardillo et al., 1989; Paderes et al., 1991). Acidic deprotection (3M HCl in MeOH), followed by cyclization under Mitsunobu conditions (Shih et al., 1987) afforded the target isoflavans **140a/b-145a/b** in excellent yields and in nearly enantiopure form (ee >96-99%).

The stereochemistry of the alkylation step is explicable in terms of the preferential formation of a Z-enolate (Evans at al., 1982). Attack of the electrophile is then directed to the face of the enolate opposite the phenyl moiety on the chiral auxiliary. The chiral auxiliary with 4S-configuration led to propanols exhibiting positive optical rotations and those from 4R-N-acyloxazolidinones showing negative values, in accordance with observations by Evans et al. (1982).

Alkylation of (4S,5R)-(+)-N-phenylacetylimidazolidinones resulted in (+)-propanols and (3S)-isoflavans and (4R,5S)-(-)-N-phenylacetylimidazolidinones in (-)-propanols and (3R)-isoflavans.

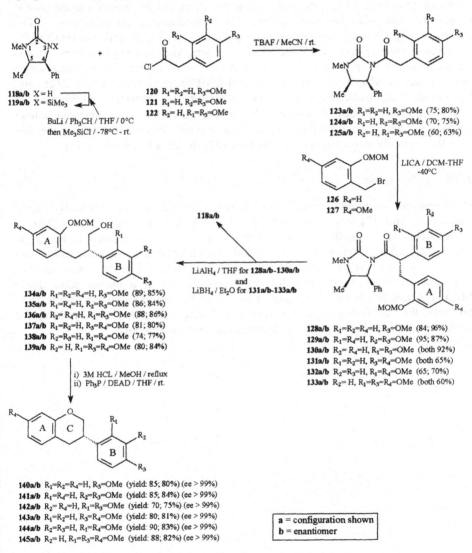

Scheme 1.9 *Stereoselective synthesis of (R)- and (S)-isoflavans.*

3.6.2. Isoflavone Epoxides

The first representatives of flavone epoxides were prepared either by alkaline hydrogen peroxide epoxidation of isoflavones or by an intramolecular Darzens reaction of α-bromo-O-acyloxyacetophenones. Lévai et al. (1998) demonstrated that dimethyldioxirane (DMDO) is a convenient and effective reagent for the epoxidation of various substituted isoflavones and subsequently prepared isoflavone glycoside epoxides in high yields by utilizing this versatile oxidizing agent. However, attempts to synthesize enantiomeric isoflavone epoxides with DMDO and a chiral auxiliary demonstrated that the sugar chiral auxiliary did not exercise enantiofacial selectivity and epoxides were isolated as 1:1 diasteromeric mixtures. The Jacobsen's Mn(III)salen complexes have proved to be highly efficient catalyst for the enantioselective epoxidation of olefins by using various oxygen donors. It was demonstrated that epoxidation of 2,2-dimethyl-2*H*-chromenes, in the presence of optically active Mn(III)salen complexes and DMDO, proceeded enantioselectively. Epoxidation of isoflavones **146–151**, utilizing the Mn(III)salen complexes (*R,R*)- and (*S,S*)-*N,N'-bis*(3,4-di-*t*-butylsalicylidene)-1,2-cyclohexanediaminomanganese chloride as catalysts and DMDO or NaOCl as oxygen donors, afforded for the first time the optically active isoflavone epoxides **152a/b–157a/b** (Scheme 1.10).

3.6.3. Isoflavanones

By employing a stereocontrolled aldol reaction as the key step, optically active isoflavones **168–171** were synthesized for the first time by Vicario et al. (2000) in good yields and excellent ee's (Scheme 1.11). This sequence included an asymmetric aldol reaction between (*S,S*)-(+)-pseudoephedrine arylacetamide and formaldehyde to introduce chirality in the isoflavanone carbon framework at C-3. This was followed by the introduction of the B-ring as a phenol ether under Mitsunobu conditions and subsequent removal of the chiral auxiliary. Acids **164– 167** were then converted by an intramolecular Friedel–Crafts acylation, yielding the isoflavanones **168–171** in good yields and essentially enantiopure.

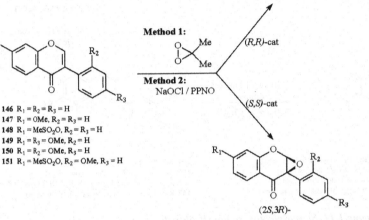

(2R,3S)-

152a $R_1 = R_2 = R_3 = H$ [method (1) yield: 31%, ee: 52%]
153a $R_1 = OMe$, $R_2 = R_3 = H$ [(1) yield: 27%, ee: 37%; (2) yield: 30%, ee: 71%]
154a $R_1 = MeSO_2O$, $R_2 = R_3 = H$ [(1) yield: 22%, ee: 21%]
155a $R_1 = R_3 = OMe$, $R_2 = H$ [(1) yield: 39%, ee: 52%]
156a $R_1 = R_2 = OMe$, $R_3 = H$ [(1) yield: 31%, ee: 82%]
157a $R_1 = MeSO_2O$, $R_2 = OMe$, $R_3 = H$ [(1) yield: 25%, ee: 72%]

Method 1:

(R,R)-cat

Method 2:
NaOCl / PPNO

(S,S)-cat

146 $R_1 = R_2 = R_3 = H$
147 $R_1 = OMe$, $R_2 = R_3 = H$
148 $R_1 = MeSO_2O$, $R_2 = R_3 = H$
149 $R_1 = R_3 = OMe$, $R_2 = H$
150 $R_1 = R_2 = OMe$, $R_3 = H$
151 $R_1 = MeSO_2O$, $R_2 = OMe$, $R_3 = H$

(2S,3R)-

152b $R_1 = R_2 = R_3 = H$ [(1) yield: 34%, ee: 56%; (2) yield: 25%, ee: 65%]
153b $R_1 = OMe$, $R_2 = R_3 = H$ [(1) yield: 36%, ee: 39%; (2) yield: 26%, ee: 77%]
154b $R_1 = MeSO_2O$, $R_2 = R_3 = H$ [(1) yield: 27%, ee: 48%; (2) yield: 23%, ee: 56%]
155b $R_1 = R_3 = OMe$, $R_2 = H$ [(1) yield: 29%, ee: 22%; (2) yield: 30%, ee: 76%]
156b $R_1 = R_2 = OMe$, $R_3 = H$ [(1) yield: 32%, ee: 86%; (2) yield: 30%, ee: 90%]
157b $R_1 = MeSO_2O$, $R_2 = OMe$, $R_3 = H$ [(1) yield: 23%, ee: 90%; (2) yield: 31%, ee: 94%]

(R,R)-cat: (R,R)-N,N'-bis(3,5-di-t-butylsalicylidene)-1,2-cyclohexanediaminomanganese chloride
(S,S)-cat: (S,S)-N,N'-bis(3,5-di-t-butylsalicylidene)-1,2-cyclohexanediaminomanganese chloride

Scheme 1.10 *Enantioselective synthesis of isoflavone epoxides **152a/b–157a/b**.*

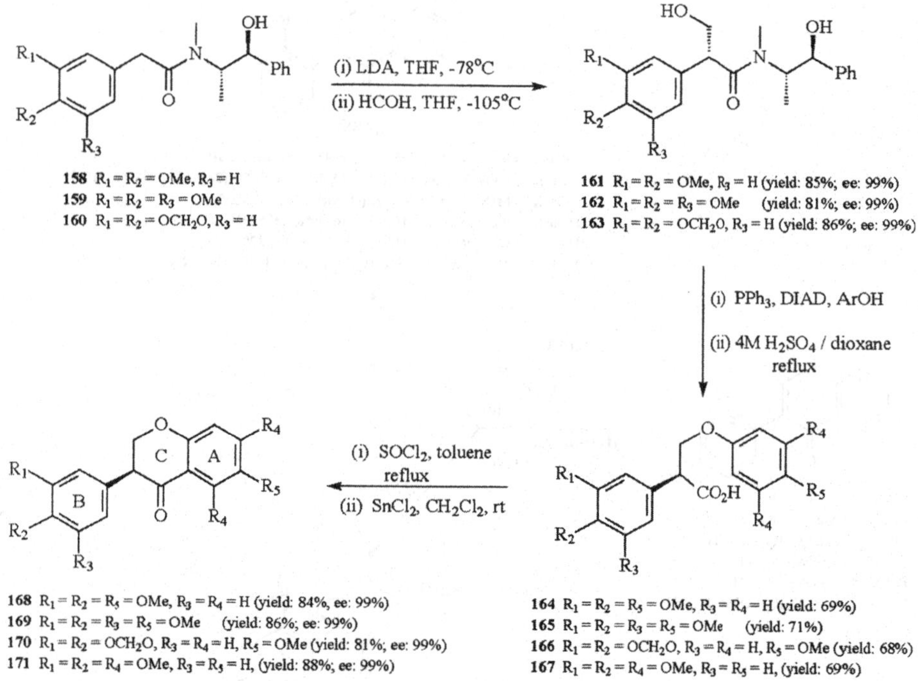

158 $R_1 = R_2 = OMe, R_3 = H$
159 $R_1 = R_2 = R_3 = OMe$
160 $R_1 = R_2 = OCH_2O, R_3 = H$

(i) LDA, THF, -78°C

(ii) HCOH, THF, -105°C

161 $R_1 = R_2 = OMe, R_3 = H$ (yield: 85%; ee: 99%)
162 $R_1 = R_2 = R_3 = OMe$ (yield: 81%; ee: 99%)
163 $R_1 = R_2 = OCH_2O, R_3 = H$ (yield: 86%; ee: 99%)

(i) PPh₃, DIAD, ArOH

(ii) 4M H_2SO_4 / dioxane
reflux

(i) SOCl₂, toluene
reflux

(ii) SnCl₂, CH₂Cl₂, rt

168 $R_1 = R_2 = R_5 = OMe, R_3 = R_4 = H$ (yield: 84%, ee: 99%)
169 $R_1 = R_2 = R_3 = R_5 = OMe$ (yield: 86%; ee: 99%)
170 $R_1 = R_2 = OCH_2O, R_3 = R_4 = H, R_5 = OMe$ (yield: 81%; ee: 99%)
171 $R_1 = R_2 = R_4 = OMe, R_3 = R_5 = H$, (yield: 88%; ee: 99%)

164 $R_1 = R_2 = R_5 = OMe, R_3 = R_4 = H$ (yield: 69%)
165 $R_1 = R_2 = R_3 = R_5 = OMe$ (yield: 71%)
166 $R_1 = R_2 = OCH_2O, R_3 = R_4 = H, R_5 = OMe$ (yield: 68%)
167 $R_1 = R_2 = R_4 = OMe, R_3 = R_5 = H$, (yield: 69%)

Scheme 1.11 *Stereoselective synthesis of isoflavanones 168–171.*

3.6.4. Pterocarpans

Despite the identification of the first 6a-hydroxypterocarpan, (+)-pisatin, in 1960 (Cruickshank and Perrin), synthetic protocols to these potent phytoalexins are limited by lengthy multistep routes and a lack of diversity as far as phenolic hydroxylation patterns are concerned. These confinements are so restrictive that only two 6a-hydroxypterocarpans, i.e., pisatin and variabilin, have been synthesized (Bevan et al., 1964; Mansfield, 1982; Pinard et al., 1998).

The results reported for the stereoselective aldol condensation between methyl ketones and aldehydes employing diisopropylethylamine and chiral boron triflates (Paterson and Goodman, 1989) prompted the investigation for a more direct synthetic approach to address the issue of stereocontrol at C-6a and C-11a of the pterocarpan framework (Van Aardt et al., 1998, 1999, 2001). Depending on the lability and/or stability of protecting groups under certain reaction conditions, this protocol included methoxymethyl protection of the benzaldehydes **181** and **182** (labile in the presence of Lewis acids such as SnCl₄) and phenylacetates **178–180** as *t*-butyldimethylsilyl (TBDMS) ethers (stable under acidic conditions). Since 2-hydroxy-, 2-hydroxy-4-methoxy- and 2-hydroxy-3,4-dimethoxyphenylacetic acids are not commercially available, the required phenylacetates **175–177** were prepared via a thallium(III)nitrate (TTN) oxidative rearrangement (McKillop et al., 1973) of 2-benzyloxyacetophenones **172–174** (Scheme 1.12). Debenzylation and silylation afforded the requisite acetates **178–180** in high yields.

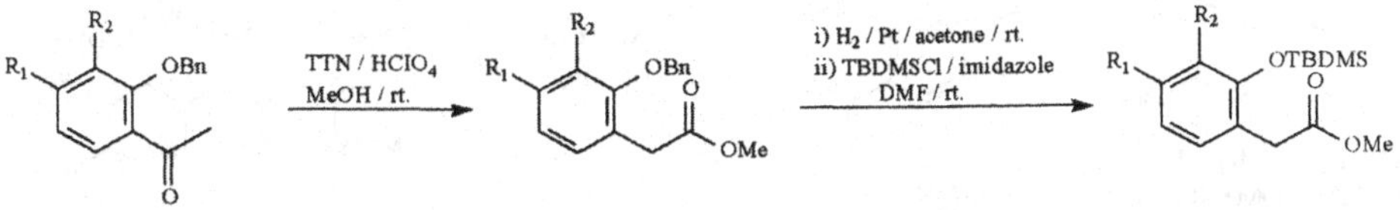

172 $R_1 = H$, $R_2 = H$
173 $R_1 = OMe$, $R_2 = H$
174 $R_1 = OMe$, $R_2 = OMe$

175 $R_1 = H$, $R_2 = H$
176 $R_1 = OMe$, $R_2 = H$
177 $R_1 = OMe$, $R_2 = OMe$

178 $R_1 = H$, $R_2 = H$
179 $R_1 = OMe$, $R_2 = H$
180 $R_1 = OMe$, $R_2 = OMe$

Scheme 1.12 The synthesis of phenylacetates 178–180.

The subsequent condensation between the ester enolates and the benzaldehydes afforded the 2,3-diaryl-3-hydroxypropanoates **183-187** in moderate to good yields (67-78%) (Scheme 1.13). Since acid deprotection of the MOM group led to decomposition (Greene and Wuts, 1991), SnCl$_4$ in the presence of PhCH$_2$SH as nucleophile was utilized as a selective deprotecting agent to afford the 2,3-diaryl-3-benzylsulfanylpropanoates **188-192** in 70-96% yield. Subsequent reduction of **188-292** with LiAlH$_4$ (yield: 77-97%) and ensuing cyclization under Mitsunobu conditions (Mitsunobu, 1981) [PPh$_3$/DEAD (diethylazodicarboxylate)] afforded the 4-benzylsulfanyliso-flavans **198-202** in good overall yields.

Cleavage of the silyl ethers using tetrabutylammonium fluoride (TBAF) on silica (Clark, 1978) gave 4-benzylsulfanyl-2′-hydroxyisoflavans **203–206**, which were converted to the 6a,11a-*cis*-pterocarpans **207–210** in yields of 39–82% using the thiophilic Lewis acids, dimethyl(methylthio)sulfonium tetra-fluoroborate (DMTSF) or silver trifluoromethanesulfonate (CF$_3$SO$_3$Ag) (Trost and Murayama, 1981; Williams et al., 1984; Trost and Sato, 1985) (Scheme 1.14).

Isoflav-3-enes **215** and **216** were obtained via periodate oxidation of the *cis*-and *trans*-4-benzyl-sulfanylisoflavans **201** and **202** followed by thermal elimination of the sulfoxides **213** and **214** (Emerson et al., 1967; Kice and Campbell, 1967; Trost et al., 1976) (Table 1.7). Owing to the instability of isoflav-3-enes **215** and **216**, swift transformation to the corresponding isoflavan-3,4-diols was essential. The commercially available AD-mix-α or -β was not reactive enough to effect asymmetric dihydroxylation. Therefore, treatment of isoflav-3-enes **215** and **216** in CH$_2$Cl$_2$ at −78°C with stoichiometric amounts of OsO$_4$ in the presence of the chiral catalyst dihydroquinine *p*-chlorobenzoate (DHQ-CLB) **211** afforded (-)-(3R,4S)-*syn*-diols **217a** and **218a** in acceptable yields (63–68%) and excellent enantiomeric excesses (>99%) (Kolb et al., 1994a; Pinard et al., 1998). The (+)-(3S,4R)-*syn*-diols **217b** and **218b** were similarly obtained by using dihydroquinidine *p*-chlorobenzoate (DHQD-CLB) **212** as chiral ligand.

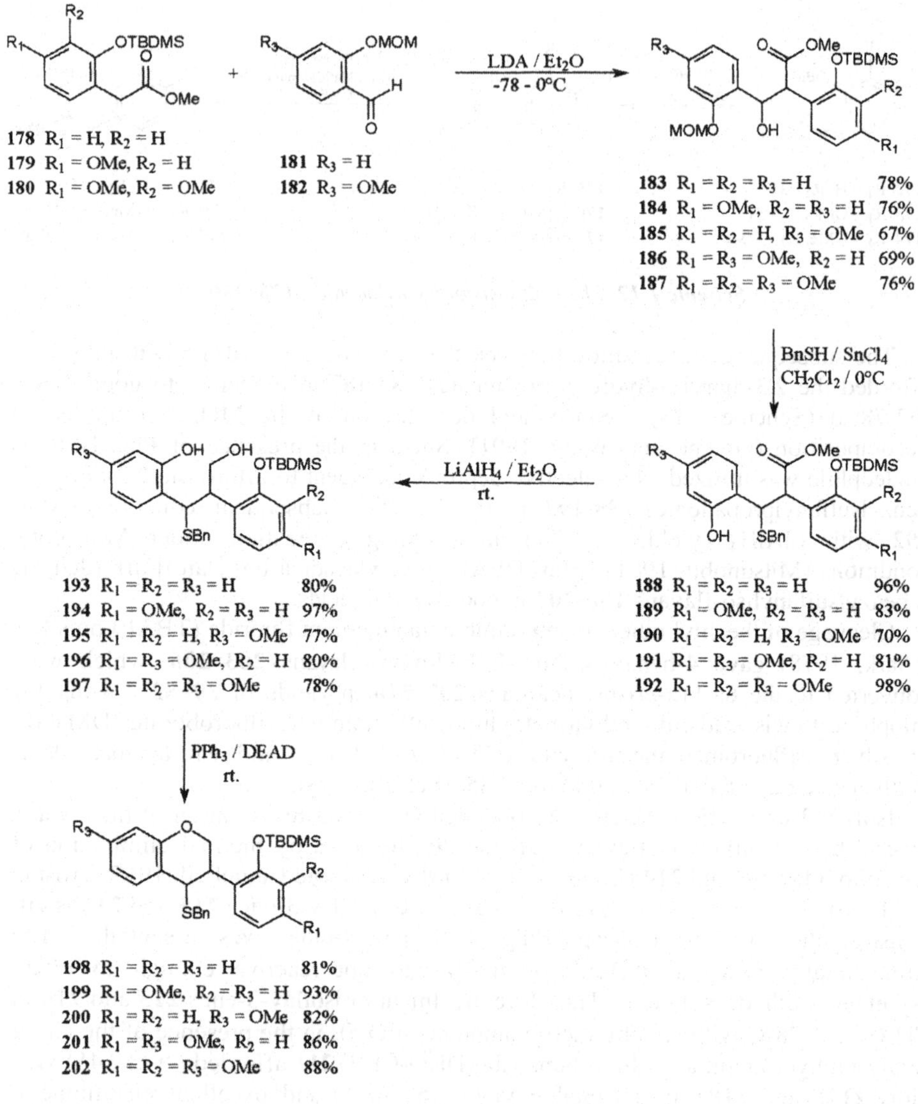

178 $R_1 = H, R_2 = H$
179 $R_1 = OMe, R_2 = H$
180 $R_1 = OMe, R_2 = OMe$

181 $R_3 = H$
182 $R_3 = OMe$

LDA / Et$_2$O
-78 - 0°C

183 $R_1 = R_2 = R_3 = H$ 78%
184 $R_1 = OMe, R_2 = R_3 = H$ 76%
185 $R_1 = R_2 = H, R_3 = OMe$ 67%
186 $R_1 = R_3 = OMe, R_2 = H$ 69%
187 $R_1 = R_2 = R_3 = OMe$ 76%

BnSH / SnCl$_4$
CH$_2$Cl$_2$ / 0°C

193 $R_1 = R_2 = R_3 = H$ 80%
194 $R_1 = OMe, R_2 = R_3 = H$ 97%
195 $R_1 = R_2 = H, R_3 = OMe$ 77%
196 $R_1 = R_3 = OMe, R_2 = H$ 80%
197 $R_1 = R_2 = R_3 = OMe$ 78%

LiAlH$_4$ / Et$_2$O
rt.

188 $R_1 = R_2 = R_3 = H$ 96%
189 $R_1 = OMe, R_2 = R_3 = H$ 83%
190 $R_1 = R_2 = H, R_3 = OMe$ 70%
191 $R_1 = R_3 = OMe, R_2 = H$ 81%
192 $R_1 = R_2 = R_3 = OMe$ 98%

PPh$_3$ / DEAD
rt.

198 $R_1 = R_2 = R_3 = H$ 81%
199 $R_1 = OMe, R_2 = R_3 = H$ 93%
200 $R_1 = R_2 = H, R_3 = OMe$ 82%
201 $R_1 = R_3 = OMe, R_2 = H$ 86%
202 $R_1 = R_2 = R_3 = OMe$ 88%

Scheme 1.13 *Direct synthesis of 4-benzylsulfanylisoflavans **198–202** via condensation of phenylacetates with bezaldehydes.*

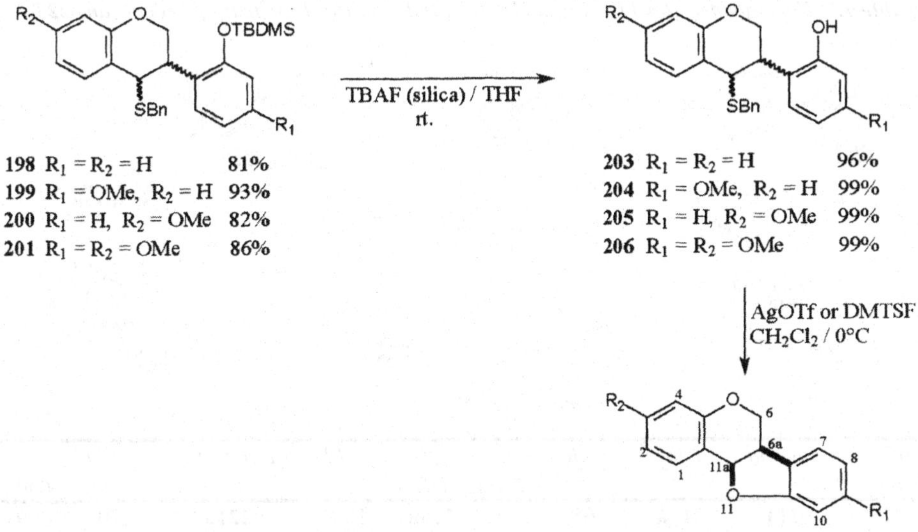

198 R$_1$ = R$_2$ = H   81%
199 R$_1$ = OMe, R$_2$ = H   93%
200 R$_1$ = H, R$_2$ = OMe   82%
201 R$_1$ = R$_2$ = OMe   86%

TBAF (silica) / THF
rt.

203 R$_1$ = R$_2$ = H   96%
204 R$_1$ = OMe, R$_2$ = H   99%
205 R$_1$ = H, R$_2$ = OMe   99%
206 R$_1$ = R$_2$ = OMe   99%

AgOTf or DMTSF
CH$_2$Cl$_2$ / 0°C

207 R$_1$ = R$_2$ = H   82%
208 R$_1$ = OMe, R$_2$ = H   39%
209 R$_1$ = H, R$_2$ = OMe   57%
210 R$_1$ = R$_2$ = OMe   50%

Scheme 1.14 *Synthesis of (6a,11a)-cis-pterocarpans 207-210.*

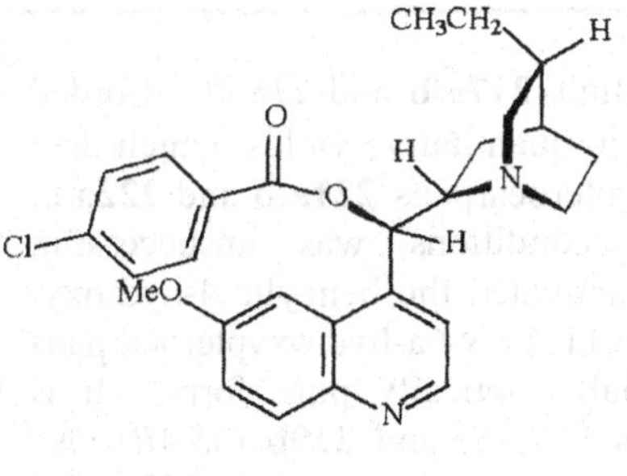

211 (DHQ-CLB)

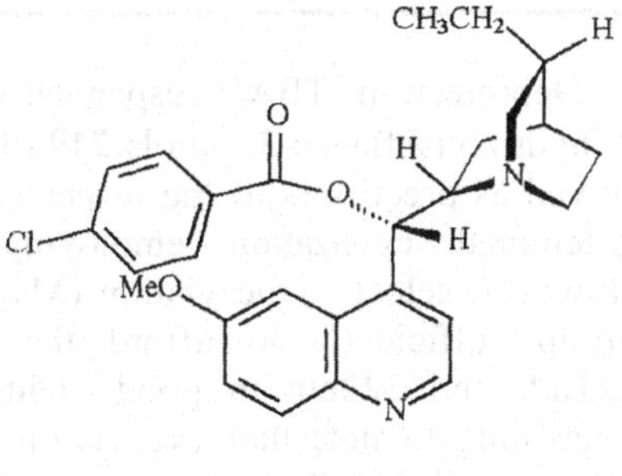

212 (DHQD-CLB)

Table 1.7 *Formation of (6a,11a)-cis- 221a/b, 222a,b and (6a,11b)-trans pterocarpans 223a*

Isoflav-3-ene	Ligand	Diol	yield (%)	ee (%)	2'-OH	yield (%)	Pterocarpan	yield (%)	ee (%)
215	211	217a (3R,4S)	65	>99	219a	100	221a (6aR,11aR)	70	>99
	212	217b (3R,4S)	68	>99	219b	100	221b (6aS,11aS)	75	>99
					219a	100	223a (6aR,11aS)	10	>99
					219b	100	223b (6aS,11aR)	9	>99
216	211	218a (3R,4S)	66	>99	220a	100	222a (6aR,11aR)	75	>99
	212	218b (3R,4S)	63	>99	220b	100	222b (6aS,11aS)	73	>99

Deprotection (TBAF suspended on silica) of diols **217a/b** and **218a/b** afforded 2'-hydroxyisoflavan-3,4-diols **219a/b** and **220a/b** in quantitative yields, which then served as precursors to the respective 6a-hydroxypterocarpans **221a/b** and **222a/b**. Attempted cyclization employing Mitsunobu conditions was unsuccessful. However, selective mesylation (Ms$_2$O, pyridine) activated the benzylic 4-hydroxyl group sufficiently to afford the requisite (6a,11a)-*cis*-6a-hydroxypterocarpans **221a/b** and **222a/b** in good yields and essentially optically pure form. It is interesting to note that cyclization of diols **219a** (3R,4S) and **219b** (3S,4R) also afforded the (6aR,11aS)- and (6aS,11aR)-*trans*-6a-hydroxyptercarpans **223a** and **223b**, respectively, as minor products (9–10% yield) and was the first report on the formation of the configurationally hindered 6a,11a-*trans*-analogues.

In all reported pterocarpan syntheses, formation of the six-membered B-ring invariably precedes closure of the five-membered C-ring. Once the B-ring is formed, Dreiding models indicate that it becomes virtually impossible to close the C-ring in a configuration other than the 6a,11a-*cis*-form. It was envisaged that the reversal of the order of cyclization, i.e., initial C-ring formation followed by B-ring closure, may provide synthetic access to the hitherto unknown 6a,11a-*trans*-pterocarpans. Thus, aldol condensation between the MOM-protected phenylacetate **224** and benzaldehyde **181**, using LDA for enolate generation, afforded the 2,3-diaryl-3-hydroxypropanoate **225** in 73% yield (Scheme 1.15).

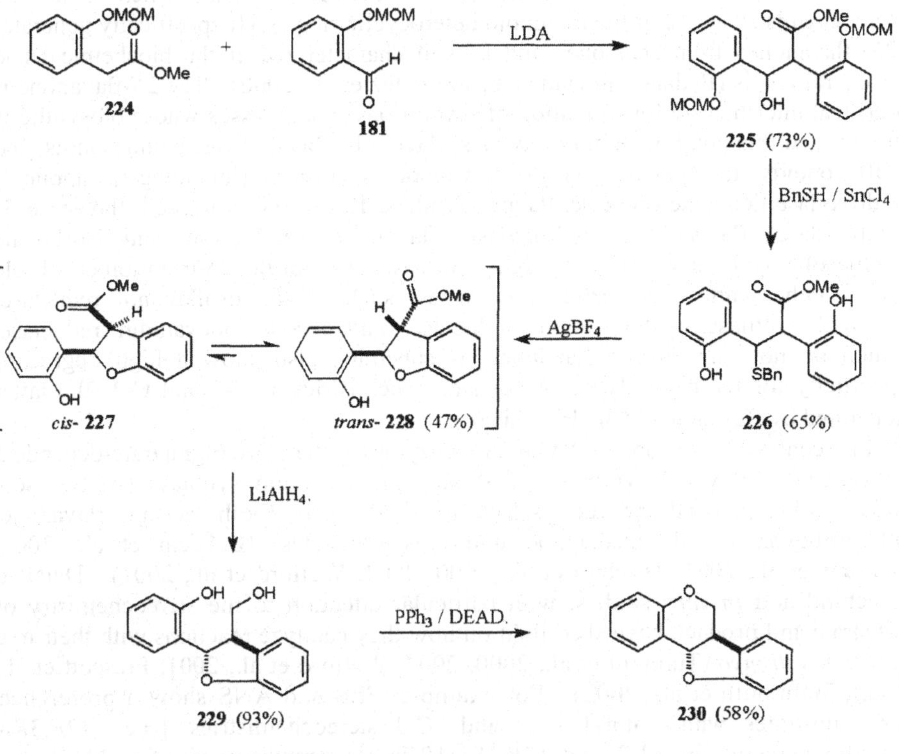

Scheme 1.15 *Synthesis of (6a,11a)-trans-pterocarpan 230.*

Deprotection of the acetal functionality of **225** using $SnCl_4/PhCH_2SH$ afforded 2,3-diaryl-3-benzylsulfanylpropanoate **226** in 65% yield. Cyclization ($AgBF_4$) of **226** to first form the pterocarpan C/D-ring system, afforded the thermodynamically more stable *trans*-fused 2,3-disubstituted dihydro-benzofuran **228** (47%; $J_{2,3} = 8.5$ Hz). Subsequent reduction ($LiAlH_4$) gave the primary alcohol **229** (93%), which was converted under Mitsunobu cyclization conditions into the 6a,11a-*trans*-pterocarpan **230** ($J_{6a,11a} = 13.5$ Hz in 58% yield).

4. ENZYMATIC STEREOSPECIFIC BIOSYNTHESIS OF FLAVONOIDS

Most enzymes of flavonoid biosynthesis are highly stereoselective and/or stereospecific; however, for many enzymes this claim rests on only one or a few published reports (Table 1.8) (Forkmann and Heller, 1999).

Flavonoids are synthesized via the phenylpropanoid pathway, beginning with the deamination of phenylalanine by the enzyme L-phenylalanine ammonia-lyase (PAL). PAL is specific for the naturally occurring L-isomer of phenylalanine; D-phenylalanine is not a substrate (Koukol and Conn, 1961).

Perhaps the most stereochemically important reaction of flavonoid biosynthesis is that catalyzed by chalcone–flavanone isomerase (CHI), which sets the stereochemistry at C-2 of the flavonoid heterocyclic ring. CHI specifically generates (2S)-flavanones from chalcones and is well characterized at the biochemical and structural levels (Bednar and Hadcock, 1988; Jez et al., 2000). The 2S-flavanone is a critical intermediate for formation of several flavonoid classes whose biosynthesis branches at this point, including flavones, flavonols, flavan-4-ols, anthocyanins, and isoflavonoids, and enzymes that use flavanone as substrate (including flavanone 2-hydroxylase/licodione synthase, flavone synthase II, flavone synthase I, flavanone 3-hydroxylase, flavonoid 3'-hydroxylase, flavanone 4-reductase, and isoflavone synthase) have been shown to be highly stereospecific for the 2S-enantiomer (Table 1.8). Other farther downstream enzymes, such as dihydroflavonol reductase, flavonol synthase, anthocyanidin reductase, and leucoanthocyanidin reductase, which do not directly use flavanone as substrate, also show a high degree of specificity for the naturally occurring stereochemistries at C-2 and C-3 (the latter generated by flavanone-3β-hydroxylase).

In contrast to most flavonoid enzymes, the 2-oxoglutarate-dependent dioxygenases flavonol synthase (FLS) and anthocyanidin synthase (ANS) show broad substrate and product selectivities *in vitro* (both accept flavanone, dihydroflavonol, and leucoanthocyanidin as substrates) (Lukacin et al., 2003; Martens et al., 2003; Turnbull et al., 2000, 2004; Welford et al., 2001). Detailed structural and *in vitro* studies, with particular attention to the stereochemistry of substrate and product, have shed light on how they catalyze reactions with their true substrates *in vivo* (Turnbull et al., 2000, 2004; Welford et al., 2001; Prescott et al., 2002; Wilmouth et al., 2002). For example, FLS and ANS show a preference for substrates with natural C-2 and C-3 stereochemistries [(i.e. (2R,3R)-dihydroquercetin for FLS and (2R,3S, 4R/S)- leucoanthocyanin for ANS], but hydroxylate both (2R)- and (2S)-naringenin equally well *in vitro*, which suggests that the C-3 hydroxyl group is important in biasing substrate selectivity (Turnbull et al., 2004).

Table 1.8 *Stereoselective and/or Specific Enzymes of Flavonoid Biosynthesis*

Enzyme	Stereoselectivity	Stereospcificity	Key references
Phenylalanine ammonia lyase	L-phenylalanine		Koukol and Conn, 1961
Chalcone isomerase		(2S)-flavanone	Bednar and Hadcock, 1988; Hahlbrock et al., 1970; Jez et al., 2000
Flavanone 2-hydroxylase (licodione synthase)	(2S)-flavanone		Akashi et al., 1998; Otani et al., 1994
Flavanone 4-reductase	(2S)-flavanone	(2S, 4R)-flavan-4-ol	Fischer et al., 1988; Stich and Forkmann, 1988
Flavone synthase	(2S)-flavanone		Britsch, 1990; Kochs et al., 1987; Martens et al., 2001; Sutter et al., 1975
Flavone 3β-hydroxylase	(2S)-flavanone	(2R,3R)-dihydroflavonol	Britsch and Grisebach, 1986; Britsch et al., 1992
Flavonoid 3'-hydroxylase	(2S)-flavanone, (2R,3R)-dihydroflavonol		Fritsch and Grisebach, 1975; Hagmann et al., 1983
Flavonol synthase	(2R,3R)-dihydroflavonol		Lukacin et al., 2003; Martens et al., 2003; Prescott et al., 2002; Turnbull et al., 2004
Dihydroflavonol 4-reductase	(2R,3R)-dihydroflavonol	(2R, 3S, 4S)-flavan-2,3-*trans*-3,4-*cis*-diol	Stafford and Lester, 1982, 1984, 1985
Leucoanthocyanidin 4-reductase	(2R, 3S, 4S)-flavan-2,3-*trans*-3,4-*cis*-diol	(2R, 3S)-flavan-3-ol	Stafford and Lester, 1984; Tanner et al., 2003
Anthocyanidin synthase	(2R, 3S, 4S)-flavan-2,3-*trans*-3,4-*cis*-diol		Turnbull et al., 2004; Wilmouth et al., 2002
Anthocyanidin reductase		(2R, 3R)-flavan-3-ol	Xie et al., 2003
Isoflavone synthase	(2S)-flavanone		Hagmann and Grisebach, 1984; Kochs and Grisebach, 1986

Table 1.8 (continued)

Enzyme	Stereoselectivity	Stereospecificity	Key references
Isoflavone reductase		(2R)-isoflavanone	Fischer et al., 1990a; Paiva et al., 1991, 1994
Vestitone reductase	(2R)-isoflavanone		Fischer et al., 1990b; Guo and Paiva, 1995
7,2'-Dihydroxy-4'-methoxyisoflavanol dehydratase		(-)-medicarpin	Guo et al., 1994
3,9-dihydroxypterocarpan 6a-hydroxylase	(6aS, 11aS)-3,9-dihydroxypterocarpan	(6aS, 11aS)-3,6a,9-trihydroxypterocarpan	Hagmann et al., 1984; Schopfer et al., 1998
6a-Hydroxymaackiain-3-O-methyltransferase	(+)-6a-hydroxymaackiain	(+)-pisatin	Preisig et al., 1989; Wu et al., 1997

The flavan-3-ols (+)-catechin and (-)-epicatechin (Figure 1.1) form the building blocks of proanthocyanidins (condensed tannins), a class of molecules of considerable interest in view of their impacts on animal health (Dixon at al., 2005). The C-2 and C-3 stereochemistries of (+)-catechin (2,3-*trans*) are identical to those of intermediates in the flavonoid pathway, and a pathway leading from (2R, 3S, 4S)-leucoanthocyanidin to (+)-catechin, catalyzed by leucoanthocyanidin reductase (LAR), has been demonstrated and confirmed by the cloning of a leucoanthocyanidin reductase from the tannin-rich forage legume *Desmodium uncinatum* (Stafford and Lester, 1984, 1985; Tanner et al., 2003). LAR is a member of the Reductase–Epimerase–Dehydrogenase family of proteins, whose members include isoflavone reductase and related homologues (Min et al., 2003).

For many years, the origin of the 2,3-*cis* (-)-epicatechin units in proanthocyanidins was a mystery. This problem was resolved by the demonstration that the pathway leading to (-)-epicatechin proceeds from leucocyanidin through cyanidin in reactions catalyzed by ANS and anthocyanidin reductase (ANR) (Xie et al., 2003). ANR, an enzyme with weak sequence homology to dihydroflavonol reductase, can introduce the 2,3-*cis* stereochemistry by acting on an achiral intermediate (anthocyanidin). Mechanisms have been proposed for this reaction, and it is possible that other ANR-like enzymes might exist with the ability to introduce alternate stereochemistries (Xie et al., 2004).

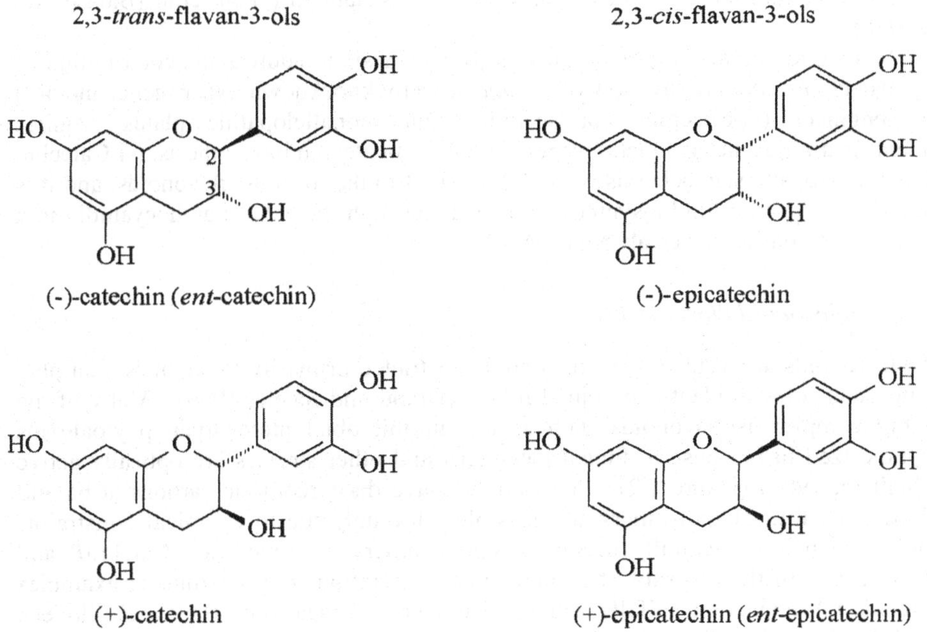

2,3-*trans*-flavan-3-ols

2,3-*cis*-flavan-3-ols

(-)-catechin (*ent*-catechin)

(-)-epicatechin

(+)-catechin

(+)-epicatechin (*ent*-epicatechin)

Figure 1.1 *Flavan-3-ol isomers.*

5. STEREOCHEMISTRY RELATED TO BIOLOGICAL ACTIVITY

5.1. Catechin

The Asian native *Centaurea maculosa* (spotted knapweed) has displaced native weeds and crops throughout the western United States. Contributing to the invasiveness of this exotic is the secretion of the phytotoxic *trans*-flavan-3-ol (-)-catechin from its roots (Bais et al., 2002) (Figure 1.1). Both enantiomers of catechin are present in root exudates of *C. maculosa*; however, only (-)-catechin had allelopathic (phytotoxic) activity. Interestingly, (+)-catechin (but not (-)-catechin) displayed antibacterial activity against several root pathogens, which suggests that secretion of a racemic mixture may simultaneously protect *C. maculosa* roots against microbial pathogens and weaken roots of neighboring plants (Bais et al., 2002). When the phytotoxicity of catechin was examined in more detail, only the (-)-enantiomer elicited generation of reactive oxygen species and calcium-signaling events in roots of susceptible species (Bais et al., 2003a). Additional studies with the *cis*-flavan-3-ols (+)-epicatechin and (-)-epicatechin showed that (+)-epicatechin, like (-)-catechin, inhibited root and shoot differentiation and seed germination of several of the plants examined, while (-)-epicatechin did not show inhibition (Bais et al., 2003b). Both (-)-catechin and (+)-epicatechin are of the 2S configuration, which suggests that the stereochemistry at C-2 is important for allelopathic activity. Interestingly, (+)-epicatechin also was effective at inhibiting *C. maculosa,* which is resistant to (-)-catechin (Bais et al., 2003b).

Metabolic engineering of (-)-catechin biosynthesis to address the contribution of (-)-catechin to the invasiveness of *C. maculosa* by knockdown experiments and also to engineer (-)-catechin root secretion into nonallelopathic plants requires identification of the enzyme(s) responsible for (-)-catechin biosynthesis. (-)-Catechin has the opposite stereochemistry at C-2 and C-3 to that of most flavonoids, and it is likely that (-)-catechin biosynthesis proceeds through the achiral anthocyanidin in a reaction similar to that catalyzed by ANR.

5.2. Isoflavonoid Phytoalexins

Isoflavonoids are a subclass of flavonoids, restricted primarily to legumes, that play important roles in plant and animal health (Dixon and Steele, 1999). Many of the more complex isoflavonoids such as the antimicrobial pterocarpan phytoalexins synthesized in response to fungal pathogens and other stresses are optically active (Ingham, 1982) (Figure 1.2). Pterocarpans have diastereomeric carbons at 6a and 11a; thus, four stereoisomers are possible, although due to chemical constraints only one pair of naturally occurring stereoisomers is found (i.e., 6aR;11aR and 6aS;11aS). In the majority of legumes the (-)-enantiomers predominate; examples include (-)-medicarpin (alfalfa, chickpea, clover), (-)-maackiain (chickpea, clover), and (-)-glycinol (soybean) (Ingham, 1982).

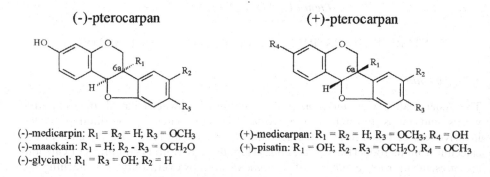

(-)-pterocarpan   (+)-pterocarpan

(-)-medicarpin: $R_1 = R_2 = H$; $R_3 = OCH_3$
(-)-maackain: $R_1 = H$; $R_2 - R_3 = OCH_2O$
(-)-glycinol: $R_1 = R_3 = OH$; $R_2 = H$

(+)-medicarpan: $R_1 = R_2 = H$; $R_3 = OCH_3$; $R_4 = OH$
(+)-pisatin: $R_1 = OH$; $R_2 - R_3 = OCH_2O$; $R_4 = OCH_3$

Figure 1.2 *Structure and stereochemistry of isoflavonoid pterocarpan phytoalexins.*

Two well-known examples of pterocarpan phytoalexins with opposite stereochemistry include (+)-medicarpin from peanut and (+)-pisatin from pea. Although molecular models of (+)- and (-)-pterocarpans are nearly superimposable (with the exception of the B ring), the absolute configuration of these phytoalexins can be an important factor in plant–pathogen interactions. Studies on the toxicity of maackiain and pisatin enantiomers to phytopathogenic fungi demonstrated that in general fungi were more sensitive to pterocarpans of the opposite stereochemistry to that found in their host plant. For example, fungal pathogens of (-)-maackiain-producing plants were more sensitive to (+)-maackiain (Delserone et al., 1992). It has been suggested that this differential sensitivity of fungal pathogens to phytoalexins of the opposite stereochemistry may be exploited for disease control by engineering plants to synthesize enantiomers of the opposite stereochemistry. Further support for this strategy comes from work on the detoxification enzymes of phytopathogenic fungi, which convert phytoalexins to less toxic forms by demethylation, hydroxylation, or reductive cleavage (VanEtten et al., 1989). These enzymes often display a high degree of stereospecificity for their host's phytoalexins. For example, an isolate of *Nectria haematococca* specifically hydroxylated (-)-maackiain and (-)-medicarpin but not their (+)-enantiomers (VanEtten et al., 1983). Similarly, a purified pterocarpan hydroxylase from the chickpea pathogen *Ascochyta rabiei* hydroxylated (-)-maackiain and (-)-medicarpin but not (+)-maackiain (Tenhaken et al., 1991). Pisatin demethylase from *N. haematococca* and *Ascochyta pisi* preferred (+)-pisatin over (-)-pisatin (53–58% of activity), although the demethylase from *Mycosphaerella pinodes* and *Phoma pinodella* had the same activity on both enantiomers (George and VanEtten, 2001). Furthermore, (+)-pisatin but not its (-)-enantiomer induced pisatin demethylase activity in *N. haematococca* (VanEtten et al., 1989).

Although the biosynthetic pathway leading to (-)-medicarpin and related compounds is well characterized, the *in vivo* enzymatic routes to (+)-pterocarpans remain unknown. In (-)-pterocarpan biosynthesis the key enzyme determining the stereochemistry of the 6a (and 11a) positions of the pterocarpan is isoflavone reductase (IFR), which stereospecifically reduces 2'-hydroxyisoflavone to (3R)-isoflavanone (Fischer et al., 1990a; Paiva et al., 1991). This (3R)-isoflavanone is further reduced to isoflavanol, then dehydrated to pterocarpan with retention of stereochemistry (Dixon, 1999). An initial hypothesis for the biosynthesis of (+)-pisatin suggested that pea IFR would specifically generate (3S)-isoflavanone. However, cloning and expression of pea IFR in *Escherichia coli* later showed that this reductase produced (3R)-isoflavanone, identical to that of alfalfa IFR (Paiva at al., 1994). The 6a-hydroxymaackiain 3-O-methyltransferase catalyzing the final step in the biosynthesis of (+)-pisatin is specific for (+)-6a-hydroxymaackiain (Preisig et al., 1989; Wu et al., 1997), suggesting that the reversal of stereochemistry occurs between reduction of isoflavanone and formation of (+)-6a-hydroxymaackiain. Possible mechanisms for the synthesis of (+)-pterocarpans include formation of an isoflav-3-ene intermediate or epimerase-mediated inversion of configuration. Support for the latter hypothesis comes from an unpublished report that in peanut, which synthesizes (+)-medicarpin, IFR produces (3R)-vestitone, but that the following enzyme vestitone reductase accepts only (3S)-vestitone (Guo and Paiva, 1995).

6. REFERENCES

Adam, W., Rao, P. B., Degen, H-G., and Saha-Möller, C. R., 2001, Asymmetric Weitz–Scheffer epoxidation of conformationally flexible and fixed enones with sterically demanding hydroperoxides mediated by optically active phase-transfer catalyst, *Tetrahedron: Asymmetry* **12**: 121-125.

Adger, B. M., Barkley, J. V., Bergeron, S., Cappi, M. W., Flowerdew, B. E., Jackson, M. P., McCague, R., Nugent, T. C., and Roberts, S. M., 1997, Improved procedure for Julia–Colonna asymmetric epoxidation of α,β-unsaturated ketones: total synthesis of diltiazem and Taxol side-chain, *J Chem Soc Perkin Trans 1* 3501-3507.

Akashi, T., Aoki, T., and Ayabe, S., 1998, Identification of a cytochrome P450 cDNA encoding (2S)-flavanone 2-hydroxylase of licorice (*Glycyrrhiza echinata* L.; Fabaceae) which represents licodione synthase and flavone synthase II, *FEBS Lett* **431**: 287-290.

Amberg, W., Bennani, Y. L., Chadha, R. K., Crispino, G. A., Davis, W. D., Hartung, J., Jeong, K-S., Ogino, Y., Shibata, T., and Sharpless, K. B., 1993, Syntheses and crystal structures of the cinchona alkaloid derivatives used as ligands in the osmium-catalyzed asymmetric dihydroxylation of olefins, *J Org Chem* **58**: 844-849.

Arai, S., Tsuge, H., Oku, M., Miura, M., and Shioiri, T., 2002, Catalytic asymmetric epoxidation of enones under phase-transfer catalyzed conditions, *Tetrahedron* **58**: 1623-1630.

Arnaudinaud, V., Nay, B., Nuhrich, A., Deffieux, G., Merillon, J-M., Monti, J-P., and Vercauteren, J., 2001a, Total synthesis of isotopically labelled flavonoids. Part 3: ^{13}C-labelled (-)-procyanidin B3 from 1-[^{13}C]-acetic acid, *Tetrahedron Lett* **42**: 1279-1281.

Arnaudinaud, V., Nay, B., Verge, S., Nuhrich, A., Deffieux, G., Merillon, J-M., Monti, J-P., and Vercauteren, J., 2001b, Total synthesis of isotopically labelled flavonoids. Part 5: Gram-scale production of ^{13}C-labelled (-)-procyanidin B3, *Tetrahedron Lett* **42**: 5669-5671.

Augustyn, J. A. N., Bezuidenhoudt, B. C. B., and Ferreira, D., 1990a, Enantioselective synthesis of flavonoids. Part I. Poly-oxygenated chalcone epoxides, *Tetrahedron* **46**: 2651-2660.

Augustyn, J. A. N., Bezoudenhoudt, B. C. B., Swanepoel, A., Ferreira, D., 1990b. Enantioselective synthesis of flavonoids, Part 2. Poly-oxygenated α-hyroxydihydrochalcones and circular dichroic assessment of their absolute configuration. *Tetrahedron* **46**, 4429-4442.

Bais, H. P., Vepachedu, R., Gilroy, S., Callaway, R. M., and Vivanco, J. M., 2003a, Allelopathy and exotic plant invasion: from molecules and genes to species interactions, *Science* **301**: 1377-1380.

Bais, H. P., Walker, T. S., Kennan, A. J., Stermitz, F. R., and Vivanco, J. M., 2003b, Structure-dependent phytotoxicity of catechins and other flavonoids: flavonoid conversions by cell-free protein extracts of *Centaurea maculosa* (spotted knapweed) roots, *J Agric Food Chem* **51**: 897-901.

Bais, H. P., Walker, T. S., Stermitz, F. R., Hufbauer, R. A., and Vivanco, J. M., 2002, Enantiomeric-dependent phytotoxic and antimicrobial activity of (±)-catechin. A rhizosecreted racemic mixture from spotted knapweed, *Plant Physiol* **128**: 1173-1179.

Bakó, P., Czinege, E., Bakó, T., Czugler, M., and Tóke, L., 1999, Asymmetric C-C bond forming reactions with chiral crown catalysts derived from D-glucose and D-galactose, *Tetrahedron: Asymmetry* **10**: 4539-4551.

Bakó, T., Bakó, P., Keglevich, G., Bombicz, P., Kubinyi, M., Pál, K., Bodor, S., Makó, A., and Tóke, L., 2004, Phase-tranfer catalyzed asymmetric epoxidation of chalcones using crown ethers derived from D-glucose, D-galactose and D-mannitol. *Tetrahedron: Asymmetry* **15**: 1589-1595.

Banfi, S., Colonna, S., Molinari, H., Julia, S., and Guixer, J., 1984, Asymmetric epoxidation of electron-poor olefins – V. Influence of stereoselectivity of the structure of poly-α-amino acids used as catalysts, *Tetrahedron* **40**: 5207-5211.

Barrett, A. G. M., Bezuidenhoudt, B. C. B., Howell, A. R., Lee, A. C., and Russell, M. A., 1989, Redox glycosidation *via* thionoester intermediates, *J Org Chem* **54**: 2275-2277.

Bednar, R. A., and Hadcock, J. R., 1988, Purification and characterization of chalcone isomerase from soybeans, *J Biol Chem*, **263**: 9582-9588.

Beltrami, E., De Bernardi, M., Fronza, G., Mellerio, G., Vidari, G., and Vita-Finzi, P., 1982, Coatline A and B, two C-glucosyl-α-hydroxydihydrochalcones from *Eysenhardtia polystachya*, *Phytochem* **21**: 2931-2933.

Bentley, P. A., Bergeron, S., Cappi, M. W., Hibbs, D. E., Hursthouse, M. B., Nugent, T. C., Pulido, R., Roberts, S. M., and Wu, L. E., 1997, Asymmetric epoxidation of enones employing polymeric α-amino acids in non-aqueous media, *Chem Commun* 739-740.

Bevan, C. W. L., Birch, A. J., Moore, B., and Mukerjee, S. K., 1964, A partial synthesis of (±)-pisatin. Some remarks on the structure and reactions of pterocarpin, *J Chem Soc Suppl* 5991-5995.

Bezuidenhoudt, B. C. B., Brandt, E. V., and Roux, D. G., 1981, A novel α-hydroxydihydrochalcone from the heartwood of *Pterocarpus angolensis* D.C.: absolute configuration, synthesis, photochemical transformations, and conversion into α-methyldeoxybenzoins, *J Chem Soc, Perkin Trans 1* 263-269.

Bezuidenhoudt, B. C. B., Swanepoel, A., Augustyn, J. A. N., and Ferreira, D., 1987, The first enantioselective synthesis of poly-oxygenated α-hydroxydihydrochalcones and circular dichroic assessment of their absolute configuration, *Tetrahedron Lett* **28**: 4857-4860.

Bhakuni, D., Bittner, M., Silva, M., and Sammes, P. G., 1973, Nubigenol. α-Hydroxydihydrochalcone from *Podocarpus nubigena*, *Phytochem*, **12**: 2777-2779.

Bougauchi, M., Watanabe, S., Arai, T., Sasai, H., and Shibasaki, M., 1997, Catalytic asymmetric epoxidation of α, β-unsaturated ketones promoted by lanthanoid complexes, *J Am Chem Soc* **119**: 2329-2330.

Britsch, L., 1990, Purification and characterization of flavone synthase I, a 2-oxoglutarate-dependent desaturase, *Arch Biochem Biophys* **282**: 152-160.

Britsch, L., and Grisebach, H., 1986, Purification and characterization of (2S)-flavanone 3-hydroxylase from *Petunia hybrida*, *Eur J Biochem* **156**: 569-577.

Britsch, L., Ruhnau-Brich, B., and Forkmann, G., 1992, Molecular cloning, sequence analysis, and *in vitro* expression of flavanone 3ß-hydroxylase from *Petunia hybrida*, *J Biol Chem* **287**: 5380-5387.

Carde, L., Davies, H., Geller, T. P., and Roberts, S. M., 1999, PaaSicats: powerful catalysts for asymmetric epoxidation of enones. Novel syntheses of α-arylpropanoic acids including (S)-fenoprofen, *Tetrahedron Lett* **40**: 5421-5424.

Cardillo, G., D'Amico, A., Orena, M., and Sandri, S., 1988, Diastereoselective alkylation of 3-acylimidazolidin-2-ones: synthesis of (R)- and (S)-lavandulol, *J Org Chem* **53**: 2354-2356.

Cardillo, G., Orena, M., Romero, M., and Sandri, S., 1989, Enantioselective synthesis of 2-benzyloxy alcohols and 1,2-diols *via* alkylation of chiral glycolate imides. A convenient approach to optically active glycerol derivatives, *Tetrahedron* **45**: 1501-1508.

Chen, R., Qian, C., and De Vries, J. G., 2001, Highly efficient enantioselective epoxidation of α, β-enones catalyzed by cheap lanthanum and gadolinium alkoxides, *Tetrahedron* **57**: 9837-9842.

Chini, M., Crotti, P., Gardelli, C., and Macchia, F., 1992, Metal salt-promoted alcoholysis of 1,2-epoxides, *Synlett* 673-676.

Claisen, L., and Claparède, A., 1881, Condensationen von ketonen mit aldehyden, *Chem Ber* **14**: 2460-2468.

Clark, J. H., 1978, Drifluor reagents: non-hygroscopic sources of the fluoride ion, *J Chem Soc, Chem Commun* 789-91.

Close, W. J., 1950, The conformation of the ephedrines, *J Org Chem* **15**: 1131-1134.

Coffey, P. E., Drauz, K-H., Roberts, S. M., Skidmore, J., and Smith, J. A., 2001, β-Peptides as catalysts: poly-β-leucine as a catalyst for the Julia–Colonna asymmetric epoxidation of enones, *Chem Commun* 2330-2331.

Colonna, S., Molinari, H., Banfi, S., Julia, S., Masana, J., and Alvarez, A., 1983, Synthetic enzymes - 4. Highly enantioselective epoxidation by means of polyamino acids in a triphase system: influence of structural variations within the catalysts, *Tetrahedron* **39**: 1635-1641.

Colonna, S., Banfi, S., and Papagni, A., 1985, Catalytic asymmetric epoxidation in the presence of cyclodextrins. Part 2, *Gazz Chim Ital* **115**: 81-83.

Colonna, S., and Manfredi, A., 1986, Catalytic asymmetric Weitz–Scheffer reaction in the presence of bovine serum albumin, *Tetrahedron Lett* **27**: 387-390.

Corey, E. J., and Zhang, F-Y., 1999, Mechanism and conditions for highly enantioselective epoxidation of α,β enones using charge accelerated catalysis by a rigid quaternary ammonium salt, *Org Lett* **1**: 1287-1290.

Cruickshank, I. A. M., and Perrin, D. R., 1960, Isolation of a phytoalexin from *Pisum sativum*, *Nature* **187**: 799-800.

Daikai, K., Kamaura, M., and Inanaga, J., 1998, Remarkable ligand effect on the enantioselectivity of the chiral lanthanum complex-catalyzed asymmetric epoxidation of enones, *Tetrahedron Lett* **39**: 7321-7322.

Dean, F. M., and Podimuang, V., 1965, The course of the Algar–Flynn–Oyamada (A.F.O.) reaction, *J Chem Soc* 3978-3987.

Delserone, L. M., Matthews, D. E., and VanEtten, H. D., 1992, Differential toxicity of enantiomers of maackiain and pisatin to phytopathogenic fungi, *Phytochem* **31**: 3813-3819.

Dixon, R. A., 1999, Isoflavonoids: biochemistry, molecular biology and biological functions, In U. Sankawa (Ed.), *Comprehensive Natural Products Chemistry* (Vol. 1, pp. 773-823), Elsevier.

Dixon, R. A., and Steele, C. L., 1999, Flavonoids and isoflavonoids—a gold mine for metabolic engineering, *Trends Plant Sci* **4**: 394-400.

Dixon, R. A., Xie, D.-Y., and Sharma, S. B., 2005, Proanthocyanidins—a final frontier in flavonoid research? *New Phytol* **165**: 9-28.

Donnelly, J. A., and Doran, H. J., 1975, Chalcone dihalides. VII. Course of the cyclization of 2'-hydroxy-6'-methoxyl derivatives, *Tetrahedron* **31**: 1565-1569.

Donnelly, J. A., Fox, M. J., and Sharma, T. C., 1979, α-Halogenoketones. XII. Extension of the Rasoda synthesis of dihydroflavonols, *Tetrahedron* **35**: 1987-1991.

Donnelly, J. A., and Emerson, G. M., 1990, Amine-effected cyclization of chalcone dihalides to aurones, *Tetrahedron* **46:** 7227-7236.

Donnelly, J. A., and Higginbotham, C. L., 1990, Flavone formation in the Wheeler aurone synthesis, *Tetrahedron* **46:** 7219-7226.

Drewes, S. E., Malissar, D. G. S., and Roos, G. H. P., 1993, Ephedrine-derived imidazolidin-2-ones. Broad utility chiral auxiliaries in asymmetric synthesis, *Chem Ber* **126:** 2663-2673.

Elston, C. L., Jackson, R. F. W., Macdonald, S. J. F., and Murray, P. J., 1997, Asymmetric epoxidation of chalcones with chirally modified lithium and magnesium *tert*-butyl peroxides, *Angew Chem* (Int. Ed. Engl.) **36:** 410-412.

Emerson, D. W., Craig, A. P., and Potts, I. W., 1967, Pyrolysis of unsymmetrical dialkyl sulfoxides. Rates of alkene formation and composition of the gaseous products, *J Org Chem* **32:** 102-105.

Enders, D., Zhu, J., and Raabe, G., 1996, Asymmetric epoxidation of enones with oxygen in the presence of diethylzinc and (*R,R*)-*N*-methylpseudoephedrine, *Angew Chem* (Int. Ed. Engl.) **35:** 1725-1728.

Enders, D., Zhu, J., and Kramps, L., 1997, Zinc-mediated asymmetric epoxidation of α-enones, *Liebigs Ann Chem* 1101-1113.

Engler, T. A., Reddy, J. P., Combrink, K. D., and Vander Velde, D., 1990, Formal [2 + 2] and [3 + 2] cycloaddition reactions of 2*H*-chromenes with 2-alkoxy-1,4-benzoquinones: regioselective synthesis of substituted pterocarpans, *J Org Chem* **55:** 1248-1254.

Engler, T. A., Letavic, M. A., and Reddy, J. P., 1991, Asymmetric induction in reactions of styrenes with 1,4-benzoquinones utilizing chiral titanium(IV) complexes, *J Am Chem Soc* **113:** 5068-5070.

Engler, T. A., Letavic, M. A., Iyengar, R., LaTessa, K. O., and Reddy, J. P., 1999, Asymmetric reactions of 2-methoxy-1,4-benzoquinones with styrenyl systems: enantioselective syntheses of 8-aryl-3-methoxybicyclo[4.2.0]oct-3-ene-2,5-diones, 7-aryl-3-hydroxybicyclo[3.2.1]oct-3-ene-2,8-diones, 2-aryl-6-methoxy-2,3-dihydrobenzofuran-5-ols, and pterocarpans, *J Org Chem* **64:** 2391-2405.

Engman, L., and Stern, D., 1994, Thiol/diselenide exchange for the generation of benzeneselenolate ion. Catalytic reductive ring-opening of α,β-epoxy ketones, *J Org Chem* **59:** 5179-5183.

Evans, D. A., Ennis, M. D., and Mathre, D. J., 1982, Asymmetric alkylation reactions of chiral imide enolates. A practical approach to the enantioselective synthesis of α-substituted carboxylic acid derivatives, *J Am Chem Soc* **104:** 1737-1739.

Evans, D. A., Britton, T. C., and Ellman, J. A., 1987, Contrasteric carboximide hydrolysis with lithium hydroperoxide, *Tetrahedron Lett* **28:** 6141-6144.

Ferrari, F., Botta, B., and Alves de Lima, R., 1983, Flavonoids and isoflavonoids from *Zollernia paraensis*, *Phytochem* **22:** 1663-1664.

Ferreira, D., Slade, D., and Marais, J.P.J., 2005, In O Andersen, K.R Markham (Eds.), *The Flavonoids: Advances in Research*, CRC Publishers, London, 2005, in press.

Ferreira, D., and Slade, S., 2002, Oligomeric proanthocyanidins: Naturally-occurring *O*-heterocycles, *Nat Prod Rep* **19:** 517-541.

Ferreira, D., and Li, X-C., 2000, Oligomeric proanthocyanidins: Naturally-occurring *O*-heterocycles, *Nat Prod Rep* **17:** 193-212.

Ferreira, D., and Bekker, R., 1996, Oligomeric proanthocyanidins: Naturally-occurring *O*-heterocycles, *Nat Prod Rep* **13:** 411-433.

Ferreira, D., Brandt, E. V., Volsteedt, F. du R., and Roux, D. G., 1975, Parameters regulating the α- and β-cyclization of chalcones, *J Chem Soc, Perkin Trans 1* 1437-1446.

Fischer, D., Ebenau-Jehle, C., and Grisebach, H., 1990a, Phytoalexin syntheis in soybean: purification and characterization of NADPH:2'-hydroxydaidzein oxidoreductase from elicitor-challenged soybean cell cultures, *Arch Biochem Biophys* **276:** 390-395.

Fischer, D., Ebenau-Jehle, C., and Grisebach, H., 1990b, Purification and characterization of pterocarpan synthase from elicitor-challenged soybean cell cultures, *Phytochem* **29:** 2879-2882.

Fischer, D., Stich, K., Britsch, L., and Grisebach, H., 1988, Purification and characterization of (+)-dihydroflavonol (3-hydroxyflavanone) 4-reductase from flowers of *Dahlia variabilis*, *Arch Biochem Biophys* **264:** 40-47.

Flood, R. W., Geller, T. P., Petty, S. A., Roberts, S. M., Skidmore, J., and Volk, M., 2001, Efficient asymmetric epoxidation of α,β-unsaturated ketones using a soluble triblock polyethylene glycol-polyamino acid catalyst, *Org Lett* **3:** 683-686.

Forkmann, G., and Heller, W., 1999, Biosynthesis of flavonoids. In U. Sankawa (Ed.), *Comprehensive Natural Products Chemistry* (Vol. 1): Elsevier.

Freudenberg, K., and Weinges, K., 1958, Leuco- und pseudoverbindungen der anthocyanidine, *Liebigs Chem Ann* **613:** 61-75.

Fritsch, H., and Grisebach, H., 1975, Biosynthesis of cyanidin in cell cultures of *Haplopappus gracilis*, *Phytochem* **14**: 2437-2442.

Geissman, T. A., and Fukushima, D. K., 1948, Flavanones and related compounds. V. The oxidation of 2'-hydroxychalcones with alkaline hydrogen peroxide, *J Am Chem Soc* **70**: 1686-1689.

Geller, T., and Roberts, S. M., 1999, A new procedure for the Julia–Colonna stereoselective epoxidation reaction under nonaqueous conditions: the development of a catalyst comprising a polyamino acid on silica (PaaSiCat), *J Chem Soc, Perkin Trans 1* 1397-1398.

George, H. L., and VanEtten, H. D., 2001, Characterization of pisatin-inducible cytochrome P450s in fungal pathogens of pea that detoxify the pea phytoalexin pisatin, *Fungal Gen Biol* **33**: 37-48.

Gobel, T., and Sharpless, K. B., 1993, Temperature effect on asymmetric dihydroxylation: proof of a stepwise mechanism, *Angew Chem* **32** (Int. Ed. Engl.): 1329-1331.

Greene, T. W., and Wuts, P. G. M., 1991, *Protective Groups in Organic Synthesis* (pp. 149-150), 2nd ed., New York: John Wiley & Sons, Inc.

Guo, L., Dixon, R. A., and Paiva, N. L., 1994, The "pterocarpan synthase" of alfalfa: association and co-induction of vesitone reductase and 7,2'-dihydroxy-4'-methoxy-isoflavonol (DMI) dehydratase, the two final enzymes in medicarpin biosynthesis, *FEBS Lett* **356**: 221-225.

Guo, L., and Paiva, N. L., 1995, Molecular cloning and expression of alfalfa (*Medicago sativa* L.) vestitone reductase, the penultimate enzyme in medicarpin biosynthesis. *Arch Biochem Biophys* **320**: 353-360.

Hagmann, M., and Grisebach, H., 1984, Enzymatic rearrangement of flavanone to isoflavone, *FEBS Lett* **175**: 199-202.

Hagmann, M. L., Heller, W., and Grisebach, H., 1983, Induction and characterization of a microsomal flavonoid 3'-hydroxylase from parsley cell cultures, *Eur J Biochem* **134**: 547-554.

Hagmann, M. L., Heller, W., and Grisebach, H., 1984, Induction of phytoalexin synthesis in soybean. Stereospecific 3,9-dihydroxypterocarpan 6a-hydroxylase from elicitor-induced soybean cell cultures, *Eur J Biochem* **142**: 127-131.

Hahlbrock, K., Zilg, H., and Grisebach, H., 1970, Stereochemistry of the enzymatic cyclisation of 4,2',4'-trihydroxychalcone to 7,4'-dihydroxyflavanone by isomerase from mung bean seedlings, *Eur J Biochem* **15**: 13-18.

Hasegawa, E., Ishiyama, K., Kato, T., Horaguchi, T., Shimizu, T., Tanaka, S., and Yamashita, Y., 1992, Photochemically and thermally induced free-radical reactions of α,β-epoxy ketones with tributyltin hydride: selective C_α-O bond cleavage of oxiranylmethyl radicals derived from α,β-epoxy ketones, *J Org Chem* **57**: 5352-5359.

Helder, R., Hummelen, J. C., Laane, R. W. P. M., Wiering, J. S., and Wynberg, H., 1976, Catalytic asymmetric induction in oxidation reactions. The synthesis of optically active epoxides, *Tetrahedron Lett* 1831-1834.

Ingham, J., 1982, Phytoalexins from the Leguminosae. In J. A. Bailey & J. W. Mansfield (Eds.), *Phytoalexins* (pp. 21-80). New York: Halsted Press.

Ishiguro, M., Tatsuoka, T., and Nakatsuka, N., 1982, Synthesis of (±)-cabenegrins A-I and A-II. *Tetrahedron Lett* **23**: 3859-3862.

Itsuno, S., Sakakura, M., and Ito, K., 1990, Polymer-supported poly(amino acids) as new asymmetric epoxidation catalyst of α,β-unsaturated ketones, *J Org Chem* **55**: 6047-6049.

Jacobson E. N., 1993, In Ojima, L., *Catalytic Asymmetric Synthesis* (pp. 159-202). New York: VCH Publishers.

Jeong, K-S., Sjo, P., and Sharpless, K. B., 1992, Asymmetric dihydroxylation of enynes, *Tetrahedron Lett* **33**: 3833-3836.

Jew, S., Kim, H., Bae, S., Kim, J., and Park H., 2000, Enantioselective synthetic method for 3-hydroxyflavanones: an approach to (2R,3R)-3',4'-O-dimethyltaxifolin, *Tetrahedron Lett* **41**: 7925-7928.

Jez, J. M., Bowman, M. E., Dixon, R. A., and Noel, J. P., 2000, Structure and mechanism of the evolutionarily unique plant enzyme chalcone isomerase, *Nat Struc Biol* **7**: 786-791.

Johnson, R. A., and Sharpless, K. B., 1993, In Ojima, L., *Catalytic Asymmetric Synthesis* (pp. 103-158). New York: VCH Publishers.

Julia, S., Masana, J., and Vega, J. C., 1980, Synthetic enzyme: highly stereoselective epoxidation of chalcone in the three-phase system toluene-water-poly-(S)-alanine, *Angew Chem* **92**: 968-969.

Julia, S., Guixer, J., Masana, J., Rocas, J., Colonna, S., Annuziata, R., and Molinari, H., 1982, Synthetic enzymes. Part 2. Catalytic asymmetric epoxidation by means of polyamino acids in a triphase system, *J Chem Soc, Perkin Trans 1* 1317-1324.

Katsuki, T., and Sharpless, K. B., 1980, The first practical method for asymmetric epoxidation, *J Am Chem Soc* **102:** 5974-5976.

Kice, J. L., and Campbell, J. D., 1967, The effect of ring size on the rate of pyrolysis of cycloalkyl phenyl sulfoxides, *J Org Chem* **32:** 1631-1633.

Klein, S., and Roberts, S. M., 2002, 2-Substituted-2,4-endo-dimethyl-8-oxabicyclo[3.2.1]octan-3-ones as catalyst for the asymmetric epoxidation of some alkenes with oxone, *J Chem Soc, Perkin Trans 1* 2686-2691.

Kochs, G., and Grisebach, H., 1986, Enzymic synthesis of isoflavones, *Eur J Biochem* **155:** 311-318.

Kochs, G., Welle, R., and Grisebach, H., 1987, Differential induction of enzyme in soybean cell cultures by elicitor or osmotic stress, *Planta* **171:** 519-524.

Kolb, H. C., Van Nieuwenhze, M. S., and Sharpless, K. B., 1994a, Catalytic asymmetric dihydroxylation, *Chem Rev* **94:** 2483-2547.

Kolb, H. C., Andersson, P. G., and Sharpless, K. B., 1994b, Toward an understanding of the high enantioselectivity in the osmium-catalyzed asymmetric dihydroxylation (AD). 1. Kinetics, *J Am Chem Soc* **116:** 1278-1291.

Koukol, J., and Conn, E. E., 1961, The metabolism of aromatic compounds in higher plants, *J Biol Chem* **136:** 2692-2698.

Krishna Prasad, A. V., Kapil, R. S., and Popli, S. P., 1986, Synthesis of (±)-isomedicarpin, (±)-homopterocarpin and tuberostan: a novel entry of hydrogenative cyclization into pterocarpans, *J Chem Soc, Perkin Trans 1* 1561-1563.

Kwong, H-L., Sorato, C., Ogino, Y., Chen, H., and Sharpless, K. B., 1990, Preclusion of the second cycle in the osmium-catalyzed asymmetric dihydroxylation of olefins leads to a superior process, *Tetrahedron Lett* **31:** 2999-3002.

Lasterra-Sanchez, M. E., Felfer, U., Mayon, P., Roberts, S. M., Thornton, S. R., and Todd, C. J., 1996, Development of the Julia asymmetric epoxidation reaction. Part 1. Application of the oxidation to enones other than chalcones, *J Chem Soc, Perkin Trans 1* 343-348.

Lévai, A., Adam, W., Fell, R.T., Gessner, R., Patonay, T., Simon, A., and Tóth, G., 1998, Enantioselective synthesis and chiroptical properties of optically active isoflavone epoxides, *Tetrahedron* **54:** 13105-13114.

Lukacin, R., Wellmann, F., Britsch, L., Martens, S., and Matern, U., 2003, Flavonol synthase from *Citrus unshiu* is a bifunctional dioxygenase, *Phytochem* **62:** 287-292.

Lygo, B., and Wainwright, P. G., 1998, Asymmetric phase-transfer mediated epoxidation of α,β-unsaturated ketone using catalysts derived from *Cinchona* alkaloids, *Tetrahedron Lett* **39:** 1599-1602.

Lygo, B., and Wainwright, P. G., 1999, Phase-transfer catalysed asymmetric epoxidation of enones using *N*-anthracenylmethyl-substituted *Cinchona* alkaloids, *Tetrahedron* **55:** 6289-6300.

Lygo, B., and To, D. C. M., 2001, Improved procedure for the room temperature asymmetric phase-transfer mediated epoxidation of α,β-unsaturated ketones, *Terahedron Lett* **42:** 1343-1346.

Mansfield, J. W., 1982, In J. A., Baily, J. W., Mansfield, *Phytoalexins* (pp. 289-312). Glasgow: Blackie & Son.

Marsman, B., Wynberg, H., 1979. Absolute configuration of chalcone epoxide. Chemical correlation. Journal of Organic Chemistry **44:** 2312-2314.

Martens, S., Forkmann, G., Britsch, L., Wellmann, F., Matern, U., and Lukacin, R., 2003, Divergent evolution of flavonoid 2-oxoglutarate-dependent dioxygenases in parsley, *FEBS Lett* **544:** 93-98.

Martens, S., Forkmann, G., Matern, U., and Lukacin, R., 2001, Cloning of parsley flavone synthase I, *Phytochem* **58:** 43-46.

McKillop, A., Swann, B. P., and Taylor, E. C., 1973, Thallium in organic synthesis. XXXIII. One-step synthesis of methyl arylacetates from acetophenones using thallium(III) nitrate (TTN), *J Am Chem Soc* **95:** 3340-3343.

Min, T., Kasahara, H., Bedgar, D. L., Youn, B., Lawrence, P. K., Gang, D. R., Halls, S. C., Park, H., Hilsenbeck, J. L., Davin L. B., Lewis, N. G., and Kang, C., 2003, Crystal structures of pinoresinol-lariciresinol and phenylcoumaran benzylic ether reductases and their relationship to isoflavone reductases, *J Biol Chem* **278:** 50714-50723.

Mitsunobu, O., 1981, The use of diethyl azodicarboxylate and triphenylphosphine in synthesis and transformation of natural products, *Synthesis* 1-28.

Molander, G. A., and Hahn, G., 1986, Lanthanides in organic synthesis. 4. Reduction of α,β-epoxy ketones with samarium diiodide. A route to chiral, nonracemic aldols, *J Org Chem* **51:** 2596-2599.

Moreno, M. J. S., Sae Melo, M. L., and Neves, C. A. S., 1993, Sonochemical reduction of α,β-epoxy ketones and α'-oxygenated analogs by aluminum amalgam, *Tetrahedron Lett* **34:** 353-356.

Narkhede, D. D., Iyer, P. R., and Iyer, C. S. R., 1990, Total synthesis of (±)-leiocarpin and (±)-isohemileiocarpin, *Tetrahedron* **46**: 2031-2034.

Nay, B., Arnaudinaud, V., Peyrat, J-F., Nuhrich, A., Deffieux, G., Mérillon, J-M., and Vercauteren, J., 2000, Total synthesis of isotopically labeled flavonoids, 2. ^{13}C-labelled (±)-catechin from potassium [^{13}C] cyanide, *Eur J Org Chem* 1279-1283.

Nay, B., Arnaudinaud, V., and Vercauteren, J., 2001, Gram-scale production and applications of optically pure ^{13}C-labeled (+)-catechin and (-)-epicatechin, *Eur J Org Chem* 2379-2384.

Nel, R. J. J., Van Heerden, P. S., Van Rensburg, H., and Ferreira, D., 1998, Enantioselective synthesis of flavonoids. Part 5. Poly-oxygenated β-hydroxydihydrochalcones, *Tetrahedron Lett* **39**: 5623-5626.

Nel, R. J. J., Van Rensburg, H., Van Heerden, P. S., Coetzee, J., and Ferreira, D., 1999a, Stereoselective synthesis of flavonoids. Part 7. Poly-oxygenated β-hydroxydihydrochalcone derivatives, *Tetrahedron* **55**: 9727-9736.

Nel, R. J. J., Mthembu, M., Coetzee, J., Van Rensburg, H., Malan, E., and Ferreira, D., 1999b, Stereoselective synthesis of flavonoids. 6. The novel flavan-3-ol, (2*R*,3*S*)-guibourtinidol and its diastereomers, *Phytochem* **52**: 1153-1158.

Nel, R. J. J., Van Rensburg, H., Van Heerden, P. S., and Ferreira, D., 1999c, Stereoselective synthesis of flavonoids. Part 8. Free phenolic flavan-3-ol diastereoisomers, *J Chem Res Synopses* 606-607 and (M), 2610-2625.

Norrby, P-O., Kolb, H. C., and Sharpless, K. B., 1994, Calculations on the reaction of ruthenium tetroxide with olefins using density functional theory (DFT). Implications for the possibility of intermediates in osmium-catalyzed asymmetric dihydroxylation, *Organometallics* **13**: 344-347.

Onda, M., Li, S., Li, X., Harigaya, Y., Takahashi, H., Kawase, H., and Kagawa, H., 1989, Heterocycles, XXIV. Synthesis of optically pure 2,3-*trans*-5,7,3',4',5'-pentahydroxyflavan-3,4-diols and comparison with naturally occurring leucodelphinidins, *J Nat Prod* **52**: 1100-1106.

Ooi, T., Ohara, D., Tamura, M., and Maruoka, K., 2004, Design of new chiral phase-transfer catalysts with dual functions for highly enantioselective epoxidation of α,β-unsaturated ketones, *J Am Chem Soc* 6844-6845.

Otani, K., Takahashi, T., Furuya, T., and Ayabe, S-I., 1994, Licodione synthase, a cytochrome P450 monooxygenase catalyzing 2-hydroxylation of 5-deoxyflavanone, in cultured *Glycyrrhiza echinata* L. cells, *Plant Physiol* **105**: 1427-1432.

Otsubo, K., Inanaga, J., and Yamaguchi, M., 1987, Samarium(II) iodide induced highly regioselective reduction of α,β-epoxy esters and γ,δ-epoxy-α,β-unsaturated esters. An efficient route to optically active β-hydroxy and δ-hydroxy esters, *Tetrahedron Lett* **28**: 4437-4440.

Ozaki, Y., Mochida, K., and Kim, S-W., 1988, Aromatic ring formation by 1,3-Michael–Claisen annulation. Total synthesis of sophorapterocarpan A, maackianin, and anhydropisatin, *J Chem Soc Chem Commun* 374-375.

Ozaki, Y., Mochida, K., and Kim, S-W., 1989, Total synthesis of sophorapterocarpan A, maackiain, and anhydropisatin: application of a 1,3-Michael–Claisen annulation to aromatic synthesis, *J Chem Soc, Perkin Trans 1* 1219-1224.

Paderes, G. D., Metivier, P., and Jorgensen, W. L., 1991, Computer-assisted mechanistic evaluation of organic reactions. 18. Reductions with hydrides, *J Org Chem* **56**: 4718-4733.

Paiva, N. L., Edwards, R., Sun, Y., Hrazdina, G., and Dixon, R. A., 1991, Stress responses in alfalfa (*Medicago sativa* L.) XI. Molecular cloning and expression of alfalfa isoflavone reductase, a key enzyme of isoflavonoid phytoalexin biosynthesis, *Plant Mol Biol* **17**: 653-667.

Paiva, N. L., Sun, Y., Dixon, R. A., VanEtten, H. D., and Hrazdina, G., 1994, Molecular cloning of isoflavone reductase from pea (*Pisum sativum* L.): Evidence for a 3R-isoflavanone intermediate in (+)-pisatin biosynthesis, *Arch Biochem Biophys* **312**: 501-510.

Paterson, I., and Goodman, J. M., 1989, Aldol reactions of methyl ketones using chiral boron reagents: a reversal in aldehyde enantioface selectivity, *Tetrahedron Lett* **30**: 997-1000.

Patonay, T., Toth, G., and Adam, W., 1993, Flavanoids. 44. A convenient and general synthesis of *trans*-3-hydroxyflavanones from chalcones by dimethyldioxirane epoxidation and subsequent base-catalyzed cyclization, *Tetrahedron* **34**: 5055-5058.

Pinard, E., Gaudry, M., Henot, F., and Thellend, A., 1998, Asymmetric total synthesis of (+)-pisatin, a phytoalexin from garden peas (*Pisum sativum* L.), *Tetrahedron Lett* **39**: 2739-2742.

Porter, L. J., 1988. In Harborne, J.B., *The Flavonoids, Advances in Research Since 1980* (pp. 27). London: Chapman & Hall.

Porter, L. J., 1994. In Harborne, J.B., *The Flavonoids, Advances in Research Since 1986* (pp. 23). London: Chapman & Hall.

Preisig, C. L., Matthews, D. E., and VanEtten, H. D., 1989, Purification and characterization of S-adenosyl-L-methionine:6a-hydroxymaackiain 3-O-methyltransferase from *Pisum sativum*, *Plant Physiol* **91**: 559-566.

Prescott, A. G., Stamford, N. P. J., Wheeler, G., and Firmin, J. L., 2002, *In vitro* properties of recombinant flavonol synthase from *Arabidopsis thaliana*, *Phytochem* **60**: 589-593.

Roder, H., Helmchen, G., Peters, E. M., Peters, K., and Von Schnering, H. G., 1984, Highly enantioselective homoaldol addition with chiral N-allylureas—use for the synthesis of optically pure γ-lactones, *Angew Chem* **96**: 895-896.

Saxena, S., Makrandi, J. K., and Grover, S. K., 1985, A facile one-step conversion of chalcones into 2,3-dihydroflavonols, *Synthesis* 110-111.

Schopfer, C. R., Kochs, G., Lottspeich, F., and Ebel, J., 1998, Molecular characterization and functional expression of dihydroxypterocarpan 6a-hydroxylase, an enzyme specific for pterocarpanoid phytoalexin biosynthesis in soybean (*Glycine max* L.), *FEBS Lett* **432**: 182-186.

Sharpless, K. B., Teranishi, A. Y., and Backvall, J. E., 1977, Chromyl chloride oxidations of olefins. Possible role of organometallic intermediates in the oxidations of olefins by oxo transition metal species, *J Am Chem Soc* **99**: 3120-3128.

Sharpless, K. B., Amberg, W., Bennani, Y. L., Crispino, G. A., Hartung, J., Jeong, K-S., Kwong, H-L., Morikawa, K., Wang, Z-M., Xu, D., and Zhang, X., 1992, The osmium-catalyzed asymmetric dihydroxylation: a new ligand class and a process improvement, *J Org Chem* **57**: 2768-2771.

Shih, T. L., Wyvratt, M. J., and Mrozik, H., 1987, Total synthesis of (±)-5-O-methyllicoricidin, *J Org Chem* **52**: 2029-2033.

Shukla, Y. N., Tandon, J. S., and Dhar, M. M., 1973, Lyonogenin, a new dihydrochalcone from *Lyonia Formosa*, *Indian J Chem* **11**: 720-722.

Singh, P., and Arora, G., 1987, Organic synthesis using phase-transfer catalysis. Part 3. Asymmetric induction in epoxidation and Michael addition reactions of chalcone under phase transfer conditions using optically active solvents, *Indian J Chem* **26B**: 1121-1123.

Stafford, H. A., and Lester, H. H., 1982, Enzymic and nonenzymic reduction of (+)-dihydroquercetin to its 3,4,-diol, *Plant Physiol* **70**: 695-698.

Stafford, H. A., and Lester, H. H., 1984, Flavan-3-ol biosynthesis. The conversion of (+)-dihydroquercetin and flavan-3,4-*cis*-diol (leucocyanidin) to (+)-catechin by reductases extracted from cell suspension cultures of Douglas fir, *Plant Physiol* **76**: 184-186.

Stafford, H. A., and Lester, H. H., 1985, Flavan-3-ol biosynthesis. The conversion of (+)-dihydromyrecetin to its flavan-3,4-diol (leucodelphinidin) and to (+)-gallocatechin by reductases extracted from tissue cultures of *Ginko biloba* and *Pseudotsuga menziesii*, *Plant Physiol* **78**: 791-794.

Stich, K., and Forkmann, G., 1988, Biosynthesis of 3-deoxyanthocyanins with flower extracts from *Sinningia cardinalis*, *Phytochem* **27**: 785-789.

Subburaj, K., Murugesh, M. G., and Trivedi, G. K., 1997, Regioselective total synthesis of edulane and its angular analog, *J Chem Soc, Perkin Trans 1* 1875-1878.

Sutter, A., Poulton, J., and Grisebach, H., 1975, Oxidation of flavanone to flavone with cell-free extracts from young parsley leaves, *Arch Biochem Biophys* **170**: 547-556.

Takahashi, H., Kubota, Y., Miyazaki, H., and Onda, M., 1984, Heterocycles. XV. Enantioselective synthesis of chiral flavanonols and flavan-3,4-diols, *Chem Pharm Bull* **32**: 4852-4857.

Tanner, G. J., Francki, K. T., Abrahams, S., Watson, J. M., Larkin, P. J., and Ashton, A. R., 2003, Proanthocyanidin biosynthesis in plants. Purification of legume leucoanthocyanidin reductase and molecular cloning of its cDNA, *J Biol Chem* **278**: 31647-31656.

Tenhaken, R., Salmen, H. C., and Barz, W., 1991, Purification and characterization of pterocarpan hydroxylase, a flavoprotein monooxygenase from the fungus *Ascochyta rabiei* involved in pterocarpan phytoalexin metabolism, *Arch Microbiol* **155**: 353-359.

Thakkar, K., and Cushman, M., 1995, A novel oxidative cyclization of 2'-hydroxychalcones to 4,5-dialkoxyaurones by thallium(III) nitrate, *J Org Chem* **60**: 6499-6510.

Trost, B. M., Salzmann, T. N., and Hiroi, K., 1976, New synthetic reactions. Sulfenylations and dehydrosulfenylations of esters and ketones, *J Am Chem Soc* **98**: 4887-4902.

Trost, B. M., and Murayama, E., 1981, Dimethyl(methylthio)sulfonium fluoroborate. A chemoselective initiator for thionium ion induced cyclizations, *J Am Chem Soc* **103**: 6529-6530.

Trost, B. M., and Sato, T., 1985, Dimethyl(methylthio)sulfonium tetrafluoroborate initiated organometallic additions to and macrocyclizations of thioketals, *J Am Chem Soc* **107**: 719-721.

Turnbull, J. J., Nakajima, J., Welford, R. W. D., Yamazaki, M., Saito, K., and Schofield, C. J., 2004, Mechanistic studies on three 2-oxoglutarate-dependent oxygenases of flavonoid biosynthesis, *J Biol Chem* **279**: 1209-1216.

Turnbull, J. J., Sobey, W. J., Aplin, R. T., Hassan, A., Firmin, J. L., Schofield, C. J., Firmin, J. L., and Prescott, A. G., 2000, Are anthocyanidins the immediate products of anthocyanidin synthase? *Chem Commun* 2473-2474.

Van Aardt, T. G., Van Heerden, P. S., and Ferreira, D., 1998, The first direct synthesis of pterocarpans *via* aldol condensation of phenylacetates with benzaldehydes, *Tetrahedron Lett* **39**: 3881-3884.

Van Aardt, T. G., Van Rensburg, H., and Ferreira, D., 1999, Direct synthesis of pterocarpans *via* aldol condensation of phenylacetates with benzaldehydes, *Tetrahedron* **55**: 11773-11786.

Van Aardt, T. G., Van Rensburg, H., and Ferreira, D., 2001, Synthesis of isoflavonoids. Enantiopure *cis*- and *trans*-6a-hydroxypterocarpans and a racemic *trans*-pterocarpan, *Tetrahedron* **57**: 7113-7126.

Van der Merwe, J. P., Ferreira, D., Brandt, E. V., and Roux, D. G., 1972, Immediate biogenetic precursors of mopanols and peltogynols, *J Chem Soc, Chem Commun* 521-522.

VanEtten, H. D., Matthews, D. E., and Matthews, P. S., 1989, Phytoalexin detoxification: Importance for pathogenicity and practical implications, *Ann Rev Phytopathol* **27**: 143-164.

VanEtten, H. D., Matthews, P. S., and Mercer, E. H., 1983, (+)-Maackiain and (+)-medicarpin as phytoalexins in *Sophora japonica* and identification of the (-) isomers by biotransformation, *Phytochem* **22**: 2291-2295.

Van Rensburg, H., Van Heerden, P. S., Bezuidenhoudt, B. C. B., and Ferreira, D., 1996, The first enantioselective synthesis of *trans*- and *cis*-dihydroflavonols, *Chem Commun* 2747-2748.

Van Rensburg, H., Van Heerden, P. S., Bezuidenhoudt, B. C. B., and Ferreira, D., 1997a, Stereoselective synthesis of flavonoids. Part 4. *Trans*- and *cis*-dihydroflavonols, *Tetrahedron* **53**: 14141-14152.

Van Rensburg, H., Van Heerden, P. S., Bezuidenhoudt, B. C. B., and Ferreira, D., 1997b, Enantioselective synthesis of the four catechin diastereomer derivatives, *Tetrahedron Lett* **38**: 3089-3092.

Van Rensburg, H., Van Heerden, P. S., and Ferreira, D., 1997c, Enantioselective synthesis of flavonoids. Part 3. *trans*- and *cis*-Flavan-3-ol methyl ether acetates, *J Chem Soc, Perkin Trans 1* 3415-3421.

Versteeg, M., Bezuidenhoudt, B. C. B., Ferreira, D., and Swart, K. J., 1995, The first enantioselective synthesis of isoflavonoids: (*R*)- and (*S*)-isoflavans, *J Chem Soc, Chem Commun* 1317-1318.

Versteeg, M., Bezuidenhoudt, B. C. B., and Ferreira, D., 1998, The direct synthesis of isoflavans *via* α-alkylation of phenylacetates, *Heterocycles* **48**: 1373-1394.

Versteeg, M., Bezuidenhoudt, B. C. B., and Ferreira, D., 1999, Stereoselective synthesis of isoflavonoids. (*R*)- and (*S*)-isoflavans, *Tetrahedron* **55**: 3365-3376.

Vicario, J.L., Badia, D., Dominguez, E., Rodriguez, M., and Carrillo, L., 2000, The first synthesis of isoflavanones, *Tetrahedron Lett* **41**: 8297-8300.

Von Konstanecki, St., and Rossbach, G., 1896, Über die Einwirkung von Benzaldehyd auf Acetophenon. *Chem Ber* **29**: 1488-1494.

Wang, Z-M., Zhang, X-L., and Sharpless, K. B., 1993, Asymmetric dihydroxylation of aryl allyl ethers, *Tetrahedron Lett* **34**: 2267-2270.

Wang, Z-X., and Shi, Y., 1997, A new type of ketone catalyst for asymmetric epoxidation, *J Org Chem* **62**: 8622-8623.

Wang, Z-X., Tu, Y., Frohn, M., Zhang, J-R., and Shi, Y., 1997, An efficient catalytic asymmetric epoxidation method, *J Am Chem Soc* **119**: 11224-11235.

Wang, Z-X., Miller, S. M., Anderson, O. P., and Shi, Y., 1999, A class of C_2 and pseudo C_2 symmetric ketone catalysts for asymmetric epoxidation conformational effect on catalysis. *J Org Chem* **64**: 6443-6458.

Weinges, K., 1958, Über catechine und ihre herstellung aus leuko-anthocyanidin-hydraten, *Liebigs Chem Ann* **615**: 203-209.

Welford, R. W. D., Turnbull, J. J., Claridge, T. D. W., Prescott, A. G., and Schofield, C. J., 2001, Evidence for oxidation at C-3 of the flavonoid C-ring during anthocyanin biosynthesis, *Chem Commun* 1828-1829.

Williams, R. M., Armstrong, R. W., and Dung, J. S., 1984, Stereocontrolled total synthesis of (±)- and (+)-bicyclomycin: new carbon-carbon bond-forming reactions on electrophilic glycine anhydride derivatives, *J Am Chem Soc* **106**: 5748-5750.

Wilmouth, R. C., Turnbull, J. J., Welford, R. W. D., Clifton, I. J., and Schofield, C. J., 2002, Structure and mechanism of anthocyanidin synthase from *Arabidopsis thaliana*, *Structure* **10**: 93-103.

Wynberg, H., and Greijdanus, B., 1978, Solvent effects in homogeneous asymmetric catalysis, *J Chem Soc, Chem Commun* 427-428.

Wu, Q., Presig, C. L., and VanEtten, H. D., 1997, Isolation of the cDNAs encoding (+)6a-hydroxymaackiain 3-*O*-methyltransferase, the terminal step for the synthesis of the phytoalexin pisatin in *Pisum satium, Plant Mol Biol* **35**: 551-560.

Xie, D., Sharma, S. B., Paiva, N. L., Ferreira, D., and Dixon, R. A., 2003, Role of anthocyanidin reductase, encoded by *BANYULS* in plant flavonoid biosynthesis, *Science* **299**: 396-399.

Xie, D.-Y., Sharma, S. B., and Dixon, R. A., 2004, Anthocyanidin reductases from *Medicago truncatula* and *Arabidopsis thaliana, Arch Biochem Biophys* **422**: 91-102.

Yu, H-B., Zheng, X-F., Lin, Z-M., Hu, Q-S., Huang, W-S., and Pu, L., 1999, Asymmetric epoxidation of α,β-unsaturated ketones catalyzed by chiral polybinaphthyl zinc complexes: great enhanced enantioselectivity by a cooperation of the catalytic sites in a polymer chain, *J Org Chem* 8149-8155.

CHAPTER 2

ISOLATION AND IDENTIFICATION OF FLAVONOIDS

MACIEJ STOBIECKI[1*] AND PIOTR KACHLICKI[2]

[1]*Institute of Bioorganic Chemistry PAS, Noskowskiego 12/14, 61-704 Poznan, Poland; [2]Institute of Plant Genetics PAS, Strzeszynska 34, 60-479 Poznan, Poland,* *corresponding author: e-mail: mackis@ibch.poznan.pl*

1. INTRODUCTION

Flavonoids and their conjugates form a very large group of natural products. They are found in many plant tissues, where they are present inside the cells or on the surfaces of different plant organs. The chemical structures of this class of compounds are based on a C_6-C_3-C_6 skeleton. They differ in the saturation of the heteroatomic ring C, in the placement of the aromatic ring B at the positions C-2 or C-3 of ring C, and in the overall hydroxylation patterns (Figure 2.1). The flavonoids may be modified by hydroxylation, methoxylation, or O-glycosylation of hydroxyl groups as well as C-glycosylation directly to carbon atom of the flavonoid skeleton. In addition, alkyl groups (often prenyls) may be covalently attached to the flavonoid moieties, and sometimes additional rings are condensed to the basic skeleton of the flavonoid core. The last modification takes place most often in the case of isoflavonoids, where the B ring is condensed to the C-3 carbon atom of the skeleton. Flavonoid glycosides are frequently acylated with aliphatic or aromatic acid molecules. These derivatives are thermally labile and their isolation and further purification without partial degradation is difficult. The multiplicity of possible modifications of flavonoids result in more than 6,000 different compounds from this class were known in the end of the last century and this number continues to increase (Harborne and Williams, 2000). Condensed tannins create a special group of flavonoid compounds formed by polymeric compounds built of flavan-3-ol units, and their molecular weights often exceeding 1,000 Da.

In the plant kingdom, different plant families have characteristic patterns of flavonoids and their conjugates. All these compounds play important biochemical and physiological roles in the various cell types or organs (seed, root, green part, fruit) where they accumulate. Different classes of flavonoids and their conjugates have numerous functions during the interactions of plant with the environment, both in biotic and abiotic stress conditions (Dixon and Paiva, 1995; Shirley, 1996). Additionally, flavonoid conjugates, because of their common presence in plants, are important components of human and animal diet. Due to the different biological activities of plant secondary metabolites, their regular consumption may have serious consequences for health, both positive and negative (Beck et al., 2003; Le March, 2002; Boue et al., 2003; Fritz et al., 2003; Nestel, 2003). For the mentioned reasons, methods for the efficient and reproducible analysis of flavonoids play a crucial role in research conducted in different fields of the biological and medical sciences.

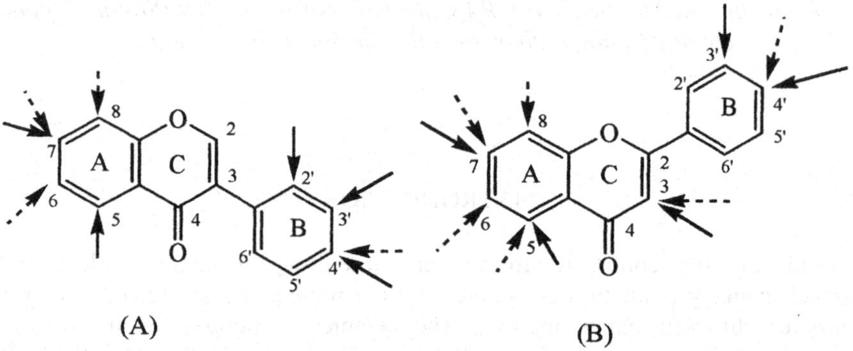

(A)   (B)

Figure 2.1 Flavonoid structures, ring labeling, and carbon atom numbering. (A) Isoflavones. (B) Flavones and flavonols. Full arrows indicate most frequent hydroxylation sites and dashed arrows indicate most frequent C- and/or O- glycosylation sites.

The identification and structural characterization of flavonoids and their conjugates isolated from plant material, as single compounds or as part of mixtures of structurally similar natural products, create some problems due to the presence of isomeric forms of flavonoid aglycones and their patterns of glycosylation. A number of analytical methods are used for the characterization of flavonoids. In many cases, nuclear magnetic resonance (NMR) analyses (^{1}H and ^{13}C) are necessary for the unambiguous identification of unknown compounds; other instrumental methods (mass spectrometry, UV and IR spectrophotometry) applied for the identification of organic compounds fail to provide the information necessary to answer all the structural questions. Utilization of standards during analyses and comparison of retention times as well as spectral properties, especially when compounds are present in a mixture, is critical. An important area of research on flavonoids is the identification of their metabolites in animal tissues and body fluids (urine, blood, spinal fluid). For this, investigators have to deal with different modifications of the flavonoid moieties, modifications often not found in plant tissues (Blaut et al., 2003). The metabolism of flavonoids in human and animal organisms, among others, is based on glucuronidation, sulfation, or methylation (Sfakianos et al., 1997;

Yasuda et al., 1994). The first two above-mentioned types of flavonoid aglycone modifications, occurring after their consumption by humans or animals, are less often found in the samples of plant origin.

One of the goals in functional genomics and systems biology studies is metabolite profiling. Qualitative and quantitative monitoring of flavonoid derivatives, together with information about the level of transcription and protein expression, enables the elucidation of gene functions (Fiehn, 2001; Hall et al., 2002; Sumner et al., 2003; Fernie et al., 2004). Another challenge in the field is to establish the flavonoid conjugate profiles in genetically modified plant lines, e.g. engineered for higher resistance against environmental conditions (pathogenic microorganisms, insects, and physical stress factors such as temperature, drought, or UV light). The plant performance in different environmental conditions and the resulting effect on crop yield may be accompanied by increased synthesis of desired and/or undesired natural products in particular plant tissues, among the biologically active flavonoids or isoflavonoids (Bovy et al., 2002; Dixon and Steel, 1999; Fisher et al., 2004; Trethewey, 2004; Verpoorte and Memelink, 2002).

2. ISOLATION OF FLAVONOIDS AND THEIR CONJUGATES FROM BIOLOGICAL MATERIALS

The analysis of flavonoids and their conjugates is one of the most important areas in the field of instrumental analytical methods, helping to solve problems in biological and medical sciences. Different methods of isolation of the natural products may be applied, and the utilization of various strategies is dependent on the origin of the biological material from which the target natural products are to be extracted (plant or animal tissue or body fluids). In the case of polyphenolic compounds, it often is important to initially determine whether the researchers are interested in the identification of individual components present in a mixture of target compounds or whether they would like to estimate the total amount of phenolic compounds in the biological material investigated. This second approach most often takes place during the nutritional studies on different foods or fodders, mainly of plant origin.

The presence of carbohydrates and/or lipophylic substances may influence the profile of the qualitative and quantitative composition of flavonoids and their derivatives in the obtained extracts. One has to consider the above-mentioned selection of the methods for sample preparation and extraction, and in many cases additional cleaning based on solid-phase extraction (SPE) of the extracted samples is required.

2.1. Preparation of Plant or Animal Tissue and Foodstuffs for Flavonoid Analysis

The utilization of dried plant material for extraction may cause a substantial decrease in the yield of flavonoid conjugates. Acylated flavonoid glycosides are especially labile at elevated temperatures and are frequently thermally degraded during the process of drying plant tissues. This is important during the profiling of this class of natural products in research directed toward the investigation of their physiological and biochemical roles in plants under the influence of environmental factors, or in studies of genetically modified plants for the elucidation of changes in metabolic pathways.

Free flavonoid aglycones exuded by plant tissues (leaf or root) may be washed from the surface with nonpolar solvents, such as methylene chloride, ethyl ether, or ethyl acetate. However, more polar glycosidic conjugates dissolve in polar solvents (methanol and ethanol), and these organic solvents are applied for extraction procedures in Soxhlet apparatus. Mixtures of alcohol and water in different ratios are applied for the extraction of flavonoids and their conjugates from solid biological material (plant or animal tissues and different food products). The extraction efficiency may be enhanced by the application of ultrasonication (Rostagno et al., 2003; Herrera et al., 2004) or pressurized liquid extraction (PLE), a procedure performed at elevated temperature ranging from 60°C to 200°C (Rostagno et al., 2004). Supercritical fluid extraction with carbon dioxide also may be used (Kaiser et al., 2004). However, the temperature conditions during the extraction procedures have to be carefully adjusted because of the possibility of thermal degradation of the flavonoid derivatives. In many cases, further purification and/or preconcentration of the target compound fraction is necessary. In these cases, liquid–liquid extraction (LLE) or SPE are most commonly used. For estimation of the extraction yield it is necessary to spike biological materials with proper internal standards. Most suitable are compounds structurally similar to the studied analytes but not present in the sample. Compounds labeled with stable isotopes (^{2}H or ^{13}C) are useful when mass spectrometric detection is applied. In the case of the extraction of flavonoids from biological materials, different classes of phenolic compounds are often added. On the other hand, quantitative analysis of consecutive components of the analyzed flavonoid mixture needs reference standard compounds necessary for preparation of calibration curves essential for a precise quantification.

The choice of the extraction procedure for obtaining flavonoid conjugates from biological material is very important and depends on the goals of the conducted research. The evaluation of the spatial distribution of target compounds on the organ, tissue, cellular, or even subcellular level is of special interest in some projects. In these situations, the amount of biological material for the isolation of natural products may be extremely small, and the application of microextraction techniques is necessary (reviewed in Vas and Veckey, 2004). In many cases, it is necessary to avoid the chemical and/or enzymatic degradation of the metabolites. This is of special importance in the profiling of flavonoid glycosides in research directed toward plant functional genomics or during physiological and biochemical studies that need information about all classes of flavonoid conjugates present, even the thermally labile acylated derivatives. On the other hand, in the phytochemical analysis of plant species or phytopharmaceutical studies of plant material, the repeatable isolation of all biologically active flavonoid aglycones with a good yield is more important. In these cases, more drastic extraction conditions are acceptable. Excellent reviews have been published on isolation strategies for the determination of active phenols in plants tissue or food and foodstuff (Naczk and Shahidi, 2004; Robards, 2003).

Robust multistep chromatographic methods are necessary for the isolation of individual components from plant extracts containing new uncharacterized compounds. Various stationary phases are used in column chromatography, including polyamide, Sephadex LH-20, and different types of silica gels (normal and reversed phase with chemically bonded functional groups). The proper choice of solvent systems is necessary, often requiring the application of gradients of more polar (normal phases) or more hydrophobic solvents (reverse phases), together with the above-mentioned chromatographic supports in different chromatography systems. The sequence and kind of separation methods used depends on the composition of the sample and the experience of the researcher. However, minor flavonoid components are difficult to obtain as pure compounds. In cases of analysis of samples containing a number of compounds present in small amounts, the application of an analytical chromatographic systems enhanced by proper detectors (UV, NMR, and/or MS) gives spectrometric information sufficient for establishing the structure of minor target components. When liquid chromatography is used for separation of compounds, multiple detector systems are available (UV diode array detector, mass spectrometers, and nuclear magnetic resonance spectrometer). It is possible to achieve complete structural information about isomeric flavonoids and their conjugates in this way.

2.2. Preparation of Body Fluids

For the isolation of flavonoids and their derivatives from liquid samples like beverages (wine or fruit juice) and physiological fluids (blood or urine), two different approaches are usually applied. The first one is based on liquid–liquid extraction and the second one on solid-phase extraction of target natural products mainly on RP C-18 silica gel cartridges. In the case of body fluids, special procedures have to be considered to avoid degradation of target compounds due to the activity of different enzymes present (Martinez-Ortega et al., 2004). However, in some cases, flavonoid conjugates can be enzymatically hydrolyzed with external glucuronidases and sulfatases prior to the isolation and analysis of products.

3. STRUCTURAL CHARACTERIZATION AND/OR IDENTIFICATION OF FLAVONOID AND THEIR CONJUGATES

All physicochemical methods applied in the field of organic chemistry are useful for structural characterization or identification of individual flavonoids and their conjugates. The separation approaches mentioned above may be considered in different ways. The first one is directed toward the analysis of single compounds obtained after exhaustive isolation and purification procedures. The method of choice in this approach is NMR of ^{1}H hydrogen and/or ^{13}C carbon isotopes, dependent on the intensity of the interactions between different atoms within a molecule placed in a high-intensity magnetic field. Different NMR experiments have been developed to achieve information concerning chemical structure of the studied molecule on this basis. Particularly useful are methods enabling recording of two-dimensional spectra showing homonuclear interactions [correlation spectroscopy (COSY) and nuclear Overhauser effect spectroscopy (NOESY)] as well as heteronuclear [heteronuclear single quantum correlation (HSQC) and heteronuclear multiple bond correlation (HMBC)] to facilitate the acquisition of all the structural information about an aglycone and the corresponding sugar substitution. In the case of diglycosides, information on the placement of the interglycosidic bonds and the possible acyl group substitutions on the sugar rings, and the position of anomeric proton(s) also can be obtained. The limitation of NMR methods is the lower sensitivity in comparison with other instrumental methods. For obtaining good quality spectra containing all the necessary structural information, relatively high amounts of purified compound (more than 1 mg) are necessary, especially when magnets of medium frequency (300 MHz) are used in the NMR spectrometer. The NMR spectrometers may be connected on line to liquid chromatographs (LC-NMR), giving a powerful tool to study mixtures of natural compounds present in complex samples. Important structural data also can be obtained from mass spectra registered on different types of mass spectrometers (MS). The application of ultraviolet and infrared spectrophotometers may give valuable information about specific compounds.

MS applied for the analysis of organic compounds utilize different ionization methods and may be equipped with different types of analyzers. In addition, these instruments may be combined with GC/LC or capillary electrophoresis (CE) apparatus. However, simple chemistry based on single reactions such as silylation, methylation, and acetylation blocking polar functional groups has to be done on the studied samples prior to GC-MS analyses. Derivatization of polar groups improves structural information obtained from MS spectra and ameliorates the volatility of analytes, decreasing the thermal degradation of compounds within the GC capillary column. The variety of MS techniques being available in laboratories is a reason that this technique has a wide range of scientific or practical applications in biological and medical disciplines. The general information about the ionization methods and types of analyzers in MS used in phytochemical analysis is presented in Table 2.1.

Analysis of natural products is possible with different types of MS available on the market. The instruments are equipped with various sample introduction systems and ionization methods, as well as diverse physical phenomena are used for separation of the created ions in MS analyzers. Positive and negative ions are analyzed in MS; the choice of the ionization mode (negative or positive) is sometimes a very important feature. The ionization methods may be divided into two groups differing with respect to the amount of energy transferred to the molecule during the ionization process. Electron ionization (EI) belongs to the first group. The transfer of energy occurs during the interaction of electrons with the molecule in the vapor state; it may cause the cleavage of chemical bonds and fragmentation of the molecule, which is characteristic for the analyzed compound. Other ionization methods deliver lower energy to the studied molecules during the protonation (positive ion mode) or deprotonation (negative ion mode) processes. In both cases, the absorbed energy is too low to cause intense fragmentation. In this situation, techniques of collision-induced dissociation with tandem MS (CID MS/MS) have to be applied for the structural characterization of compounds.

Different designs of tandem analyzers in MS may be used (McLafferty, 1983; Jennings, 1996). An example of such instruments enabling multistage tandem MS (MS^n) are instruments with ion trap analyzers. In these analyzers, the fragment ions created with CID may be further studied using additional MS^n stages (March, 1997).

Table 2.1 Ionization methods and types of analyzers for phytochemical and clinical analysis of flavonoids and their conjugates using mass spectrometry

Sample introduction system	Ionization method	Analyzer type	Comments
Direct insertion or infusion of single compound	Electron ionization – EI Chemical ionization – CI Liquid secondary ion mass spectrometry – LSIMS[b] Fast atom bombardment – FAB[b] Atmospheric pressure chemical ionization – APCI[b] Electrospray ionization – ESI[b] Matrix assisted laser desorption – MALDI[b]	Electromagnetic – B/E or E/B[a] Quadrupole – Q Ion trap – IT Time of flight – ToF[a] Fourier transform ion cyclotron resonance – FT ICR[a]	When soft ionization methods are applied, MS/MS techniques Increase the amount of structural information.
Gas chromatography – GC[c]	Electron ionization – EI Chemical ionization – CI	Electromagnetic – B/E or E/B[a] Quadrupole – Q Ion trap – IT Time of flight – ToF[a]	Need of derivatization, only analysis of free aglycones is possible.
Liquid chromatography – LC or capillary electrophoresis – CE	Continuous flow fast atom bombardment – CF FAB[b] Atmospheric pressure chemical ionization – APCI[b] Electrospray ionization – ESI[b]	Quadrupole – Q Ion trap – IT Time of flight – ToF[a] Fourier transform ion cyclotron resonance – FT ICR[a]	Only soft ionization methods are applied, MS/MS techniques increase the amount of structural information

[a] High resolution analyzer, ions with mass difference of 0.1 Da may be resolved.
[b] Soft ionization method.
[c] Possible thermal degradation of analyte on GC column.

However, the fragmentation mechanisms are different during high- and low-energy collisions used in electromagnetic or quadrupole and ion trap analyzers. It has been shown that low-energy CID MS/MS spectra of *C*-glycosidic flavonoids differ from those obtained when high-energy measurements are performed. In tandem electromagnetic analyzers, high-energy collisions are performed, while other types of tandem MS require low-energy collisions. Mass spectra registered with these two approaches may give slightly different fragmentation pathways (Waridel et al., 2001). During the analysis of flavonoid glycosides in negative ion CID

MS/MS experiments, the cleavage of aglycone fragment ions and intensities of the product ions in registered mass spectra are dependent on the ions chosen for the fragmentation. March and coworkers (2004) registered CID MS/MS product ion mass spectra of the fragment ions at 268 m/z $(Y_0 - H)^{\bullet-}$ and at 269 m/z (Y_0^-), created after cleavage of the glycosidic bond of genistein 7-O-glucoside and demonstrated that both spectra differed substantially.

In the last few years some review articles have been published describing methods of MS analysis of flavonoids and their conjugates including information on fragmentation pathways of flavonoid aglycones and their glycosidic conjugates (Justesen et al., 1998; Stobiecki, 2000, 2004; Careri et al., 2002; Cuyckens and Claeys, 2004; Prasain et al., 2004).

3.1. Structural Characterization of Individual Flavonoids and Their Conjugates

3.1.1. Nuclear Magnetic Resonance

NMR is a well-established and the most commonly used method for natural product structure analysis. The studies of flavonoid structures using ^{1}H-NMR were initiated in 1960s (Markham and Mabry, 1975) and along with ^{13}C-NMR have became the method of choice for the structure elucidation of these compounds. The chemical shifts and multiplicity of signals corresponding to particular atoms and their coupling with other atoms within the molecule allow for easy identification of the aglycone structure, the pattern of glycosylation, and the identity of the sugar moieties present. The literature of this topic is abundant and rapidly growing (Markham and Geiger, 1994; Albach et al., 2003; Kazuma et al., 2003; Francis et al., 2004).

3.1.2. Mass Spectrometry

MS is a very sensitive analytical method used to identify flavonoid conjugates or to perform partial structural characterization using microgram amounts of sample (Cuyckens and Claeys, 2004). Indeed, significant structural data can be obtained from less than 1 mg of the analyzed compound when different MS techniques are used in combination with chemical derivatization of the characterized compounds (Franski et al., 1999, 2002, 2003). A strategy for the combined application of different MS techniques and chemical derivatization are presented in Figure 2.2.

Flavonoid glycosides are thermally labile compounds and the evaporation without decomposition of the analyte is impossible, even in the ion source of a MS, where high vacuum exists (about 3×10^{-5} torr). In this situation, soft ionization methods need to be applied for the analysis of this group of compounds, and the analyte molecules are ionized without evaporation in high vacuum (FAB or LSIMS, MALDI) or under atmospheric pressure (ESI, APCI). From normal mass spectra, information can be obtained about the molecular weight of the whole conjugate, the size of the sugar moieties attached to the aglycone, and the molecular weight of the aglycone (Stobiecki, 2000; Cuyckens and Claeys, 2004).

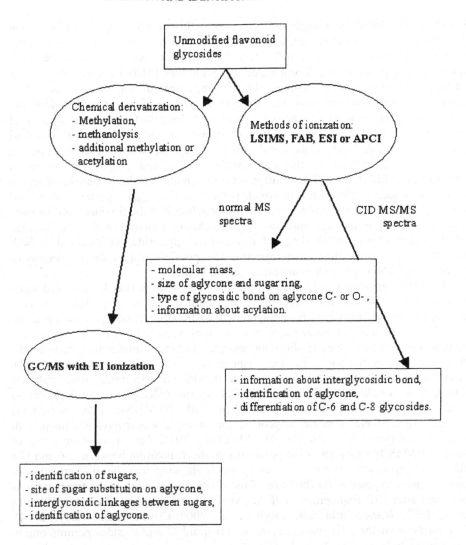

Figure 2.2 *Structural information obtainable with different mass spectrometric methods.*

With FAB or LSIMS ionization, the desorption of the analyte molecule ions from the liquid matrix may be improved when the interactions of the polar groups of the analyte with the matrix decrease. Improved efficiency of ion desorption may be further achieved after the methylation of the analyzed compounds. In addition, the methylation of a flavonoid glycoside may help to elucidate the glycosylation pattern of the aglycone hydroxyl groups (Stobiecki et al., 1988).

The *O*-glycosides of flavonoids give positive ion mass spectra containing intense [M+H]$^+$ ions as well as fragment ions created after the cleavage of glycosidic bonds between sugar moieties or sugar and aglycone, in this case Y_n^+ type ions (for the ion nomenclature, see Domon and Costello, 1988; Claeys et al., 1996). A slightly different situation is observed in the negative ion mass spectra mode, where much lower fragmentation of the deprotonated molecule ions $[M-H]^-$ occurs. In the case of flavonoid O-glycosyl-glycosides, the rearrangement of sugars may take place during the fragmentation process, and the sequence of sugar losses does not correspond to the sequence of the sugar moieties in the intact molecule (Cuyckens et al., 2001, 2002). This situation was observed for the first time in flavonoid rutinosides $\alpha(1\text{-}6)$ and neohesperidosides $\alpha(1\text{-}2)$, in which rhamnose occurs on the nonreducing end of the diglucoside moiety and glucose is bound to the aglycone (Ma et al., 2001). A different fragmentation pathway may be observed in the flavonoid C-glycosides, in which rupture of the sugar ring takes place and strong $[M+H\text{-}120]^+$ ions are created. The nomenclature of the fragment ions in mass spectra of C-glycosides was introduced by Becchi and Fraisse (1989). The cleavage of the sugar moiety in all types of flavonoid C-glycosides is observed in both positive and negative mode mass spectra. The possible fragmentation patterns of flavonoid O- and C-glycosides are presented on Figure 2.3.

Structural information obtainable from the mass spectra may be achieved when tandem mass spectrometric techniques are applied. It is possible to increase the degree of fragmentation by applying CID MS/MS technologies; however, in LC/MS instruments with atmospheric pressure ionization (APCI or ESI) the increase of potential between the entry slit to the analyzer and the skimmer also promotes the fragmentation of the molecule ions, especially in the positive ion mode. The analysis of metastable ions is also possible in MS with electromagnetic analyzers equipped with collision cells in the field-free regions (McLafferty, 1983; Stobiecki et al., 1988; Becchi and Fraisse, 1989). When a full CID MS/MS system is used, the identification of the isomeric aglycones, for example kaempferol and luteolin or apigenin and genistein, is also possible (Ma et al., 1997). An important advantage of some CID MS/MS systems is the possibility of discrimination between C-6 and C-8 flavonoid glycosides. There are several reports showing that the differentiation of both isomers is possible on the basis of the relative intensities of the fragment ions obtained after CID fragmentation of the $[M+H]^+$ ions (Becchi and Fraisse, 1989; Li et al., 1992; Waridel et al., 2001; Bylka et al., 2002; Cuyckens and Claeys, 2004). The analysis of the CID mass spectra of flavonoid O-diglucosides permits one to distinguish between $\alpha(1\text{-}2)$- and $\alpha(1\text{-}6)$-linked sugars in the investigated molecules (Cuyckens et al., 2000; Ma et al., 2001; Franski et al., 2002).

Mass spectrometers with laser desorption ionization combined with the time of flight analyzer (MALDI ToF) also were used for the analysis of isoflavone glycosides in soy products (Wang and Sporns, 2000a), and $[M+H]^+$ and Y_n^+ type ions were observed in the mass spectra. The same authors tried to apply this MS technique for the analysis of flavonol glycosides in food (Wang and Sporns, 2000b).

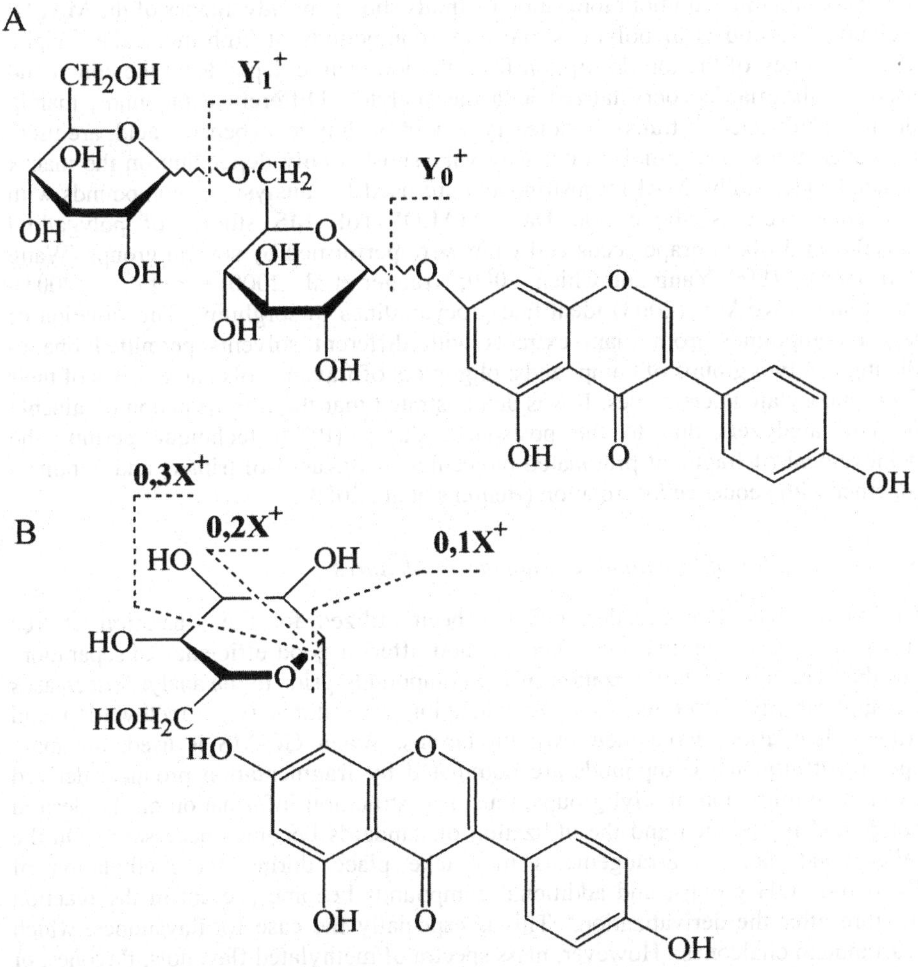

Figure 2.3 Fragmentation pattern and daughter ion nomenclature of different glycosides of isoflavonoids. (A) Genistein-7-O-glucosyl-glucoside. (B) Genistein-8-C-glucoside.

Studies on the application of the MALDI ToF technique for the quantitative analysis of flavonoid glycosides also were performed in the same laboratory; in this case the proper choice of the matrix and the deposition of samples in target wells was very important (Frison and Sporns, 2002). MALDI ToF also has been used for the analysis of condensed tannins isolated from plant material. In the first paper on this subject (Ohnishi-Kameyama et al., 1997), the MALDI and FAB ionization for the identification of flavan-3-ol oligomers in apples was compared. The presence of up to pentadecamer polymers composed of catechins was reported. However, only tetramers were identified in the same plant source when FAB ionization was used in earlier experiments (Self et al., 1986). This comparison of results obtainable with both ionization methods shows the advantages of the MALDI technique in studies of polymers and other compounds of high molecular weight. The efficiency of the ion desorption from the ion source depends on the compound used for the matrix cocrystalized with the analyte. Different compounds, mainly organic acids such as trans-3-indoleacrylic acid or dihydroxybenzoic acid, are used. However, the signal-to-noise ratio may vary substantially depending on the matrix utilized. Generally, MALDI instruments are used for analysis of compounds with molecular weights above 500 Da. MALDI ToF MS studies of polygalloyl polyflavan-3-ols in grape seeds and wine were performed by several groups (Wang and Sporns, 1999; Yang and Chien, 2000; Krueger et al., 2000; Perret et al., 2003). Reed and coworkers (2003) identified procyanidines in sorghum. The isolation of target compounds from plant extracts with different solvents permitted one to distinguish two groups of compounds: oligomers of flavan-3-ols and a series of their 5-O-glucosylated derivatives. It was demonstrated that the fragmentation obtainable in ToF analyzers due to the postsource decay (PSD) technique permits the achievement of fragment-protonated molecules of flavan-3-ol trimers and tetramers together with sequence information (Behrens et al., 2003).

3.2 Identification of Flavonoid Conjugates in Mixtures

During the past few decades, GC has been utilized for the separation of free flavonoids present in mixtures. This method offers a good efficiency of separation, but the necessity of derivatization of the compounds prior to the analysis decreases its applicability. Moreover, the derivatization procedures (e.g., methylation and trimethylsilylation) have their own limitations. When GC-MS is used, the mass spectra of silylated compounds are dominated by fragmentation products derived from the elimination of silyl groups; thus, less structural information on the desired compound is obtained and the utilization of standards becomes necessary. On the other hand, some rearrangements may take place during the methylation of flavonoid –OH groups, and additional compounds become present in the reaction mixture after the derivatization. This is especially the case for flavanones, which rearrange to chalcones. However, mass spectra of methylated flavonols, flavones, or isoflavones give good structural information about the corresponding isomers. The comparison of mass spectra of silylated and methylated isoflavones from extracts obtained from lupine roots was previously demonstrated (Stobiecki and Wojtaszek, 1990). It is possible to obtain additional information when CD_3I is used for the methylation of flavonoids, where the placement of native methyl group on the aglycone can be established (Bednarek et al., 2001). The chromatographic separation of the trimethylsilyl derivatives of components of a given mixture was better than that of the methylated compounds, but the mass spectra of methylated isoflavones provided more structural information.

    The two other chromatographic techniques, LC and CE, mentioned before, allow one to work without the chemical derivatization of analytes and permit the simultaneous separation of flavonoid aglycones and glycosidic conjugates. The combination of LC or CE with MS permits the use of different soft ionization methods. The most important are atmospheric pressure ionization, ESI and APCI, used in negative or positive ion mode. Other ionization methods, such as thermospray (TSP) and continuous flow fast atom bombardment (CF FAB), are no longer in common use due to their technical complications. However, significant descriptions of the use of LC/TSP/MS or LC/CF FAB/MS for the identification (profiling) of secondary metabolites were published in the last decade of the twentieth century (Stobiecki, 2000, 2001). Various analyzers are widely used in the MS, both high resolution (time of flight) and low resolution (quadrupole and ion trap). All the mentioned analyzers may be utilized in CID MS/MS mode. For the detection of flavonoid conjugates after their separation with LC or CE systems, other instruments besides MS are also commonly applied as detectors, including UV detectors [single wavelength or diode array (UV DAD)], electrochemical, laser-induced fluorescence (LIF), and NMR (Sumner et al., 2003). The sensitivity of LC-MS or CE-MS systems depends on the detector applied. The highest sensitivity achievable is estimated to approach 10^{-21} mole in the case of CE combined with a LIF detector. The detection limit of LC-MS or LC-UV instruments is in the range of 10^{-12} to 10^{-15} mole, whereas the LC-NMR sensitivity is much lower and the minimal amount of compound that can be detected is around 10^{-5} mole.

Many reports describe the combination of different detectors coupled in line. The most promising system, permitting in many cases an unambiguous structural characterization of components in extract from unknown source, consists of UV DAD detector, NMR, and mass spectrometer in line. The analysis of plant extracts using this system was demonstrated for the first time almost 10 years ago (Wolfender et al., 1998).

The structural elucidation of compounds using NMR experiments and their separation by HPLC are well-established analytical methods. This is why the connection of LC and NMR spectroscopy for analysis of mixtures of unknown compounds was originally devised. The first such experiment was performed in 1978 (Watanabe and Niki, 1978). However, until the mid-1990s, not much progress was achieved in the development of this technique. Several reviews discussed serious obstacles, which had to be overcome to develop LC-NMR as a useful method (Albert, 1999; Wolfender et al., 2003). The major problems concerned a low sensitivity of NMR, the presence of protonated solvents used in the reversed phase HPLC covering some signals of protons of analytes, and the possibility of running only ^{1}H-NMR experiments.

Most of these problems have been solved partially within the last decade. The low sensitivity of the NMR techniques has been addressed in several ways. Using high-field magnets (at least 500 and up to 750–900 MHz) in spectrometers connected to HPLC gives satisfactory results (Wolfender et al., 2001). Another method for increasing the sensitivity is a "stop-flow" mode of the LC-NMR system operation, allowing for a prolonged time of acquisition of NMR spectra, of up to several days. In this method, the software controlling the whole system switches off the HPLC pump in the moment when a chromatographic peak, registered by another detector (most frequently UV), reaches the NMR probe (Holt et al., 1998). Different NMR experiments then may be performed, including two-dimensional ^{1}H-^{13}C

HSQC and HMBC spectra that could be obtained using even 10 μg of sample (Wolfender et al., 2001). Preconcentration of the sample, using higher-field strength of the NMR instrument and optimization of the column parameters, further improve the detection limit by an order of magnitude (Hansen et al., 1999). Signals of solvents most frequently used in the reversed-phase chromatographic systems (CH_3CN-D_2O and CH_3OD-D_2O) may be suppressed using the WET (water suppression-enhanced through t_1 effects) technique (Smallcombe et al., 1995).

The LC-NMR technique has been applied successfully in several laboratories for the studies of flavonoid compounds from different plant species (Wolfender et al., 1997; Hansen et al., 1999; Lommen et al., 2000; Vilegas et al., 2000; Andrade et al., 2002; Queiroz et al., 2002; Le Gall et al., 2003; de Rijke et al., 2004a; Waridel et al., 2004). In most cases, the LC-NMR data are supported by results obtained with different types of LC-MS experiments. However, even though the direct connection of an HPLC with both spectrometers in one system is possible, it is not a frequent arrangement. The reason for this is the need of use of deuterated water as a solvent for NMR. In these conditions, some protons in the flavonoid molecules (especially hydroxyl groups of the aglycone as well as those of the glycoside moieties) are exchanged easily for deuterium atoms, causing the increase of the apparent molecular weights of analytes and the eventual differences in the MS spectra. In these cases, running additional LC-MS with H_2O instead of D_2O in the mobile phase is recommended (Hansen et al., 1999).

The application of LC-NMR provides the most valuable structural information in studies of mixtures of flavonoids of natural origin and allows for a proper identification of compounds indistinguishable using other methods. Hansen et al. (1999) have shown in *Hypericum perforatum* that two different glycosides of quercitin exist, which coelute from the RP C-18 column and have the same MS and MS/MS spectra. However, according to the differences in the NMR spectra, they could be identified as quercitin 3-β-D-glucoside and 3-β-D-galactoside. The same compounds were found and similarly identified in the apple peel by Lommen et al. (2000). However, in this material, in addition there were three quercitin pentosides present, which could be identified only using LC-NMR as 3-α-L-arabinofuranoside, 3-β-D-xyloside, and most probably 3-β-L-arbinopyranoside. Similarly to the above examples, two different kempferol glycosides, obtained from transgenic tomato fruits, eluted from the column very close to each other and had almost identical MS spectra (LeGall et al., 2003). These compounds could be differentiated only on the basis of the spectra acquired in the LC-NMR system. The isoflavones formononetin and biochanin A along with their 7-O-β-D-glucosides and malonyl-glucosides were found in clover (*Trifolium pratense*) leaves (de Rijke et al., 2004b). Isomeric forms of malonyl-glucosides of both aglycones were present, and according to the LC-NMR analysis were identified to differ in the 4″ or 6″ esterification of the glucose ring with malonic acid.

Successful identification and quantification of components in analyzed samples is not only dependent on the applied detectors. The influence of the composition of the liquid phase on results obtained with mass spectrometry detectors was studied for different classes of compounds (Rauha et al., 2001; Zhao et al., 2002; Huck et al., 2002). In addition, different interfaces applied for the coupling of LC with MS were described (Gelpi, 2002; Niessen, 1999). The type of LC columns used also influences the quality of data collected. The application of columns with SiO_2 particles of 3 μm diameter as well as the application of RP C-30 stationary phase (Vilegas et al., 2000) may improve the resolution and further reduce the time of analysis.

Most of the described LC separations of flavonoid mixtures obtained from different biological materials were performed on reversed-phase C-18 columns with acidified mobile phases, in most cases with acetic or formic acids. A strong suppression of cationized molecules ($[M+Na]^+$ and $[M+K]^+$ ions) was observed in mass spectra due to the acid addition to the phase (Stobiecki et al., 1997). An increase in sensitivity may be achieved this way owing to the presence of single $[M+H]^+$ ions. Comparable effects may be achieved after the addition of ammonium acetate to the mobile phase; in this case strong ammonium adduct ions $[M+NH_4]^+$ are observed (de Rijke et al., 2003). The flow rate of the mobile phase is an important parameter when API methods are applied. In LC-MS systems the column flow varies between 0.2 ml/min to 1 ml/min, for column diameters of 2 mm and 4 mm, respectively. The decrease of the column diameter may improve the sensitivity due to a lower dilution of the analyte in the liquid phase. The accessible sensitivity during the mass spectrometric analysis of flavonoids is in the low nanogram level of a single compound injected onto the column. However, in the case of extracts from biological origin, the sensitivity may decrease due to the coelution of different natural products and the suppression of the ionization efficiency of the target compounds. Extracts from 200 mg of fresh weight of plant tissue material (root or green parts) give reasonable amounts of natural products to obtain good quality single-ion chromatograms and mass spectra. The additional purification and/or preconcentration of target flavonoid compounds is not necessary; however, solid-phase extraction often was used to improve the quality of obtained results. On the other hand, many efforts have been undertaken to improve the sample preparation procedures and instrumental identification methods in order to achieve the sensitivity sufficient for secondary metabolites analysis in single cells. At present the utilization of micro- and capillary LC or CE allows one to approach this limit of sensitivity.

During LC/MS system utilization, increased structural information may be obtained when tandem mass spectrometric approaches are utilized. Different types of analyzers can be applied including triple quadrupole, quadrupole and time of flight, or ion traps. The last system makes it possible to perform multiple MS/MS experiments (March, 1997). From the MS^n product ion mass spectra one can achieve data about structure of an aglycone (Sanchez-Rabaneda et al., 2003; de Rijke et al., 2003) and differentiation of C-6 and C-8 substituted glycosides (Sikorska et al., 2003; Waridel et al., 2001) and can draw conclusions about patterns of O-glycosylation (Cuyckens and Claeys, 2004). During MS^n experiments it is important to establish proper conditions of mass spectrometric analysis. These include the voltage between the entrance slit and skimmer or choice of parent ions. Both characteristic fragmentation pathways of parent ions and their relative intensities bring information about the structure of flavonoid glycosides present in the mixture. Comparison of positive ion mass spectra registered during LC/MS and LC/MS/MS analyses with ESI ionization is presented on Figures 2.4 and 2.5. In normal mass spectra, protonated molecules $[M+H]^+$ ions and Y_n^+ fragment ions are present, created after the cleavage of glycosidic bonds. Tandem experiments after further fragmentation of Y_0^+ ions allow the identification of the structure of an aglycone in O-glycosides. This approach creates possibilities for establishing mass spectral data bases of flavonoid glycosides or free aglycones (Bristow et al., 2004). However, during analyses of crude extracts, the coelution of different secondary metabolites is possible and in this situation simultaneous ionization of compounds in the column eluate occurs; sometimes ratios of ion intensities of target compound with these, originating from other substances, are very low. In this situation, the application of MS/MS systems is useful, enabling the identification of natural product in extracts of complex composition.

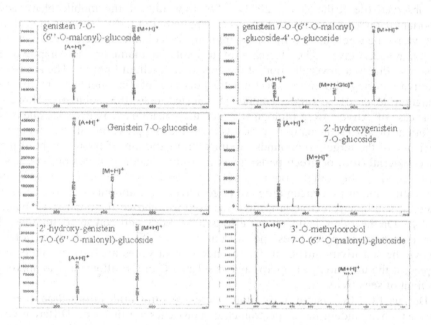

Figure 2.4 *Mass spectra of different isoflavonoid glycosides obtained using ESI HPLC-MS system.*

Metabolite profiling of flavonoids and isoflavonoids and their conjugates gained a wide interest in last few years. Many papers presenting applications of LC-UV or LC-MS systems for the analysis of the above-mentioned natural products in tissues of different plant species have been presented. Roles of isoflavones under the influence of abiotic stress or as phytoalexins or phytoanticipins during interaction of plants with pathogenic microorganisms were studied (Graham, 1991a, 1991b; Lozovaya et al., 2004; Bednarek et al., 2001, 2003). The same group of compounds was analyzed in different plant tissues (Gu and Gu, 2001; de Rijke et al., 2001, 2004a, 2004b; Klejdus et al., 2001; Lin et al., 2000; Sumner et al., 1996). Numerous conjugates of flavones and flavonols were identified in extracts obtained from herbs, vegetables, fruits, and berries or juices. In the last few years, numerous papers related to the analysis of anthocyanins in wine and grape or catechin derivatives in green tea also were published (reviewed in Justesen et al., 1998; Stobiecki, 2000, 2001; Careri et al., 1999, 2002; Prasain et al., 2004).

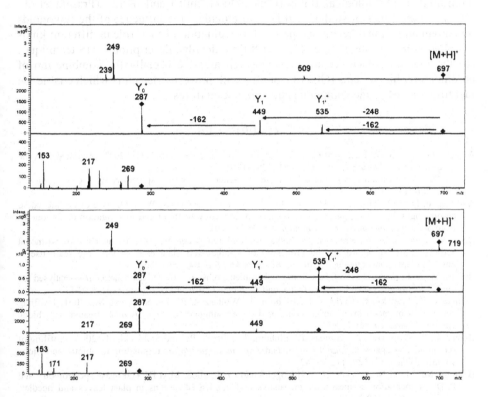

Figure 2.5 MS^n spectra of different malonyl-diglucosides of 2'-hydroxy-genistein obtained using ESI HPLC/MS ion trap system.

Different methods of CE also can be also applied for the analysis of flavonoids and their conjugates in various biological materials (Mellenthin and Gallensa, 1999; Urbanek et al., 2002). The comparison of LC and CE chromatographic systems for flavonoids analysis also has been published (Tomas-Barberan and Garcia-Viguera, 1997; Vanhoenacker et al., 2001; Smyth and Brooks, 2004; Wang and Huang, 2004).

4. CONCLUSIONS

Due to a wide range of biological activities of flavonoids consumed by humans and animals, there is a high interest in the metabolism of these compounds. The most important groups of this class of natural products are the phytoestrogenes (isoflavones: genistein derivatives) and antioxidants (anthocyanins: flavonols and flavones); their interactions with proteins (tannins) and their metabolites are monitored in physiological fluids (urine, blood, milk) and tissues (Prasain et al., 2004). These kinds of studies will help to elucidate the influence of the flavonoids on human and animal health and permit the evaluation of their role in different kinds of epidemiological studies (see Chapter 8 for a detailed description). MS techniques are the methods of choice in these research areas, especially the combination of chromatographic systems (GC, LC, or CE) with powerful detectors which enable the identification of single compounds in complex mixtures.

5. REFERENCES

Albach, D.C., Grayer, R.J., Jensen, S.R., Özgökce F., and Veitch, N.C., 2003, Acylated flavone glycosides from *Veronica*, *Phytochem* **64**: 1295-1301.

Albert, K., 1999, Liquid chromatography–nuclear magnetic resonance spectroscopy, *J Chromatogr A* **856**: 199-211.

Andrade, F. D. P., Santos, L. C., Datchler, M., Albert, K., and Vilegas, W., 2002, Use of on-line liquid chromatography–nuclear magnetic resonance spectroscopy for the rapid investigation of flavonoids from *Sorocea bomplandii*, *J Chromatogr A* **953**: 287-291.

Beck, V., Unterrieder, E., Krenn, L., Kubelka, W., and Jungbauer A., 2003, Comparison of hormonal activity (estrogen, androgen and progestin) of standardized plant extracts for large scale use in hormone replacement therapy, *J Steroid Biochem Mol Biol* **84**: 259-268.

Becchi, M., and Fraisse, D., 1989, Fast atom bombardment collision-activated dissociation/mass-analysed ion kinetic energy analysis of C-glycosidic flavonoids, *Biomed Environ Mass Spectrom* **18**: 122-130.

Bednarek, P., Franski, R., Kerhoas, L., Einhorn, J., Wojtaszek, P., and Stobiecki, M., 2001, Profiling changes in metabolism of isoflavonoids and their conjugates in *Lupinus albus* treated with biotic elicitor, *Phytochem* **56**: 77-85.

Bednarek, P., Wojtaszek, P., Kerhoas, L., Einhorn, J., Franski, R., and Stobiecki, M., 2003, Profiling of flavonoid conjugates in *Lupinus albus* and *Lupinus angustifolius* responding to abiotic and biotic stimuli, *J Chem Ecol* **29**: 1127-1142.

Behrens, M., Maie, N., Knicker, H., and Kogel-Knabner, I., 2003, MALDI TOF mass spectrometry and PSD fragmentation as means for the analysis of condensed tannins in plant leaves and needles, *Phytochem* **62**: 1159-1170.

Blaut, M., Schoefer, L., and Braune, A., 2003, Transformation of flavonoids by intestinal microorganisms, *Int J Vitamin Nutr Res* **73**: 79-87.

Boue, S. M., Wiese, T. E., Nehls, S., Burow, M.E., Elliott, S., Carter-Wientjes, C.H., Shih, B. Y., Mclachlan, J. A., and Cleveland, T. E., 2003, Evaluation of the estrogenic effects of legume extracts containing phytoestrogens, *J Agric Food Chem* **51**: 2193-2199.

Bristow, A. W. T., Webb, K. S., Lubben, A. T., and Halket, J., 2004, Reproducible product-ion tandem mass spectra on various liquid chromatography/mass spectrometry instruments for development of spectral libraries, *Rapid Comm Mass Spectrom* **18**: 1447-1454.

Bovy, A., de Vos, R., Kemper, M., Schijlen, E., Pertejo, M. A., Muir, S., Collins, G., Robinson, S., Verhoyen, M., Hughes, S., Santos-Buelga, C., and van Tunen, A., 2002, High-flavonol tomatoes resulting from the heterologous expression of the maize transcription factors genes *Lc* and *C1*, *Plant Cell* **14**: 2509-2526.

Bylka, W., Franski, R., and Stobiecki, M., 2002, Differentiation between isomeric acacetin-6-*C*-(6"-*O*-malonyl)-glucoside and acacetin-8-*C*-(6"-*O*-malonyl)-glucoside by using low-energy CID mass spectra, *J Mass Spectrom* **37**: 648-650.

Careri, M., Bianchi, F., and Corradini C., 2002, Recent advances in the application of mass spectrometry in food related analysis, *J Chromatogr A* **970:** 3-64.

Careri, M., Elviri, L., and Mangia, A., 1999, Validation of liquid chromatography ionspray mass spectrometry method for analysis of flavanones, flavones and flavonols, *Rapid Comm Mass Spectrom* **13:** 2399-2405.

Claeys, M., Li, Q., van den Heuvel, H. and Dillen, L., 1996, Mass spectrometric studies on flavonoid glucosides. In: Applications of Modern Mass Spectrometry in Plant Sciences. R.P Newton and T.J. Walton (Eds.), Clarendon Press, Oxford, 182-194.

Cuyckens, F., Ma, Y. L., Pocsfalvi, G., and Claeys, M., 2000, Tandem mass spectral strategies for the structural characterization of flavonoid glycosides, *Analusis* **28:** 888-895.

Cuyckens, F., Rozenberg, R., de Hoffman, E., and Claeys, M., 2001, Structure characterization of flavonoid O-glycosides by positive and negative nano-electrospray ionization ion trap mass spectrometry, *J Mass Spectrom* **36:** 1203-1210.

Cuyckens, F., Shahat, A. A., Pieters, L., and Claeys, M., 2002, Direct stereochemical assignement of hexose and pentose residues in flavonoid O-glycosides by fast atom bombardment and electrospray ionisation mass spectrometry, *J Mass Spectrom* **37:** 1272-1279.

Cuyckens, F., and Claeys, M., 2004, Mass spectrometry in structural analysis of flavonoids, *J Mass Spectrom* **39:** 1-15.

de Rijke, E., Zafra-Gomez, A., Ariese, F., Brinkman, U. A. T., and Gooijer, C., 2001, Determination of isoflavone glucoside malonates in *Trifolium pratense* L. (red clover) extracts: Quantification and stability studies, *J Chromatogr A* **932:** 55-64.

de Rijke, E., Zappey, H., Ariese, F., Gooijer, C., and Brinkman, U. A. T., 2003, Liquid chromatography with atmospheric pressure chemical ionization and electrospray ionization mass spectrometry of flavonoids with triple-quadrupole and ion-trap instruments, *J Chromatogr A* **984:** 45-58.

de Rijke E., de Kanter, F., Ariese, F., Brinkman, U. A. T., and Gooijer, C., 2004a, Liquid chromatography coupled to nuclear magnetic resonance spectroscopy for the identification of isoflavone glucoside malonates in *T. pretense* L. leaves, *J Sep Sci* **27:** 1061-1070.

de Rijke, E., Zappey, H., Ariese, F., Gooijer, C., and Brinkman, U. A. T., 2004b, Flavonoids in Leguminosae: Analysis of extracts of *T. pratense* L., *T. dubium* L., *T. repens* L., and *L corniculatus* L. leaves using liquid chromatography with UV, mass spectrometric and fluorescence detection, *Anal Bioanal Chem* **378:** 995-1006.

Dixon, R. A., and Paiva, N. L., 1995, Stress-induced phenylpropanoid metabolism, *Plant Cell* **7:** 1085-1097.

Dixon, R. A., and Steel, C. L., 1999, Flavonoids and isoflavonoids—a gold mine for metabolic engineering, *Trends Plant Sci* **4:** 394-400.

Domon, B., and Costello, C. E., 1988, A systematic nomenclature for carbohydrate fragmentations in FAB MS/MS spectra of glycoconjugates, *Glycoconj J* **5:** 397-409.

Fernie, A. R., Trethewey, R. N., Krotzky, A. J., and Willmitzer, L., 2004, Metabolite profiling: from diagnostic to systems biology. *Nat Rev Mol Cell Biol* **5:** 763-769.

Fiehn, O.. 2001, Combining genomics, metabolome analysis, and biochemical modeling to understand metabolic networks, *Comp Functional Genomics* **2:** 155-168.

Fisher, R., Stoger, E., Schillberg, S., Christou, P., and Twyman, R. M., 2004, Plant-based production of biopharmaceuticals, *Curr Opin Plant Biol* **7:** 152-158.

Francis, J.A., Rumbeiha, W., and Nair, M. G., 2004, Constituents in Easter lily flowers with medicinal activity, *Life Sci* **76:** 671-683.

Franski, R., Bednarek, P., Wojtaszek, P., and Stobiecki, M., 1999, Identification of flavonoid diglycosides in yellow lupin (*Lupinus luteus* L.) with mass spectrometric techniques, *J Mass Spectrom* **34:** 486-495.

Franski, R., Matlawska, I., Bylka, W., Sikorska, M., Fiedorow P., and Stobiecki, M., 2002, Differentiation of interglycosidic linkages in permethylated flavonoid glycosides from linked-scan mass spectra (B/E), *J Agric Food Chem* **50:** 976-982.

Franski, R., Eitner, K., Sikorska, M., Matlawska, I., and Stobiecki, M., 2003, Electrospray mass spectrometric decomposition of some glucuronic acid containing flavonoid diglycosides, *Phytochem Anal* **14:** 170-175.

Friedrich, W., Eberhardt, A., and Galensa, R., 2000, Investigation of proanthocyanidins by HPLC with electrospray ionization mass spectrometry, *Eur Food Res Technol* **211:** 56-64.

Frison, S., and Sporns, P., 2002, Variation in the flavonol glycosides composition of almond seedcoats as determined by Maldi-Tof mass spectrometry, *J Agric Food Chem* **50**: 6818-6822.

Fritz, K. L., Seppanen, C. M., Kurzer, M. S., and Csallany, A. S., 2003, The *in vivo* antioxidant activity of soybean isoflavones in human subjects, *Nutr Res* **23**: 479-487.

Gelpi, E., 2002, Interfaces for coupled liquid-phase separation/mass spectrometry techniques. An update on recent developments, *J Mass Spectrom* **37**: 241-253.

Graham, T. L., 1991a, A rapid, high-resolution high performance liquid chromatography profiling procedure for plant and microbial aromatic secondary metabolites, *Plant Physiol* **95**: 584-593.

Graham, T. L., 1991b, Flavonoid and isoflavonoid distribution in developing soybean seedlings tissues and in seed and root exudates, *Plant Physiol* **95**: 594-603.

Gu, L., and Gu, W., 2001, Characterisation of soy isoflavones and screening for novel malonyl glycosides using high performance liquid chromatography electrospray ionisation-mass spectrometry, *Phytochem Anal* **12**: 377-382.

Hall, R., Beale, M., Fiehn, O., Hardy, N., Sumner, L. W., and Bino, R., 2002, Plant metabolomics: the missing link in functional genomics strategies, *Plant Cell* **14**: 1437-1440.

Hansen, S. H., Jensen, A. G., Cornett, C., Bjørnsdottir, I., Taylor, S., Wright, B., and Wilson, I. D., 1999, High-performance liquid chromatography on-line coupled to high-field NMR and mass spectrometry for structure elucidation of constituents of *Hypericum perforatum* L., *Anal Chem* **71**: 5235-5241.

Harborne, J. B., and Williams, C. A., 2000, Advances in flavonoid research since 1992, *Phytochem* **55**: 481–504.

Herrera, M. C., Lugue, M. D., and de Castro, L., 2004, Ultrasound assisted extraction for analysis of phenolic compounds in strawberries, *Anal Bioanal Chem* **379**: 1106-1112.

Holt, R. M., Newman, M. J., Pullen, F. S., Richards, D. S. and Swanson A. G., 1998, High-performance liquid chromatography-NMR spectrometry-mass spectrometry: further advances in hyphenated technology, *J Mass Spectrom* **32**: 64-70.

Huck, C. W., Buchmeister, M. R. and Bonn, G. K., 2002, Fast analysis of flavonoids in plant extracts by liquid chromatography—ultraviolet absorbance detection on poly(carboxylic acid)–coated silica and electrospray ionization tandem mass spectrometric detection, *J Chromatography A* **943**: 33-38.

Jennings, K. R., 1996, MS/MS instrumentation. In: *Applications of Modern Mass Spectrometry in Plant Sciences*, R.P. Newton and T.J. Walton (Eds.), Clarendon Press, Oxford, 25-43.

Justesen, U., Knuthesen, P. and Leth, T., 1998, Quantitative analysis of flavonols, flavones and flavanones in fruits vegetables and beverages by high-performance liquid chromatography with photodiode array and mass spectrometric detection, *J Chromatogr A* **799**: 101-110.

Kaiser, C. S., Rompp, H. and Schmidt, P. C., 2004, Supercritical carbon dioxide extraction of chamomile flowers: extraction efficiency, stability, and on-line inclusion of chamomile-carbon dioxide extract in beta cyclodextrin, *Phytochem Anal* **15**: 249-256.

Kazuma, K., Noda, N. and Suzuki, M., 2003, Malonylated flavonol glycosides from the petals of *Clitoria ternatea*, *Phytochem* **62**: 229–237.

Klejdus, B., Vitamvasova-Sterbova, D. and Kuban, V., 2001, Identification of isoflavone conjugates in red clover (*Trifolium pratense*) by liquid chromatography-mass spectrometry after two-dimensional solid phase extraction, *Anal Chim Acta* **450**: 81–97.

Krueger, C. G., Dopke, N. C., Treichel, P. M., Folts, J. and Reed, J. D., 2000, Matrix assisted laser desorption/ionization time of flight mass spectrometry of polygalloyl polyflavan-3-ols in grape seed extract, *J Agric Food Chem* **48**: 1663-1667.

Krueger, C. G., Vestling, M. M. and Reed, J. D., 2003, Matrix-assisted laser desorption/ionization time-of-flight mass spectrometry of heteropolyflavan-3-ols and glucosylated heteropolyflavans in sorghum [*Sorghum bicolor* (L.) Moench], *J Agric Food Chem* **51**: 538-543.

Le Gall, G., DuPont, M. S., Mellon, F. A., Davis, A. L., Collins, G. J., Verhoyen, M. E. and Colquhoun, I. J., 2003, Characterization and content of flavonoid glycosides in genetically modified tomato (*Lycopersicon esculentum*) fruits, *J Agric Food Chem* **51**: 2438-2446.

Le March, L., 2002, Cancer preventive effects of flavonoids—a review, *Biomed Pharmacother* **56**: 78-83.

Li, C. A., Meng, X. F., Winnik, B., Lee, M. J., Lu, H., Sheng, S. Q., Buckley, B. and Yang, C. S., 2001, Analysis of urinary metabolites of tea catechins by liquid chromatography/electrospray ionization mass spectrometry, *Chem Res Toxicol* **14**: 702-707.

Li, Q. M., van den Heuvel, H., Dillen, L. and Clayes, M., 1992, Differentiation of C-6- and C-8-glycosidic flavonoids by positive ion fast atom bombardment and tandem mass spectrometry, *Biol Mass Spectrom* **21**: 213-221.

Lin, L. Z., He, X. G., Lindenmaier, M., Yang, J., Cleary, M., Qiu, S.X. and Cordell, G. A., 2000, LC/ESI/MS study of the flavonoid glycoside malonates of red clover (*Trifolium pratense*), *J Agric Food Chem* **48**: 354-356.

Lommen, A., Godejohann, M., Venema, D. P., Hollman, P. C. H. and Spraul, M., 2000, Application of directly coupled HPLC-NMR-MS to the identification and confirmation of quercitin glycosides and phloretin glycosides in apple peel, *Anal Chem* **72**: 1793-1797.

Lozovaya, V. V., Lygin, A. V., Zernova, O. V., Li, S. X., Hartman, G. L. and Widholm, J. M., 2004, Isoflavonoid accumulation in soybean hairy roots upon treatment with *Fusarium solani, Plant Physiol Biochem* **42**: 671-679.

Ma, Y. L., Li, Q. M., van den Huevel, H. and Clayes, M., 1997, Characterization of flavone and flavonol aglycones by collision induced dissociation tandem mass spectrometry, *Rapid Commun Mass Sp* **11**: 136-144.

Ma, Y. L., Cuyckens, F., van den Heuvel, H. and Claeys, M., 2001, Mass spectrometric methods for the characterisation and differentiation of isomeric *O*-diglycosyl flavonoids. *Phytochem Anal* **12**: 159-165.

March, R. E., 1997, An introduction to quadrupole ion trap mass spectrometry, *J Mass Spectrom* **32**: 351-369.

March, R. E., Miao, X-S., Metcalfe, C. D., Stobiecki, M. and Marczak, L., 2004, A fragmentation study of an isoflavone glycoside, genistein-7-*O*-glucoside, using quadrupole time of flight mass spectrometry at high mass resolution, *Intern J Mass Spectrom* **232**: 171-183.

Markham, K. R. and Geiger, H., 1994, ^{1}H nuclear magnetic resonance spectroscopy of flavonoids and other their glycosides in hexadeuterodimethylsulfoxide. In: Harborne, J. B. (Eds.), *The Flavonoids, Advances in Research Since 1986.* Chapman Hill, London, 452.

Markham, K. R. and Mabry, T. J., 1975, Ultraviolet-visible and proton magnetic resonance spectroscopy of flavonoids. In: *The Flavonoids,* Harborne J. B., Mabry T. J. and Mabry H. (Eds), Chapman and Hall, London, 45-77.

Martinez-Ortega, M. V., Garcia-Parilla, M. C. and Troncoso, A. M., 2004, Comparison of different sample preparation treatments for the analysis of wine phenolic compounds in human plasma be reversed phase high-performance liquid chromatography, *Anal Chim Acta* **502**: 49-55.

McLafferty, F. W., Ed., 1983, *Tandem Mass Spectrometry*. Wiley-Interscience, New York.

Mellenthin, O. and Galensa, R., 1999, Analysis of polyphenols using capillary zone electrophoresis and HPLC: Detection of soy, lupin, and pea protein in meat products, *J Agric Food Chem* **47**: 594-602.

Naczk, M. and Shahidi, F., 2004, Extraction and analysis of phenolics in food, *J Chromatogr A* **1054**: 95-111.

Nestel, P., 2003, Isoflavones: their effects on cardiovascular risk and function, *Curr Opin Lipidol* **14**: 3-8.

Niessen, W. M. A., 1999, *Liquid Chromatography-Mass Spectrometry,* Second Edition, Marcel Dekker Inc., New York.

Ohnishi-Kameyama, M., Yanagida, A., Kanda, T. and Nagata, T., 1997, Identification of catechin oligomers from apple (*Malus pumila* cv. Fuji) in matrix assisted laser desorption/ionization time of flight mass spectrometry and fast atom bombardment mass spectrometry, *Rapid Commun Mass Spectrom* **11**: 31-36.

Perret, C., Pezet, R. and Tabacchi, R., 2003, Fractionation of grape tannins and analysis by matrix-assisted laser desorption/ionization time of flight mass spectrometry, *Phytochem Anal* **14**: 202-208.

Prasain, J. K., Wang, C-C., and Barnes, S., 2004, Mass spectrometric methods for the determination of flavonoids in biological samples, *Free Radical Bio Medicine* **37**: 1324-1350.

Queiroz, E. F., Wolfender, J. L., Atindehou, K. K., Traore, D. and Hostetmann, K., 2002, On-line identification of the antifungal constituents of *Erythrina vogelii* by liquid chromatography with tandem mass spectrometry, ultraviolet absorbance detection and nuclear magnetic resonance spectrometry combined with liquid chromatography micro-fractionation, *J Chromatogr A* **974**: 123-134.

Rauha, J-P., Vuorela, H. and Kostiainen, R., 2001, Effect of eluent on the ionization efficiency of flavonoids by ion spray atmospheric pressure chemical ionization, and atmospheric pressure photoionization mass spectrometry, *J Mass Spectrom* **36**: 1269-1280.

Robards, K., 2003, Strategies for the determination of bioactive phenols in plants, fruit and vegetables, *J Chromatogr A* **1000**: 657-691.

Rostagno, M. A., Palma, M. and Barroso, C. G., 2003, Ultrasound assisted extraction of soy isoflavones, *J Chromatogr A* **1012**: 119-128.

Rostagno, M. A., Palma, M. and Barroso, C.G., 2004, Pressurized liquid extraction of isoflavones from soybeans, *Anal Chim Acta* **522:** 169-177.

Sanchez-Rabaneda, F., Jauregui, O., Casals, I., Andres-Lacueva, C., Izquierdo-Pulido, M. and Lamuela-Raventos, R. M., 2003, Liquid chromatographic/electrospray ionization tandem mass spectrometric study of the phenolic composition of cocoa (*Theobroma cacao*), *J Mass Spectrom* **38:** 35-42.

Self, R., Eagels, J., Galletti, G. C., Mueller-Harvey, I., Hartley, R. D., Lea, A. G. H., Magnolato, D., Richli, U., Gajer, R. and Haslam, E., 1986, Fast atom bombardment mass spectrometry of polyphenols (*syn.* vegetable tannins), *Biomed Environ Mass Spectrom* **13:** 449-468.

Sfakiaos, J., Coward, L., Kirk, M. C. and Barnes, S., 1997, Intestinal uptake and biliary excretion of the isoflavone genistein in rats, *J Nutr* **127:** 1260-1269.

Shirley, B. W., 1996, Flavonoids biosynthesis "new" function for an "old" pathway, *Trends Plant Sci* **1:** 377-382.

Sikorska, M., Matlawska, I., Franski, R. and Stobiecki, M., 2003, Application of mass spectrometric techniques for structural analysis of apigenin 8-*C*-(6-*O*-glucopyranosyl) glucopyranoside—a novel flavonoid *C*-diglycoside, *Rapid Comm Mass Spectrom* **17:** 1380-1382.

Smallcombe, S. H., Patt, S. L. and Keiffer, P. A., 1995, WET solvent suppression and its application to LC-NMR and high-resolution NMR spectroscopy, *J Magn Res Ser A* **117:** 295-303.

Smyth, W. F. and Brooks, P., 2004, A critical evaluation of high performance liquid chromatography-electrospray ionization-mass spectrometry and capillary electrophoresis-electrospray-mass spectrometry for the detection and determination of small molecules of significance in clinical and forensic studies, *Electrophoresis* **25:** 1413-1446.

Stobiecki M. and Makkar H.P.S., 2004, Recent advances in analytical methods for identification of phenolic compounds, In: Proceedings of Fourth Workshop on Antinutritional Factors in Legume Seeds and Oilseeds, 8-10.03.2004 Toledo, Spain, eds. M. Muzquiz, G.D. Hill, C. Cuadado, M.M. Pedrosa, C. Burbano, Wageningen Academic Publisher, 11-28.

Stobiecki, M., Ollechnowicz-Stepien, W., Rzadkowska-Bodalska, H., Cisowski, W. and Budko, E., 1988, Identification of flavonoid glycosides isolated from plants by Fast Atom Bombardment mass spectrometry and gas chromatography/mass spectrometry, *Biomed Environm Mass Spectrom* **15:** 589-594.

Stobiecki, M. and Wojtaszek, P., 1990, Application of gas chromatography-mass spectrometry for identification of isoflavonoids in lupin root extracts, *J Chromatogr* **508:** 391-398.

Stobiecki, M., 2000, Review—Application of mass spectrometry for identification and structural studies of flavonoid glycosides, *Phytochem* **54:** 237-256.

Stobiecki, M., 2001, Applications of separation techniques hyphenated to mass spectrometer for metabolic profiling, *Curr Opin Chem* **5:** 89-111.

Stobiecki, M., Wojtaszek, P. and Gulewicz, K., 1997, Application of solid phase extraction for profiling quinolizidine alkaloids and phenolic compounds in *Lupinus albus*, *Phytochem Anal* **8:** 153-158.

Stobiecki, M., Carr, S. A., Seviert, H-J., Reinhold, V. N. and Goldman, P., 1984, The contribution of bacterial microflora in metabolic profiling, *32nd ASMS Annual Conference on Mass Spectrometry and Allied Topics. San Antonio, Texas 27.05.-1.06. 1984.* Abstracts, 580-581.

Sumner, L. W., Mendes, P. and Dixon, R. A., 2003, Plant metabolomics: large-scale phytochemistry in the functional genomics era, *Phytochem* **62:** 817-836.

Sumner, L. W., Paiva, N. L., Dixon, R. A. and Geno, P. W., 1996, High performance liquid chromatography/continuous-flow liquid secondary ions mass spectrometry of flavonoid glycosides in leguminous plant extracts, *J Mass Spectrom* **31:** 472-485.

Tomas-Barberan, F. A. and Garcia-Viguera, C., 1997, Capillary electrophoresis versus HPLC in plant polyphenol analysis, *Analusis* **25:** M23-M25.

Tomas-Barberan, F. A., Gil, M. I., Cremin, P., Waterhouse, A. L., Hess-Pierce, B. and Kader, A. A., 2001, HPLC-DAD-ESIMS analysis of phenolic compounds in nectarines, peaches, and plums, *J Agric Food Chem* **49:** 4748-4760.

Trethewey, R. N., 2004, Metabolite profiling as an aid to metabolic engineering in plants. *Curr Opin Plant Biol* **7:** 196-201.

Urbanek, M., Blechtova, L., Pospisilova, M. and Polasek, M., 2002, On-line coupling of capillary isotachophoresis and capillary zone electrophoresis for the determination of flavonoids in methanolic extracts of *Hypericum perforatum* leaves or flowers, *J Chromatogr A* **958:** 261-271.

Vanhoenacker, G., de Villiers, A., Lazou, K., de Keukeleire, D. and Sandra, P., 2001, Comparison of high-performance liquid chromatography-mass spectroscopy and capillary electrophoresis-mass spectroscopy for the analysis of phenolic compounds in diethyl ether extracts of red wines, *Chromatographia* **54**: 309-315.

Vas, G. and Veckey, K., 2004, Solid-phase microextraction: a powerful sample preparation tool prior to mass spectrometric analysis, *J Mass Spectrom* **39**: 233-254.

Verpoorte, R. and Memelink, J., 2002, Engineering secondary metabolite production in plants, *Curr Opin Biotechnol* **13**: 181-187.

Vilegas, W., Vilegas, J. H. Y., Dachtler, M., Glaser, T. and Albert, K., 2000, Application of on-line C_{30} RP-HPLC-NMR for the analysis of flavonoids from leaf extract of *Mayatenus aquifolium*, *Phytochem Anal* **11**: 317-321.

Wang, J. and Sporns, P., 1999, Analysis of anthocyanins in red wine and fruit juice using Maldi-MS, *J Agric Food Chem* **47**: 2009-2015.

Wang, J. and Sporns, P., 2000a, MALDI-ToF MS analysis of isoflavones in soy products, *J Agric Food Chem* **48**: 5887-5892.

Wang, J. and Sporns, P., 2000b, MALDI-ToF MS analysis of food flavonol glycosides, *J Agric Food Chem* **48**: 1657-1662.

Wang, S. P. and Huang, K. J., 2004, Determination of flavonoids by high performance liquid chromatography and capillary electrophoresis, *J Chromatogr A* **1032**: 273-279.

Waridel, P., Wolfender, J. L., Lachavanne, J. B. and Hostettmann, K., 2004, Identification of the polar constituents of *Potamogeton* species by HPLC-UV with post column derivatization, HPLC-MS^n and HPLC-NMR, and isolation of a new *ent*-labdane diglycoside, *Phytochemistry* **65**: 2401-2410.

Waridel, P., Wolfender, J-L., Ndjoko, K., Hobby, K. R., Major, H. J. and Hostettmann, K., 2001, Evaluation of quadrupole time of flight tandem mass spectrometry and ion trap multiple-stage mass spectrometry for differentiation of *C*-glycosidic flavonoid isomers, *J Chromatogr A* **926**: 29-41.

Watanabe, N. and Niki, E., 1978, Direct coupling of FT-NMR to high performance liquid chromatography, *Proc Jpn Acad Ser B* **54**: 194-199.

Wolfender, J. L., Ndjoko, K. and Hostettmann, K., 2001, The potential of LC-NMR in phytochemical analysis, *Phytochem Anal* **12**: 2-22.

Wolfender, J. L., Ndjoko, K. and Hostettmann, K., 2003, Liquid chromatography with ultraviolet absorbance–mass spectrometric detection and with nuclear magnetic resonance spectroscopy: a powerful combination for the on-line structural investigation of plant metabolites, *J Chromatogr A* **1000**: 437-455.

Wolfender, J. L., Rodriguez, S., Hostettmann, K. and Hiller, W., 1997, Liquid chromatography/ultra violet/mass spectrometric and liquid chromatography/nuclear magnetic resonance spectroscopic analysis of crude extracts of Gentianaceae species, *Phytochem Anal* **8**: 97-104.

Wolfender, J-L., Rodriguez, S. and Hostettmann, K., 1998, Liquid chromatography coupled to mass spectrometry and nuclear magnetic resonance spectroscopy for screening of plant constituents, *J Chromatogr A* **794**: 299-316.

Yasuda, T., Kano, Y., Saito, K. and Ohsawa, K., 1994, Urinary and biliary metabolites of daidzin and daidzein in rats, *Biol Pharm Bull* **17**: 1369-1374.

Yang, Y. and Chien, M., 2000, Characterization of grape procyanidins using high performance liquid chromatography/mass spectrometry and matrix assisted laser desorption/ionization time of flight mass spectrometry, *J Agric Food Chem* **48**: 3990-3996.

Zhao, J. J., Yang, A. Y. and Rogers, D. 2002, Effects of liquid chromatography mobile phase buffer contents on the ionisation and fragmentation of analytes in liquid chromatography/ion spray tandem mass spectrometric determination, *J Mass Spectrom* **37**: 421-433.

CHAPTER 3

THE BIOSYNTHESIS OF FLAVONOIDS

BRENDA S.J. WINKEL

Department of Biological Sciences and Fralin Center for Biotechnology, Virginia Tech, Blacksburg, VA 24061-0346 USA

1. INTRODUCTION

Flavonoids have long sparked the interest of scientists and nonscientists alike, largely because these metabolites account for much of the red, blue, and purple pigmentation found in plants and increasingly for their association with the health benefits of wine, chocolate, and generally with diets rich in fruits and vegetables. The flavonoid pathway, illustrated in Figure 3.1, therefore has become one of the most well-studied of the many unique secondary metabolic systems that characterize the plant kingdom. There is good evidence that these systems are derived from primary metabolism, with a variety of enzymes, including members of the cytochrome P450 hydroxylase, 2-oxoglutarate-dependent dioxygenase (2-ODDs), short-chain dehydrogenase/reductase (SDR), *O*-methyltransferase (OMT), and glycosyltransferase (GT) families, having been recruited into new functions during the rapid evolution that accompanied the movement of plants onto land (Stafford, 1991). In the case of flavonoids, rudimentary forms of these compounds may have played early roles as signaling molecules and then evolved functions in processes as diverse as UV protection, growth and development, defense against herbivores and pathogens, and recruitment of pollinators and seed dispersers. The remarkable diversity of form and function of flavonoids in present-day plants has provided a rich foundation for research in areas ranging from genetics and biochemistry to chemical ecology and evolution to human health and nutrition. To date, more than 6,400 different flavonoid compounds have been described in the literature (Harborne and Baxter, 1999) and the pathways responsible for their synthesis have been characterized in detail in numerous plant species (Dixon and Steele, 1999; Harborne and Williams, 2000, 2001; Winkel-Shirley, 2001a; Springob et al., 2003).

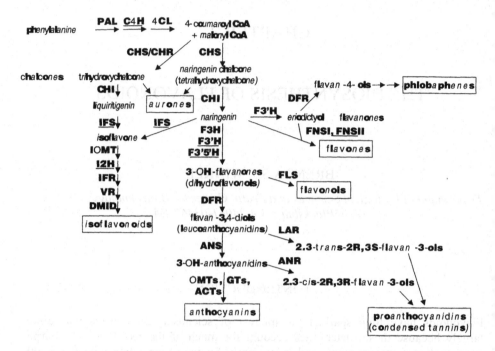

Figure 3.1 *Schematic of the flavonoid pathway showing the enzymatic steps leading to the major classes of end products, flavonols, anthocyanins, proanthocyanidins, phlobaphenes, aurones, flavones, and isoflavonoids, which are identified with boxes. Names of the major classes of intermediates are given, with names of specific compounds in italics. Enzymes are indicated with standard abbreviations in bold; names of cytochrome P450 monooxygenases that may function as membrane anchors for other flavonoid enzymes are underlined. Abbreviations: ACTs, acetyl transferases; ANR, anthocyanidin reductase; ANS, anthocyanidin synthase (also known as leucoanthocyanidin dioxygenase); C4H, cinnamate-4-hydroxylase; CHI, chalcone isomerase; CHR, chalcone reductase; CHS, chalcone synthase; 4CL, 4-coumaroyl:CoA-ligase; DFR, dihydroflavonol 4-reductase; DMID, 7,2'-dihydroxy, 4'-methoxyisoflavanol dehydratase; F3H, flavanone 3-hydroxylase; FNSI and FNSII, flavone synthase I and II; F3'H and F3'5'H, flavonoid 3' and 3'5' hydroxylase; IOMT, isoflavone O-methyltransferase; IFR, isoflavone reductase; I2'H, isoflavone 2'-hydroxylase; IFS, isoflavone synthase; LAR, leucoanthocyanidin reductase; OMTs, O-methyltransferase; PAL, phenylalanine ammonia-lyase; GTs, glucosyl transferases; VR, vestitone reductase.*

There also is a growing understanding of the diverse physiological functions of these compounds in plants and their effects, both beneficial and detrimental, when consumed by mammals. The aim of this chapter is to provide a brief historical account of the work that has led to our current understanding of the flavonoid pathway and then discuss recent advances in elucidating the biochemistry and organization of this intriguing metabolic system.

2. A BRIEF HISTORY OF RESEARCH ON FLAVONOID BIOSYNTHESIS

In-depth historical accounts of research leading to our current understanding of the structure and synthesis of flavonoid pigments can be found in two classic texts, Muriel Wheldale Onslow's *The Anthocyanin Pigments of Plants*, 2nd ed. (Onslow, 1925) and Helen Stafford's *Flavonoid Metabolism* (Stafford, 1990) (Figure 3.2), highlights of which are recounted here. Among the earliest documented studies of flavonoid pigments is Robert Boyle's 1964 *Experiments and Considerations Touching Colours* (Boyle, 1664), which describes the effects of acids and bases on the color of extracts from plant flowers and other pigmented tissues. The chemical compositions of flavonoids, specifically the blue pigment of cornflower and the red pigment of wine, was described in the mid-1800s as involving carbon, hydrogen, and oxygen (Morot, 1849; Glénard, 1858). Colorless substances also began to be recognized as being related to anthocyanin pigments during this period (Filhol, 1854; Wigand, 1862).

The first major progress toward understanding the biochemistry and genetics of flavonoid metabolism came from studies of inheritance, starting with Mendel's use of flower color in pea to study the segregation of visible traits (Mendel, 1865), and with renewed focus in the early 1900s in studies with common stock (a close relative of *Arabidopsis thaliana*) and sweet pea by Bateson, Saunders, and Punnett (1904, 1905) (Figure 3.2). The finding that purple-flowered progeny could be produced by crossing different white-flowered lines gave rise to the hypothesis that two genetic factors, *C* and *R*, were required for the production of red pigments in sweet pea, which were further modified by a third factor, *B*, producing blue or purple pigments. These critical insights opened the door to elucidating the chemical processes that underlie flavonoid biosynthesis. Palladin's theory of "Atmuhngspigmente," which proposed that plants contain chromogens that are oxidized by enzymes to pigments, including anthocyanins (Palladin, 1908), together with chemical and genetic studies in *Antirrhinum*, led Wheldale to suggest that anthocyanins are formed from a flavanone by the action of oxidase and reductase enzymes (Wheldale, 1909a, 1909b). Substantial progress was made soon thereafter on the chemical structures of flavonoids by Willstätter and colleagues, who showed that pigments from a wide range of plant species all derived from the three anthocyanins: pelargonidin, cyanidin, and delphinidin (e.g., Willstätter and Everest, 1913; Willstätter and Weil, 1916). This group also described the chemical relationship of anthocyanins with the flavonols, quercetin, kaempferol, and myricetin, as well as the presence of sugar and methoxy groups on these compounds.

C. Eric Conn lecturing on nitrogen metabolism in the General Biochemistry course at Davis.

A. Reginald Punnett (1875-1967) and William Bateson (1861-1926) in 1907.

B. Helen Stafford in her laboratory at Reed College, 1960.

D. Hans Grisebach (1926-1990).

E. Geza Hrazdina and George Wagner, early 1990s.

F. Klaus Hahlbrock with Fritz Kreuzaler at his Ph.D. examination, ~1974.

Figure 3.2 Some of the key contributors to research on flavonoid metabolism. Photographs reprinted with permission from Mary Catherine Bateson, Institute for Intercultural Studies (A); courtesy of Special Collections, Eric V. Hauser Memorial Library, Reed College (B); and with permission from the Phytochemical Society (D). Additional photographs were provided by Dr. Eric Conn (C), Dr. Geza Hrazdina and Dr. George Wagner (E) and Dr. Klaus Hahlbrock (F).

However, it was not until the 1950s that the availability of radioisotope tracer molecules fueled major new progress toward elucidating the flavonoid biosynthetic pathway (reviewed in Stafford, 1990). These studies also generated the first evidence for channeling of intermediates in phenylpropanoid biosynthesis (reviewed in Stafford, 1981; Hrazdina and Jensen, 1992). Integration of the resulting biochemical information with a wealth of accumulated genetic data led to a description of the sequence of genes involved in the production of anthocyanins in the maize aleurone in 1962 (Reddy and Coe, 1962). Enzymological information was also appearing during this time, with the first enzyme of the phenylpropanoid pathway, phenylalanine ammonia-lyase (PAL), described in 1961, followed by cinnamate 4-hydroxylase (C4H) and the first flavonoid enzyme, chalcone isomerase (CHI) from soybean in 1967 (Koukol and Conn, 1961; Moustafa and Wong, 1967; Russell and Conn, 1967) (Figure 3.2). Eight years later, Kreuzaler and Hahlbrock (1975) (Figure 3.2) described the purification from parsley of a second enzyme, chalcone synthase (CHS), which functions at the entry point into the pathway. Soon thereafter, Styles and Ceska (1977) published a scheme for flavonoid biosynthesis that incorporated some 25 structural and regulatory genes functioning in diverse tissues of maize. This was followed by the isolation of numerous additional flavonoid enzymes and the corresponding genes (reviewed in Winkel-Shirley, 2001b), the first gene, for parsley CHS, being described in 1983 by Hahlbrock's group (Kreuzaler et al., 1983). During this period, Stafford raised the possibility that the enzymes of flavonoid metabolism, like those of the phenylpropanoid pathway, might be organized as one or more enzyme complexes (Stafford, 1974), and substantial experimental evidence in support of this idea was published in the mid-1980s by Hrazdina and his colleagues (Wagner and Hrazdina, 1984; Hrazdina and Wagner, 1985) (Figure 3.2).

Numerous recent reviews have summarized the extensive progress made in the flavonoid field since 1990, including advances in deciphering the signaling pathways that regulate expression of flavonoid genes as well as mechanisms controlling the intracellular distribution of flavonoid end products (Winkel-Shirley, 2001b; Vom Endt et al., 2002; Marles et al., 2003; Springob et al., 2003; Schijlen et al., 2004; Dixon et al., 2005). In addition, the first three-dimensional structures of several flavonoid enzymes were solved in the past few years, first for CHS and CHI from *Medicago truncatula*, solved by Noel's group (Ferrer et al., 1999; Jez et al., 2000), and then for anthocyanidin synthase [(ANS) also known as leucoanthocyanidin dioxygenase (LDOX)] from *Arabidopsis* by Wilmouth et al. (2002).

Still, major gaps remain in our understanding of flavonoid biochemistry, including the structural and biochemical basis of substrate- and stereospecificity for most flavonoid enzymes and the mechanisms by which flux is distributed among the various branch pathways. Unanswered questions also remain regarding the identity of several enzymes and precise sequence of events that lead to the major classes of flavonoid endproducts.

3. RECENT ADVANCES IN IDENTIFICATION AND CHARACTERIZATION OF THE PRIMARY FLAVONOID ENZYMES

3.1. Proanthocyanidin Biosynthesis

Despite extensive biochemical and genetic efforts over the past several decades to identify the enzymes that mediate flavonoid biosynthesis, a number of key steps have resisted elucidation.   An important branch pathway that has long posed challenges in terms of defining the relevant biochemical steps is the one leading to the proanthocyanidins (a.k.a. condensed tannins) (reviewed in Dixon et al., 2005). This pathway is of substantial interest as a potential target for genetic modification to improve the yield, forage traits, and nutritional properties of crop species.   A major recent breakthrough has been the isolation of genes encoding two key enzymes—anthocyanidin reductase (ANR) and leucoanthocyanidin reductase (LAR)—that provide the initiating and extension units for proanthocyanidin biosynthesis (Figure 3.1).

ANR was initially identified through characterization of the *banyuls* locus in *Arabidopsis* (Devic et al., 1999).   The gene and enzyme now have been studied in detail in both *Arabidopsis* and *Medicago* (Xie et al., 2003, 2004a) and the enzyme activity has been described in *Camellia sinensis* and a number of other plant species (Punyasiri et al., 2004; Dixon et al., 2005).   ANR, which is closely related in sequence to dihydroflavonol reductase (DFR) as well as the phenylpropanoid and isoflavonoid enzymes, cinnamoyl-CoA reductase, cinnamoyl alcohol dehydrogenase, and vestitone reductase, converts the product of ANS/LDOX, an anthocyanidin, to a 2,3-*cis*-2R,3R-flavan 3-ol (Figure 3.3).   This discovery provided the first indication that ANS/LDOX has a role in both anthocyanin and proanthocyanidin biosynthesis and offered an explanation for the difference in stereochemistry between 2,3-*trans* isomers for flavonols and anthocyanins, and 2,3-*cis* isomers for the proanthocyanidin extension units in most plants.   This is further supported by the isolation of two mutant alleles of *Arabidopsis* ANS/LDOX, *tt18* and *tds4*, in screens for plants deficient in proanthocyanidin biosynthesis, which also provided genetic evidence that ANS/LDOX precedes ANR in the pathway (Abrahams et al., 2003; Shikazono et al., 2003).   A gene encoding another key enzyme of proanthocyanidin biosynthesis, LAR, was cloned from the legume *Desmodium uncinatum* and found to be a member of the plant reductase-epimerase-dehydrogenase (RED) superfamily, which also includes the gene for isoflavone reductase (Tanner et al., 2003).   LAR generates (2,3-*trans*) catechin from leucoanthocyanidin, competing with ANS/LDOX for substrate to produce an alternative initiating unit for proanthocyanidin biosynthesis.   Interestingly, LAR does not appear to exist in *Arabidopsis*, consistent with the presence of only 2,3-*cis* initiating and extension units in this species (Abrahams et al., 2003; Tanner et al., 2003).   Although we have moved much closer to an understanding of this pathway with the cloning and characterization of ANR and LAR, the mechanism by which the extension units are incorporated into proanthocyanidin polymers remains unknown (Dixon et al., 2005).

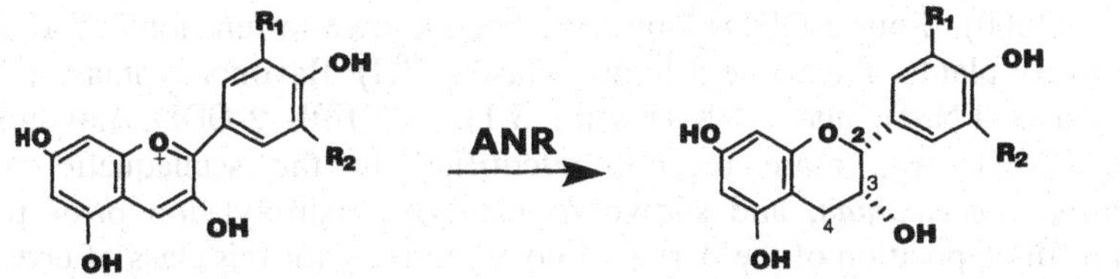

ANTHOCYANIDIN:
$R_1=R_2=H$, pelargonidin
$R_1=OH$, $R_2=H$, cyanidin
$R_1=R_2=OH$, delphinidin

2,3-cis-2R,3R-FLAVAN-3-OL:
$R_1=R_2=H$, (-)-epiafzelechin
$R_1=OH$, $R_2=H$, (-)-epicatechin
$R_1=R_2=OH$, (-)-epigallocatechin

Figure 3.3 Reaction catalyzed by anthocyanidin reductase (ANR) (Xie et al., 2003).

3.2. Dihydroflavonol Reductase

In contrast to ANR, DFR has long been known to contribute to both anthocyanidin and proanthocyanidin biosynthesis (Heller et al., 1985; Reddy et al., 1987). Recently, Shimada et al. (2004) reported the characterization of DFR enzymes from *Spinachia oleracea* and *Phytolacca americana*. As with most other members of the order Caryophyllales, anthocyanins are entirely replaced in these plants by betalain pigments, which are derived from tyrosine via dihydroxyphenylalanine by a pathway unrelated to flavonoid metabolism (Strack et al., 2003). Recombinant DFR enzymes from these species are functional *in vitro*, producing cyanidin from dihydroquercetin; however, no such activity could be detected in extracts from seedlings or mature leaves of either species. This could be due to the low expression levels of the corresponding genes, which were assessed by Northern analysis and semiquantitative RT-PCR in various tissues. It also was suggested that these enzymes participate in proanthocyanidin biosynthesis, although these compounds were not present in the vegetative tissues examined in this study. Another possibility is that the DFR enzymes in these plants have lost the ability to interact with other components of a flavonoid enzyme complex, so that only flavonols are synthesized in tissues where enzymes of the proanthocyanidin pathway, such as ANS/LDOX, ANR, or LAR, are not present. More work is needed on this interesting system to assess the specific role(s) of the DFR enzymes in these species and how the flavonoid and betalain pathways have evolved in the Caryophyllales.

3.3. The 2-ODD Enzymes

Advances in recombinant protein expression technology and substrate/product detection methods have fueled rapid advances in the biochemical characterization of numerous key flavonoid enzymes in recent years. This is particularly true of members of the 2-ODD enzyme family for which conventional biochemical analysis of plant extracts had been limited by the oxygen sensitivity of the enzyme and its susceptibility to cleavage by endogenous enzymes (Britsch and Grisebach, 1986; Lukacin et al., 2000). Four 2-ODDs long have been known to function in flavonoid metabolism in all plants: flavanone 3-hydroxylase (F3H), flavonol synthase (FLS), flavone synthase (FNSI), and ANS (Figure 3.1). A fifth 2-ODD, flavonol 6-hydroxylase (F6H), was more recently identified in the semiaquatic weed, *Chrysosplenium americanum*, and shown to catalyze hydroxylation prior to O-methylation at the 6 position of the A ring, a novel activity for this class of enzymes (Anzellotti and Ibrahim, 2000, 2004) (Figure 3.4).

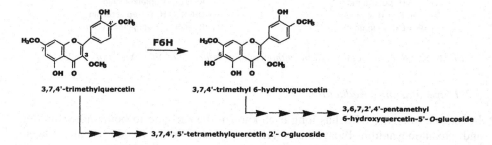

3,7,4'-trimethylquercetin

3,7,4'-trimethyl 6-hydroxyquercetin

3,6,7,2',4'-pentamethyl 6-hydroxyquercetin-5'- O-glucoside

3,7,4', 5'-tetramethylquercetin 2'- O-glucoside

Figure 3.4 *The reaction catalyzed by the 2-ODD enzyme, flavonol 6-hydroxylase (F6H) from Chrysosplenium americanum (Anzellotti and Ibrahim, 2000).*

A great deal of biochemical information on the flavonoid 2-ODD enzymes has emerged since the mid-1980s, starting with the seminal work of Lothar Britsch and colleagues (reviewed in Prescott and John, 1996). Recent studies incorporating HPLC analysis into enzyme assays using conventional radiolabeled substrates, or increasingly unlabeled substrates, have provided evidence for overlapping substrate and catalytic specificities among these enzymes. The first indications for this overlap came from studies on FLS enzymes from *Arabidopsis* and *Citrus unshiu,* which were found to exhibit substantial F3H activity in addition to their primary enzymatic activity in the terminal step of flavonol production (Prescott et al., 2002; Lukacin et al., 2003). In addition, *in vitro* experiments with recombinant ANS suggested that the enzyme could produce not only its natural product, cyanidin, but depending on the C-4 stereochemistry of the leucocyanidin substrate possibly also dihydroquercetin and quercetin (Turnbull et al., 2003). Turnbull et al. (2004) subsequently compared the substrate preferences and stereospecificities of F3H, FLS, and ANS enzymes from *Arabidopsis* and *Petunia hybrida.* The results of this study indicated that the reactions catalyzed by the four enzymes all proceed via an initial oxidation at the C-3 position. However, it also grouped F3H/FNSI and FLS/ANS into two distinct subfamilies. The former have a very narrow substrate preference, recognize only a single stereoisomer of their natural substrate, naringenin, and are classified as C-3 β-face oxygenases. In contrast, FLS and ANS are C-3 α-face oxygenases that efficiently recognize different stereoisomers of their natural, as well as several unnatural, substrates. The flavonoid 2-ODDs clearly offer an excellent experimental system for defining the molecular determinants of the substrate and stereoselectivities of this important class of enzymes. Domain swapping experiments between *Petunia* F3H and *Citrus* FLS already have provided evidence that the selectivity of F3H is not conferred by the 52 C-terminal amino acids (Wellmann et al., 2004). Further enhancement of methods for biochemical characterization of 2-ODDs, such as a recently described fluorescence-based assay (McNeill et al., 2004), should help facilitate these types of studies in the future. Enormous promise also lies in combining biochemical analyses with structural approaches to understanding 2-ODD enzyme function. A high-resolution crystal structure for *Arabidopsis* ANS, including one for the enzyme in complex with enantiomerically pure dihydroquercetin, already is providing new insights into the mechanism of stereoselective C-3 hydroxylation by this enzyme that are likely also to apply to other members of this class of enzymes (Wilmouth et al., 2002).

3.4. Common Activities Specified by Members of Different Enzyme Classes

There are now three different examples of the overlapping contributions of 2-ODDs and cytochrome P450s in the evolution of flavonoid metabolism. Work by Shin-ichi Ayabe's group (Sawada et al., 2002; Sawada and Ayabe, 2005) involving site-directed mutagenesis of the P450 enzyme, isoflavone synthase (IFS), suggests that this enzyme might have been derived from an ancestral CYP93 enzyme that also gave rise to flavanone 2-hydroxylase and flavone synthase II (FNSII) (Figure 3.1). These experiments indicate that IFS could have evolved via an intermediate enzyme having flavanone 3-hydroxylase activity. Interestingly, FNS activity is specified by either a P450 (FNSII) or a 2-ODD (FNSI) enzyme in different plant species (Britsch, 1990; Martens and Forkmann, 1999). It is conceivable that F3H was represented by members of these two different enzyme classes in ancestral plant species. There is now also evidence that F6H activity, first described as a 2-ODD enzyme in *Chrysosplenium americanum* (Anzellotti and Ibrahim, 2000, 2004), is specified by a P450 monooxygenase in *Tagetes* species (Halbwirth et al., 2004). The activities differ in that the enzyme from *Chrysosplenium americanum* accepts methylated flavonols, while the *Tagetes* enzyme does not. These findings underscore the remarkable evolutionary plasticity of plant secondary metabolism and the potential to achieve similar ends through entirely unrelated means. It could be interesting to consider whether similar evolutionary processes may have occurred in other metabolic systems. Further evidence for the biochemical plasticity of the flavonoid pathway comes from analyses of the substrate and product specificities of FLS and DFR. These enzymes carry out unrelated reactions in flavonoid biosynthesis (Figure 3.1) and belong to entirely different enzyme classes: the 2-ODD and SDR superfamilies. However, both enzymes utilize primarily dihydroflavonols, but also have been shown to convert flavanones (e.g., naringenin and eriodictyol) to dihydroflavonols in the case of *Arabidopsis* and *Citrus* FLS, an activity traditionally associated with F3H (Prescott et al., 2002; Lukacin et al., 2003), or to flavan 4-ols in the case of DFR from *Dahlia variabilis, Gerbera hybrida, Zea mays,* and *Medicago,* an activity previously named flavanone 4-reductase (FNR) (Fischer et al., 1988, 2003; Halbwirth et al., 2003; Xie et al., 2004b) (Figure 3.5).

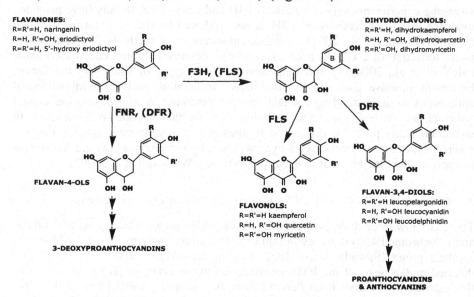

Figure 3.5 *Flexible substrate recognition of flavonol synthase (FLS) and dihydroflavonol 4-reductase (DFR) (Prescott et al., 2002; Lukacin et al., 2003; Xie et al., 2004b). Abbreviations: F3H, flavanone 3-hydroxylase; FNR, flavanone 4-reductase.*

The broad substrate specificity of DFR has led to the suggestion that it is the level of F3H activity that ultimately determines the type of anthocyanin that accumulates in maize silks (Halbwirth et al., 2003). Although FLS and DFR do recognize a similar set of substrates, a difference between these enzymes is that only DFR exhibits a strong stereospecificity. For example, two recently characterized DFR enzymes from *Medicago* convert only the 2R,3R form of dihydroquercetin (Xie et al., 2004b), consistent with the stereochemistry of naturally occurring flavonols and anthocyanins. These enzymes therefore are likely to provide a useful experimental system for understanding the determinants of substrate and stereospecificity in individual enzymes.

Efforts also continue to be focused on elucidating how the preference of DFR for its natural dihydroflavonol substrates is determined (Punyasiri et al., 2004; Xie et al., 2004b). DFR enzymes from different plant species vary in acceptance of dihydroflavonols with different B-ring oxidation states (Figure 3.5), a characteristic that has been of long-standing interest with regard to engineering flower color (Meyer et al., 1987). There long has been evidence for the role of a 26 amino acid region, and more recently for one residue in particular, in determining DFR substrate specificity (Beld et al., 1989; Johnson et al., 2001). However, the recent analysis of the *Medicago* enzymes shows that the molecular basis of this specificity, including the roles of other active site residues, remains to be elucidated (Xie et al., 2004b). Accumulating information on members of the SDR class of enzymes in a wide range of species, including more than 40 three-dimensional structures now deposited in the protein database, suggests that new insights will soon be forthcoming (Oppermann et al., 2003).

4. ADVANCES IN ELUCIDATING FLAVONOID MODIFICATION REACTIONS

In addition to the ongoing interest in the primary enzymes of flavonoid metabolism, there has been a growing focus on enzymes that perform the substitution reactions responsible for the tremendous structural and functional diversity of flavonoids found in nature. This is driven in part by opportunities for metabolic engineering of flavonoid biosynthesis for agronomic and nutritional improvement of plants. Methylases, acetyltransferases, and GTs confer many of the ultimate chemical and bioactive properties of flavonoids, for example, modulating flower color or enhancing the activity of these compounds for use in defense against herbivores and pathogens. These modifications also may play an important part in controlling the distribution of flux across branch pathways by altering intermediates to favor utilization by one or more competing enzymes at critical branch points. It is noteworthy that many of the substitution reactions are quite species-specific and most are still poorly understood both in terms of enzymology and their biochemical or physiological significance.

4.1. Flavonoid Methyltransferases

A variety of *O*-methylated flavonoids have been described, involving substitutions at the 3, 5, 6, 7, 4' or 5' positions (for example, Figure 3.4). Huang et al. (2004) have recently described the purification and biochemical characterization of a 3-*O*-methyltransferase from *Serratula tinctoria* that exhibits a preference for quercetin over other flavonols. This enzyme is of particular interest because 3-*O*-methylation confers a number of distinct activities to quercetin, including antiinflammatory and antiviral properties. Thus, the enzyme clearly has potential metabolic engineering applications and cloning of the corresponding gene is underway. Other studies have focused on how the stereo- and regiospecific addition of sugars and other groups is controlled. Comparison of the substrate preferences and kinetic characteristics of variants of a class I OMT from the ice plant, *Mesembryanthemum crystallinum*, indicates that the N-terminus of this enzyme is an important determinant of these properties (Vogt, 2004). Deletion of 11 N-terminal residues allows the enzyme to add sugars to not only the 3' and 6 positions observed for the native plant protein, but also to the 5 position (Figure 3.6). However, it also reduces the overall catalytic efficiency of the enzyme, apparently affecting turnover rate rather than substrate binding affinity. Efforts are underway to solve the crystal structure of this enzyme in order to understand these observations at a molecular level. Interestingly, characterization of a novel 4'-O-methyltransferase from *Catharanthus roseus* indicates that, at least for the four major 2-ODDs that function in flavonoid synthesis, methylation at the 4' position did not reduce substrate utilization (Schröder et al., 2004). This suggests that, although this modification may alter the biological activity of flavonoid end products, it does not by itself control flux into different branch pathways.

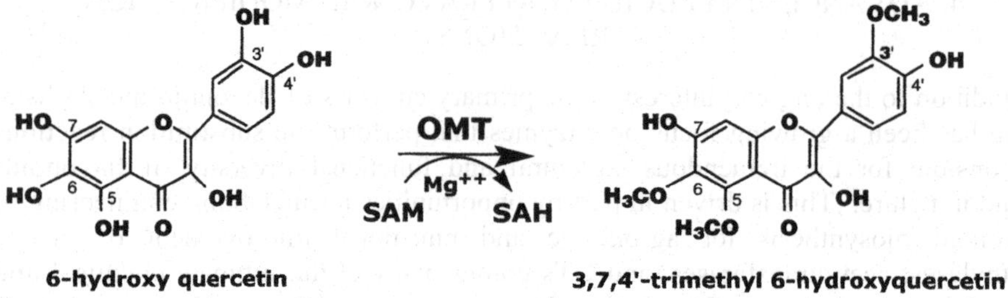

6-hydroxy quercetin   **3,7,4'-trimethyl 6-hydroxyquercetin**

Figure 3.6 Regiospecific methylation of 6-hydroxy quercetin catalyzed by a phenylpropanoid and flavonoid O-methyltransferase from the ice plant, Mesembryanthemum crystallinum (Vogt, 2004).

4.2. Flavonoid Glycosyltransferases

The addition of sugar groups has been well documented to enhance the solubility of flavonoids, as well as many other metabolites, and is likely to be critical for the transport and storage of these compounds at the final destinations in the vacuole or cell wall.

Most flavonoid end products exhibit complex glycosylation patterns involving the addition of one or more glucose, rhamnose, or other sugars. Recent progress in understanding how this occurs includes elucidation of the terminal steps in the biosynthesis of maysin, a flavone produced in maize that provides resistance against the corn earworm, *Helicoverpa zea*. Quantitative trait locus and metabolite analysis of the maize *salmon silk* (sm) phenotype by McMullen and colleagues (2004) has shown that the *sm1* and *sm2* genes encode or control a glucose modification enzyme and a rhamnosyl-transferase activity, respectively, in the maysin pathway. This analysis has made possible the ordering of the final intermediates in the pathway from the flavone, luteolin, to isoorientin, to rhamnosylisoorientin, and then to the

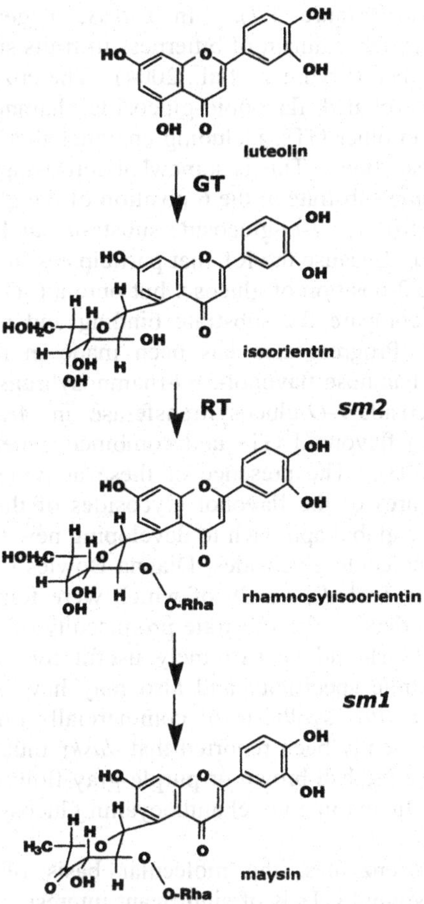

Figure 3.7 *Pathway for the synthesis of maysin from the flavone luteolin. The sm1 and sm2 loci are now known to encode or control the terminal steps (Vogt, 2004). Abbreviations: GT, glucosyl transferase; RT, rhamnosyl transferase.*

bioactive product, maysin (Figure 3.7). In *Citrus*, a gene encoding a 1,2 rhamnosyltransferase, a key determinant of bitterness in fruits such as pummelo and grapefruit, has been described (Frydman et al., 2004). The enzyme is more closely related to the only other cloned flavonoid-glucoside rhamnosyl transferase, an enzyme from *Petunia*, than other GTs, including enzymes that conjugate rhamnose directly to the flavonoid backbone. This is somewhat surprising as the *Petunia* adds the sugar to a 3-*O*-glucoside substrate at the 6 position of the glucose moiety, while the *Citrus* enzyme is specific for 7-*O*-glucoside substrates and places rhamnose at the 2 position (Figure 3.8). Because the RT that participates in maysin biosynthesis also adds rhamnose at the 2 position of glucose, but onto a 6-*O*-glucoside substrate, it will be interesting to compare the substrate binding and reaction mechanisms for these three enzymes. Progress also has been made in the identification of genes encoding UDP-rhamnose:flavonol-3-*O*-rhamnosyltransferase and UDP-glucose:flavonol-3-*O*-glycoside-7-*O*-glucosyltransferase in *Arabidopsis* based on homology to other known flavonoid GTs and combined genetic and biochemical analyses (Jones et al., 2003). The presence of these activities in *Arabidopsis* is consistent with the structures of the flavonol glycosides of this species (Veit and Pauli, 1999). Using a more global approach to developing new tools for engineering the production of small molecule glycosides, Dianna Bowles's group has surveyed 91 recombinant GTs from *Arabidopsis*, 29 of which were found to be capable of glycosylating quercetin, reflecting the substrate promiscuity of these enzymes (Lim et al., 2004). These GTs should be extremely useful for efforts to define the determinants of GT substrate specificity and also may have substantial practical value for directing the *in vitro* synthesis of commercially important mono- and diglucosides. It also has recently been reported that *dusky* mutants of the Japanese morning glory, which have reddish-brown or purple-gray flowers rather than bright red or blue flowers, are deficient in a novel anthocyanin glucosyltransferase (Morita et al., 2005).

As with many other enzymes, the molecular basis of the substrate and regioselectivity of the flavonoid GTs is of significant interest. A single amino acid change recently has been shown to increase the sugar donor specificity of a UDP-galactose:anthocyanin galactosyltransferase from *Aralia cordata* to include UDP-glucose (Kubo et al., 2004). However, UDP-galactose specificity did not appear to be associated with the same residue, indicating that much remains to be learned about the catalytic mechanism of this group of enzymes. Cloning and characterization of a novel glucuronyltransferase (UGAT) from red daisy (*Bellis perennis*) that is believed to confer solubility and stability to the pigment in flowers suggests that this enzyme has a relatively broad substrate specificity (Sawada et al., 2005). This may allow the enzyme to participate in a metabolic grid at the end of pigment biosynthesis in which either malonylation at the 6 position or glucuronosylation at the 2 position of glucose may take place first, similar to what has been proposed for the late steps in pigment biosynthesis in red *Perilla* leaves (Yamazaki et al., 1999). Together, these findings are opening new doors in understanding substrate binding and regiospecificity of the various modification

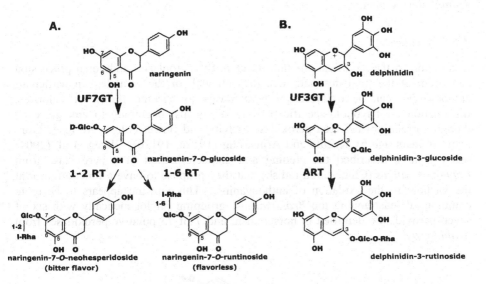

Figure 3.8 *Glycosylation of the flavanone, naringenin, in* Citrus (Frydman et al., 2004) *and the anthocyanidin, delphinidin, in* Petunia (Brugliera et al., 1994). *Abbreviations: UDPG flavonoid 3 or 7-glucosyltransferase (UF3GT or UF7GT); anthocyanin rhamnosyltransferase (RT).*

enzymes that function in secondary pathways, information that is clearly important for efforts to engineer the chemical and biological characteristics of flavonoids and other metabolites.

4.3. Flavonoid Acetyltransferases

Although acetylation of flavonoids has been known for some time, it is less common and not as well understood as the other modifications. It has been suggested that acyl groups are effective competitors for addition of water at the 2 and 4 position and may serve to stabilize the anthocyanin chromophore (Stafford, 1990). Insights on substrate selection of flavonoid acetylases recently has come from isolation and characterization of a novel anthocyanin malonyltransferase from scarlet sage (*Salvia splendens*) (Suzuki et al., 2004). This enzyme catalyzes the second malonylation reaction of anthocyanins in this species, adding malonyl at the 4 position of the 5-glucosyl moiety to produce salvianin flower pigments (Figure 3.9). Sequence comparison with the enzyme that catalyzes the previous step, malonylation at the 6 position of this glucose moiety, indicates that even these two enzymes have very similar substrate specificities; they evolved from different branches of the BAHD acyltransferase family.

4.4. Peroxidases

In addition to the modifications that occur in the cytoplasm, additional processing can occur at the final destination in the cell wall or the vacuole. In particular, anthocyanins can be oxidized by peroxidases to become brown or colorless. Information on flavonoid-specific peroxidases is just beginning to emerge, even though correlations between peroxidase activity and flower color were made more than 90 years ago by Keeble and Armstrong (1912a, 1912b). Wang et al. (2004) recently have described the cloning and characterization of a peroxidase from *Raphanus sativus* (Chinese red radish) that they propose to have a function in or at the vacuole in the oxidation of anthocyanins. This enzyme appears to be quite common, at least among the *Brassicaceae*, grouping phylogenetically with six of seven peroxidases identified in horseradish, and 7 of 73 putative peroxidases from *Arabidopsis*.

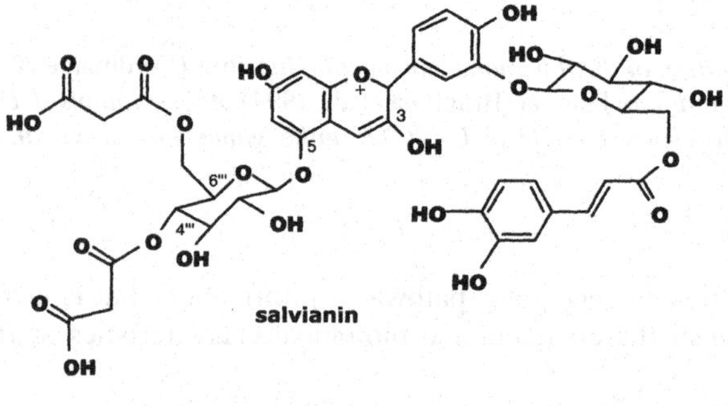

Figure 3.9 *Structure of the anthocyanin pigment, salvianin, from* Salvia splendens. *The terminal steps in the synthesis of this compound involve addition of malonyl to the 5-glycosyl moiety, first at the 6 and then at the 4 position (Suzuki et al., 2004).*

5. ENGINEERING OF FLAVONOID METABOLISM

The wealth of information that has resulted from decades of biochemical and molecular genetic characterization of flavonoid metabolism increasingly seems to underscore just how much remains to be learned about this system. Efforts to engineer flavonoid metabolism is one of the best illustrations of this, with early work resulting in some striking success stories, such as the use of a maize *DFR* gene to produce a new flower color in *Petunia* (Meyer et al., 1987). In other well-known cases, introduction of flavonoid transgenes has resulted in unstable phenotypes and in fact provided one of the first examples of cosuppression (Jorgensen, 1995). Gene downregulation using RNAi technology at least now appears to provide a way to target genes in a highly specific manner, as illustrated in modulation of flower color in *Torenia hybrida* using RNAi against CHS (Fukusaki et al., 2004).

At the same time, the unexpected outcomes of attempts to engineer flavonoid biosynthesis continue to generate new insights into how metabolism really operates in cells. For example, two recent engineering efforts have uncovered new clues as to how flux through competing branch pathways is controlled. Dixon's laboratory has shown that the bottleneck for engineering isoflavonoid synthesis in *Arabidopsis* involves competition for flavanone between the introduced enzyme, isoflavone synthase (IFS), and the endogenous FLS1 enzyme (Figure 3.1) (Liu et al., 2002). Similarly, anthocyanin production in a white-flowered *Petunia* line has been achieved, not just by enhancing DFR expression, but also by downregulating expression of FLS; the highest levels of anthocyanin production were achieved by using both approaches together (Davies et al., 2003). These findings are building a case for the FLS reaction as a key point in the regulation of flux into the branch pathways of flavonoid metabolism. There also are examples of the effects of engineering competition with the entry point enzyme, CHS, including reduced coloration of *Petunia* and *Peuraria montana* flowers upon introduction of chalcone reductase, which in legumes shifts flux into the isoflavonoid pathway (Davies et al., 1998; Joung et al., 2003). Similarly, reduced flower color and fertility were reported in tobacco upon introduction of a stilbene synthase gene, which normally sends flux into the stilbene branch pathway (Fischer et al., 1997). Some of these negative effects can be ameliorated by selecting lines with altered expression of the endogenous pathway or with moderate levels of transgene expression (for example, Fettig and Hess, 1999). All these observations indicate that a great deal remains to be learned about the mechanisms normally used by cells to distribute flux among competing branch pathways.

Reconstitution in yeast of parts of the flavonoid and isoflavonoid pathways, as well as early steps in phenylpropanoid metabolism, is another application of metabolic engineering that may provide insights into the cellular organization and regulation of these enzyme systems. Yu's group has demonstrated the feasibility of expressing CHI, F3H, and FNSII in yeast cells to produce both the natural products from supplied chalcone precursors as well as novel compounds when uncommon substrates were provided (Ralston et al., 2005). This system is proposed to offer a unique platform for functional analysis of these enzymes, including their physical and genetic interactions. However, a similar study of the first enzymes of general phenylpropanoid metabolism from poplar (two isoforms of PAL, cinnamate 4-hydroxylase, and a cytochrome P450 reductase), failed to detect channeling of *trans*-cinnamic acid, even though this has been demonstrated for the corresponding enzymes in *Nicotiana tabacum* (tobacco) cells (Rasmussen and Dixon, 1999). Individual isoforms of tobacco PAL behave very differently in terms of association with the ER and presumably other phenylpropanoid enzymes (Achnine et al., 2004), which suggests that it could be the isoform used in these experiments rather than the yeast that is responsible for the lack of corroboration of the results obtained in tobacco.

6. FRONTIERS IN THE STUDY OF FLAVONOID METABOLISM

The localization of flavonoid metabolism and how this impacts the deposition of end products is another area with a long history of study where new information and new questions continue to arise. Our laboratory recently has demonstrated the presence of CHS and CHI in the nucleus of *Arabidopsis* cells (Saslowsky et al., 2005). This finding is consistent with several reports from other groups of the presence of flavonoids in nuclei in such diverse species as *Arabidopsis*, *Brassica napus*, *Flaveria chloraefolia*, *Picea abies*, *Tsuga Canadensis*, and *Taxus baccata* (Grandmaison and Ibrahim, 1996; Hutzler et al., 1998; Kuras et al., 1999; Peer et al., 2001; Buer and Muday, 2004; Feucht et al., 2004). These studies have raised interesting questions regarding the physiological roles of flavonoids, speculated to include protecting DNA from UV and oxidative damage (Feucht et al., 2004) or controlling the transcription of genes required for growth and development such as auxin transport proteins (Grandmaison and Ibrahim, 1996; Kuras et al., 1999; Buer and Muday, 2004). However, it also begs the question of how intracellular distribution of the enzymes is controlled. There is good evidence that flavonoid synthesis in the cytoplasm proceeds via an enzyme complex, tethered to the endoplasmic reticulum by the cytochrome P450 enzyme, flavonoid 3'-hydroxylase. However, this enzyme cannot be the assembly point for an enzyme complex in the nucleus; also, only CHI is sufficiently small enough to enter the nucleus on it own, but only CHS contains a predicted nuclear localization signal (NLS). Thus, either the other enzymes required to synthesize flavonoids such as quercetin 3-sulfate in *Flaveria chloraefolia* (Grandmaison and Ibrahim, 1996) are cotransported with CHS, or NLSs are formed upon association of these enzymes. At present the mechanisms by which cells regulate the distribution of flavonoid metabolism to these different compartments remains entirely unknown.

As already alluded to, one of the most important recent developments in the effort to understand the organization and operation of flavonoid metabolism has been the application of the tools of structural biology. The release over the past 5 years of the three-dimensional structures of CHS and CHI from *Medicago truncatula* and for ANS for *Arabidopsis* has enhanced dramatically our understanding of several of the key reactions of flavonoid metabolism (Jez et al., 2000, 2002; Wilmouth et al., 2002; Austin and Noel, 2003). Homology-based modeling is proving useful also in the analysis of numerous other enzymes where structures are available for even distantly related enzymes, including the 2-ODD (F3H, FLS, FNSI, and F6H), cytochrome P450 (FNSII, IFS, and flavonoid 3'-hydroxylase, F3'H), and SDR (DFR and ANR) enzymes that constitute the flavonoid pathway. This holds true for several modifying enzymes, with substantial structural information already available for OMTs (*O*-methyltransferases) and evidence that the Family I class of GT enzymes will be similar in structure to the GT-B enzymes, for which several crystal structures have been solved (Lim et al., 2004). In addition to the examples cited above, Yang et al. (2004) have described the use of docking simulations to model substrate in the active site of a flavonoid OMT from *Arabidopsis* in order to develop a mechanistic explanation for differences between the substrate preferences of this enzyme and that of caffeic acid OMT, the enzyme used as the template for homology modeling. Molecular modeling also is being used as part of a functional genomics study of the *Arabidopsis* cytochrome P450 enzymes by Schuler's laboratory. Initial efforts have focused on four enzymes of phenylpropanoid metabolism, including F3'H (Figure 3.10), the structures of which were modeled on the crystal structures of four bacterial and one mammalian P450 enzyme (Rupasinghe et al., 2003). Analysis of these models indicated that, despite as little as 13% sequence identity among these proteins, the structural cores and several loop regions are highly conserved. Substrate docking simulations suggested that the enzymes use a similar substrate recognition mechanism and are informing mutagenesis experiments aimed at clarifying the roles of specific residues in defining the substrate preferences of these enzymes.

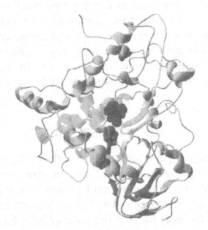

Figure 3.10 *Homology model of Arabidopsis F3'H developed by the Schuler laboratory* (Rupasinghe et al., 2003). *The structure was visualized using Swiss-PdbViewer 3.7. Side chains of residues corresponding to the P450 signature motif, which includes the heme-binding cysteine (C445, red) and the DT pair that mediates dioxygen activation (D305 and T306, blue), are shown. The six regions that determine substrate specificity (SRS1-6) are highlighted in green. See Color Section for figure in color.*

The availability of authentic structures and homology models already has enhanced radically our understanding of reaction mechanisms. It will be interesting to see how these types of modeling experiments perform overall as predictive tools for enzyme function. The promise of this approach is underscored by the fact that the substrate and product specificity of a Gerbera CHS-like protein were accurately predicted by modeling the active site architecture based on the *Medicago* CHS crystal structure (Ferrer et al., 1999). Structural information also will be essential for understanding the organization and intracellular distribution of the flavonoid pathway. For example, our laboratory has begun fitting docking simulations of homology models of *Arabidopsis* CHS and CHI to data generated by small-angle neutron-scattering experiments to explore the structural basis of the interaction of these two enzymes (Dana, Martins, Kreuger, and Winkel, unpublished data). The fact that many other metabolic systems are composed of members of the same classes of enzymes suggests that what is learned from this highly tractable experimental system will have broad applicability in understanding the organization of cellular metabolism.

7. CONCLUSIONS

Studies of flavonoid metabolism are increasingly intertwined with efforts to understand a wide array of other primary and secondary metabolic systems. The growing accessibility of sophisticated tools for determining the biochemical and structural characteristics of enzymes is driving this work to new levels, even as it adds to our appreciation for the true complexity of cellular metabolism. The practical implications of new knowledge for engineering enhanced agronomic and nutritional traits in plants also are more and more evident. It seems likely that flavonoid metabolism will continue to serve as an important and tractable experimental model for efforts to understand cellular metabolism for some time to come.

8. ACKNOWLEDGMENTS

The author is indebted to Eric Conn, Klaus Hahlbrock, Geza Hrazdina, George Wagner, and Mary Catherine Bateson of the Institute for Intercultural Studies; Donald Forsdyke of Queens University; and Laurie Lindquist of Reed College for generous contributions of photographs for Figure 3.2 and to Sanjeewa Rupasinghe and Mary Schuler for the pdb file for the *Arabidopsis* F3'H structure. She also gratefully acknowledges the National Science Foundation (grants MCB 0131010 and 0445878) and the U.S. Department of Agriculture (grant 2001–03,371) for support of her laboratory's current work on flavonoid metabolism.

9. REFERENCES

Abrahams, S., Lee, E., Walker, A. R., Tanner, G. J., Larkin, P. J., and Ashton, A. R., 2003, The *Arabidopsis* TDS4 gene encodes leucoanthocyanidin dioxygenase (LDOX) and is essential for proanthocyanidin synthesis and vacuole development, *Plant J* **35**: 624-636.

Achnine, L., Blancaflor, E. B., Rasmussen, S., and Dixon, R. A., 2004, Colocalization of L-phenylalanine ammonia-lyase and cinnamate 4-hydroxylase for metabolic channeling in phenylpropanoid biosynthesis, *Plant Cell* **16**: 3098-3109.

Anzellotti, D., and Ibrahim, R. K., 2000, Novel flavonol 2-oxoglutarate dependent dioxygenase: affinity purification, characterization, and kinetic properties, *Arch Biochem Biophys* **382**: 161-172.

Anzellotti, D., and Ibrahim, R. K., 2004, Molecular characterization and functional expression of flavonol 6-hydroxylase, *BMC Plant Biol* **4**: 20.

Austin, M. B., and Noel, J. P., 2003, The chalcone synthase superfamily of type III polyketide synthases, *Nat Prod Rep* **20**: 79-110.

Bateson, W., Saunders, E. R., and Punnett, R. C., 1904, Report II. Experimental studies in the physiology of heredity, *Rep Evol Com Roy Soc* 1-131.

Bateson, W., Saunders, E. R., and Punnett, R. C., 1905, Further experiments on inheritance in sweet peas and stocks, *Proc R Soc* **LXXVIIB**: 236-238.

Beld, M., Martin, C., Huits, H., Stuitje, A. R., and Gerats, A. G. M., 1989, Flavonoid synthesis in *Petunia:* partial characterization of dihydroflavonol 4-reductase genes, *Plant Mol Biol* **13:** 491-502.

Boyle, R., 1664, *Experiments and Considerations Touching Colours*. London.

Britsch, L., 1990, Purification and characterization of flavone synthase I, a 2-oxoglutarate-dependent desaturase, *Arch Biochem Biophys* **282:** 152-160.

Britsch, L., and Grisebach, H., 1986, Purification and characterization of (2*S*)-flavanone 3-hydroxylase from *Petunia hybrida*, *Eur J Biochem* **156:** 569-577.

Brugliera, F., Holton, T. A., Stevenson, T. W., Farcy, E., Lu, C. Y., and Cornish, E. C., 1994, Isolation and characterization of a cDNA clone corresponding to the *Rt* locus of *Petunia hybrida*, *Plant J* **5:** 81-92.

Buer, C.S., and Muday, G. K., 2004, The *transparent testa4* mutation prevents flavonoid synthesis and alters auxin transport and the response of *Arabidopsis* roots to gravity and light. *Plant Cell* **16:** 1191-1205.

Davies, K. M., Bloor, S. J., Spiller, G. B., and Deroles, S. C., 1998, Production of yellow color in flowers: redirection of flavonoid biosynthesis in *Petunia*, *Plant J* **13:** 259-266.

Davies, K. M., Schwinn, K. E., Deroles, S. C., Manson, D. G., Lewis, D. H., Bloor, S. J., and Bradley, J. M., 2003, Enhancing anthocyanin production by altering competition for substrate between flavonol synthase and dihydroflavonol 4-reductase, *Euphytica* **131:** 259-268.

Devic, M., Guilleminot, J., Debeaujon, I., Bechtold, N., Bensaude, E., Koornneef, M., Pelletier, G., and Delseny, M., 1999, The *BANYULS* gene encodes a DFR-like protein and is a marker of early seed coat development, *Plant J* **19:** 387-398.

Dixon, R. A., and Steele, C. L., 1999, Flavonoids and isoflavonoids—a gold mine for metabolic engineering, *Trends Plant Sci* **4:** 394-400.

Dixon, R. A., Xie, D. Y., and Sharma, S. B., 2005, Proanthocyanidins—a final frontier in flavonoid research? *New Phytol* **165:** 9-28.

Ferrer, J-L., Jez, J. M., Bowman, M. E., Dixon, R. A., and Noel, J. P., 1999, Structure of chalcone synthase and the molecular basis of plant polyketide biosynthesis, *Nature Struc Biol* **6:** 775-784.

Fettig, S., and Hess, D., 1999, Expression of a chimeric stilbene synthase gene in transgenic wheat lines, *Transgenic Res* **8:** 179-189.

Feucht, W., Treutter, D., and Polster, J., 2004, Flavanol binding of nuclei from tree species, *Plant Cell Rep* **22:** 430-436.

Filhol, E., 1854, Observations sur les matières colorantes des fleurs, *C R Acad Sci* **39:** 194-198.

Fischer, D., Stich, K., Britsch, L., and Grisebach, H., 1988, Purification and characterization of (+)dihydroflavonol (3-hydroxyflavanone) 4-reductase from flowers of *Dahlia variabilis*, *Arch Biochem Biophys* **264:** 40-47.

Fischer, R., Budde, I., and Hain, R., 1997, Stilbene synthase gene expression causes changes in flower colour and male sterility in tobacco, *Plant J* **11:** 489-498.

Fischer, T.C., Halbwirth, H., Meisel, B., Stich, K., and Forkmann, G., 2003, Molecular cloning, substrate specificity of the functionally expressed dihydroflavonol 4-reductases from *Malus domestica* and *Pyrus communis* cultivars and the consequences for flavonoid metabolism, *Arch Biochem Biophys* **412:** 223-230.

Frydman, A., Weisshaus, O., Bar-Peled, M., Huhman, D. V., Sumner, L.W., Marin, F. R., Lewinsohn, E., Fluhr, R., Gressel, J., and Eyal, Y., 2004, Citrus fruit bitter flavors: isolation and functional characterization of the gene *Cm1,2RhaT* encoding a 1,2 rhamnosyltransferase, a key enzyme in the biosynthesis of the bitter flavonoids of citrus, *Plant J* **40:** 88-100.

Fukusaki, E., Kawasaki, K., Kajiyama, S., An, C. I., Suzuki, K., Tanaka, Y., and Kobayashi, A., 2004, Flower color modulations of *Torenia hybrida* by downregulation of chalcone synthase genes with RNA interference, *J Biotechnol* **111:** 229-240.

Glénard, A.,1858). Recherches sur la matière colorante du vin. *Ann Chim Phys* (Paris) 366-376.

Grandmaison, J., and Ibrahim, R. K., 1996, Evidence for nuclear binding of flavonol sulphate esters in *Flaveria chloraefolia*, *J Plant Physiol* **147:** 653-660.

Halbwirth, H., Forkmann, G., and Stich, K., 2004, The A-ring specific hydroxylation of flavonols in position 6 in *Tagetes* sp is catalyzed by a cytochrome P450 dependent monooxygenase, *Plant Science* **167:** 129-135.

Halbwirth, H., Martens, S., Wienand, U., Forkmann, G., and Stich, K., 2003, Biochemical formation of anthocyanins in silk tissue of *Zea mays*, *Plant Science* **164**: 489-495.

Harborne, J. B., and Baxter, H., 1999, *Handbook of Natural Flavonoids* Chichester: Wiley.

Harborne, J. B., and Williams, C. A., 2000, Advances in flavonoid research since 1992, *Phytochem* **55**: 481-504.

Harborne, J. B., and Williams, C. A., 2001, Anthocyanins and other flavonoids, *Nat Prod Rep* **18**: 310-333.

Heller, W., Forkmann, G., Britsch, L., and Griseback, H., 1985, Enzymatic reduction of (+)-dihydroflavonols to flavan-3,4-*cis*-diols with flower extracts from *Matthiola incana* and its role in anthocyanin biosynthesis, *Planta* **165**: 284-287.

Hrazdina, G., and Wagner, G. J., 1985, Metabolic pathways as enzyme complexes: evidence for the synthesis of phenylpropanoids and flavonoids on membrane associated enzyme complexes, *Arch Biochem Biophys* **237**: 88-100.

Hrazdina, G., and Jensen, R. A., 1992, Spatial organization of enzymes in plant metabolic pathways, *Ann Rev Plant Physiol Plant Mol Biol* **43**: 241-267.

Huang, T. S., Anzellotti, D., Dedaldechamp, F., and Ibrahim, R. K., 2004, Partial purification, kinetic analysis, and amino acid sequence information of a flavonol 3-*O*-methyltransferase from *Serratula tinctoria*, *Plant Physiol* **134**: 1366-1376.

Hutzler, P., Rischbach, R., Heller, W., Jungblut, T. P., Reuber, S., Schmitz, R., Veit, M., Weissenbck, G., and Schmitzler, J-P., 1998, Tissue localization of phenolic compounds in plants by confocal laser scanning microscopy, *J Exp Bot* **49**: 953-965.

Jez, J. M., Bowman, M. E., and Noel, J. P., 2002, Role of hydrogen bonds in the reaction mechanism of chalcone isomerase, *Biochemistry* **41**: 5168-5176.

Jez, J. M., Bowman, M. E., Dixon, R. A., and Noel, J. P., 2000, Structure and mechanism of chalcone isomerase: an evolutionarily unique enzyme in plants, *Nature Struct Biol* **7**: 786-791.

Johnson, E. T., Ryu, S., Yi, H., Shin, B., Cheong, H., and Choi, G., 2001, Alteration of a single amino acid changes the substrate specificity of dihydroflavonol 4-reductase, *Plant J* **25**: 325-333.

Jones, P., Messner, B., Nakajima, J., Schaffner, A. R., and Saito, K., 2003, UGT73C6 and UGT78D1, glycosyltransferases involved in flavonol glycoside biosynthesis in *Arabidopsis thaliana*, *J Biol Chem* **278**: 43910-43918.

Jorgensen, R. A., 1995, Cosuppression, flower color patterns, and metastable gene expression studies, *Science* **268**: 686-691.

Joung, J-Y., Kasthuri, M., Park, J-Y., Kang, W-J., Kim, H-S., Yoon, B-S., Joung, H., and Jeon, J-H., 2003, An overexpression of chalcone reductase of *Pueraria montana* va. *Iobata* alters biosynthesis of anthocyanins and 5'-deoxyflavonoids in transgenic tobacco, *Biochem Biophys Res Commun* **3003**: 326-331.

Keeble, F., and Armstrong, E. F., 1912a, The distribution of oxydases in plants and their role in the formation of pigments, *Proc R Soc Lon, Series B* **85**: 214-218.

Keeble, F., and Armstrong, E. F., 1912b, The oxydases of *Cytisus adami*, *Proc R Soc Lon, Series B* **85**: 460-465.

Koukol, J., and Conn, E. E., 1961, The metabolism of aromatic compounds in higher plans. IV. Purification and properties of the phenylalanine deaminase of *Hordeum vulgare*, *J Biol Chem* **236**: 2692-2698.

Kreuzaler, F., and Hahlbrock, K., 1975, Enzymic synthesis of an aromatic ring from acetate units. Partial purification and some properties of flavanone synthase from cell-suspension cultures of *Petroselinum hortense*, *Eur J Biochem* **56**: 205-213.

Kreuzaler, F., Ragg, H., Fautz, E., Kuhn, D. N., and Hahlbrock, K., 1983, UV-induction of chalcone synthase mRNA in cell suspension cultures of *Petroselinum hortense*, *Proc Natl Acad Sci USA* **80**: 2591-2593.

Kubo, A., Arai, Y., Nagashima, S., and Yoshikawa, T., 2004, Alteration of sugar donor specificities of plant glycosyltransferases by a single point mutation, *Arch Biochem Biophys* **429**: 198-203.

Kuras, M., Stefanowska-Wronka, M., Lynch, J. M., and Zobel, A. M., 1999, Cytochemical localization of phenolic compounds in columella cells of the root cap in seeds of *Brassica napus*—Changes in the localization of phenolic compounds during germination, *Ann Bot* **84**: 135-143.

Lim, E. K., Ashford, D. A., Hou, B., Jackson, R. G., and Bowles, D. J., 2004, *Arabidopsis* glycosyltransferases as biocatalysts in fermentation for regioselective synthesis of diverse quercetin glucosides, *Biotechnol Bioeng* **87**: 623-631.

Liu, C. J., Blount, J. W., Steele, C. L., and Dixon, R. A., 2002, Bottlenecks for metabolic engineering of isoflavone glycoconjugates in *Arabidopsis, Proc Natl Acad Sci USA* **99:** 14578-14583.

Lukacin, R., Groning, I., Schiltz, E., Britsch, L., and Matern, U., 2000, Purification of recombinant flavanone 3ß-hydroxylase from *Petunia hybrida* and assignment of the primary site of proteolytic degradation, *Arch Biochem Biophys* **375:** 364-370.

Lukacin, R., Wellmann, F., Britsch, L., Martens, S., and Matern, U., 2003, Flavonol synthase from *Citrus unshiu* is a bifunctional dioxygenase, *Phytochem* **62:** 287-292.

Marles, M. A., Ray, H., and Gruber, M. Y., 2003, New perspectives on proanthocyanidin biochemistry and molecular regulationm, *Phytochem* **64:** 367-383.

Martens, S., and Forkmann, G., 1999, Cloning and expression of flavone synthase II from *Gerbera* hybrids, *Plant J* **20:** 611-618.

McMullen, M. D., Kross, H., Snook, M. E., Cortes-Cruz, M., Houchins, K. E., Musket, T. A., and Coe, E. H., Jr., 2004, Salmon silk genes contribute to the elucidation of the flavone pathway in maize (*Zea mays* L.), *J Hered* **95:** 225-233.

McNeill, L. A., Bethge, L., Hewitson, K. S., and Schofield, C. J., 2004, A fluorescence-based assay for 2-oxoglutarate-dependent oxygenases, *Anal Chem* **336:** 125-131.

Mendel, G., 1865, Versuche uber Pflanzen-Hybriden, *Verh naturforschenden Vereines Brunn* **4:** 3-47.

Meyer, P., Heidmann, I., Forkmann, G., and Saedler, H., 1987, A new petunia flower color generated by transformation of a mutant with a maize gene, *Nature* **330:** 677-678.

Morita, Y., Hoshino, A., Kikuchi, Y., Okuhara, H., Ono, E., Tanaka, Y., Fukui, Y., Saito, N., Nitasaka, E., Noguchi, H., and Iida, S., 2005, Japanese morning glory *dusky* mutants displaying reddish-brown or purplish-gray flowers are deficient in a novel glycosylation enzyme for anthocyanin biosynthesis, UDP-glucose:anthocyanidin 3-*O*-glucoside-2"-*O*-glucosyltransferase, due to 4-bp insertions in the gene, *Plant J* **42:** 353-363.

Morot, F. S., 1849, Recherches sur la coloration des végétaux. *Ann Sci Nat* (Bot.) (Paris), pp. 160-235.

Moustafa, E., and Wong, E., 1967, Purification and properties of chalcone-flavanone isomerase from soya bean seed, *Phytochem,* **6:** 625-632.

Onslow, M. W., 1925, *The Anthocyanin Pigments of Plants.* Cambridge: Cambridge University Press.

Oppermann, U., Filling, C., Hult, M., Shafqat, N., Wu, X., Lindh, M., Shafqat, J., Nordling, E., Kallberg, Y., Persson, B., and Jornvall, H., 2003, Short-chain dehydrogenases/reductases (SDR): the 2002 update, *Chem Biol Interact* **143-144:** 247-253.

Palladin, W., 1908, Ueber die Bildung der Atmungschromogene in den Pflanzen. *Ber D Bot Ges* **26a:** 389-394.

Peer, W. A., Brown, D. E., Tague, B. W., Muday, G. K., Taiz, L., and Murphy, A. S., 2001, Flavonoid accumulation patterns of *transparent testa* mutants of *Arabidopsis thaliana, Plant Physiol* **126:** 536-548.

Prescott, A. G., and John, P., 1996, Dioxygenases: Molecular structure and role in plant metabolism, *Ann Rev Plant Physiol Plant Mol Biol* **47:** 245-271.

Prescott, A. G., Stamford, N. P., Wheeler, G., and Firmin, J. L., 2002, *In vitro* properties of a recombinant flavonol synthase from *Arabidopsis thaliana, Phytochem* **60:** 589-593.

Punyasiri, P. A., Abeysinghe, I. S., Kumar, V., Treutter, D., Duy, D., Gosch, C., Martens, S., Forkmann, G., and Fischer, T. C., 2004, Flavonoid biosynthesis in the tea plant *Camellia sinensis*: properties of enzymes of the prominent epicatechin and catechin pathways, *Arch Biochem Biophys* **431:** 22-30.

Ralston, L., Subramanian, S., Matsuno, M., and Yu, O., 2005, Partial reconstruction of flavonoid and isoflavonoid biosynthesis in yeast using soybean type I and type II chalcone isomerases, *Plant Physiol* **137:** 1375-1388.

Rasmussen, S., and Dixon, R. A., 1999, Transgene-mediated and elicitor-induced perturbation of metabolic channeling at the entry point into the phenylpropanoid pathway, *Plant Cell* **11:** 1537-1551.

Reddy, A. R., Britsch, L., Salamini, F., Saedler, H., and Rohde, W., 1987, The *A1* (anthocyanin-1) locus in *Zea mays* encodes dihydroquercitin reductase, *Plant Sci* **52:** 7-13.

Reddy, G. M., and Coe, E. H., Jr., 1962, Inter-tissue complemention: a simple technique for direct analysis of gene-action sequence, *Science* **138:** 149-150.

Rupasinghe, S., Baudry, J., and Schuler, M. A., 2003, Common active site architecture and binding strategy of four phenylpropanoid P450s from *Arabidopsis thaliana* as revealed by molecular modeling, *Protein Eng* **16:** 721-731.

Russell, D. W., and Conn, E. E., 1967, The cinnamic acid 4-hydroxylase of pea seedlings, *Arch Biochem Biophys* **122**: 256-258.

Saslowsky, D., Warek, U., and Winkel, B. S. J., 2005, Nuclear localization of flavonoid metabolism in *Arabidopsis thaliana*, *J Biol Chem* **280**: 23735-23740.

Sawada, S., Suzuki, H., Ichimaida, F., Yamaguchi, M. A., Iwashita, T., Fukui, Y., Hemmi, H., Nishino, T., and Nakayama, T., 2005, UDP-glucuronic acid:anthocyanin glucuronosyltransferase from red daisy (*Bellis perennis*) flowers. Enzymology and phylogenetics of a novel glucuronosyltransferase involved in flower pigment biosynthesis, *J Biol Chem* **280**: 899-906.

Sawada, Y., and Ayabe, S., 2005, Multiple mutagenesis of P450 isoflavonoid synthase reveals a key active-site residue, *Biochem Biophys Res Commun* **330**: 907-913.

Sawada, Y., Kinoshita, K., Akashi, T., Aoki, T., and Ayabe, S., 2002, Key amino acid residues required for aryl migration catalysed by the cytochrome P450 2-hydroxyisoflavanone synthase, *Plant J* **31**: 555-564.

Schijlen, E. G., Ric de Vos, C. H., van Tunen, A. J., and Bovy, A. G., 2004, Modification of flavonoid biosynthesis in crop plants, *Phytochem* **65**: 2631-26348.

Schröder, G., Wehinger, E., Lukacin, R., Wellmann, F., Seefelder, W., Schwab, W., and Schröder, J., 2004, Flavonoid methylation: a novel 4'-*O*-methyltransferase from *Catharanthus roseus*, and evidence that partially methylated flavanones are substrates of four different flavonoid dioxygenases, *Phytochem* **65**: 1085-1094.

Shikazono, N., Yokota, Y., Kitamura, S., Suzuki, C., Watanabe, H., Tano, S., and Tanaka, A., 2003, Mutation rate and novel *tt* mutants of *Arabidopsis thaliana* induced by carbon ions, *Genetics* **163**: 1449-1455.

Shimada, S., Takahashi, K., Sato, Y., and Sakuta, M., 2004, Dihydroflavonol 4-reductase cDNA from non-anthocyanin-producing species in the Caryophyllales, *Plant Cell Physiol* **45**: 1290-1298.

Springob, K., Nakajima, J., Yamazaki, M., and Saito, K., 2003, Recent advances in the biosynthesis and accumulation of anthocyanins, *Nat Prod Rep* **20**: 288-303.

Stafford, H. A., 1974, Possible multi-enzyme complexes regulating the formation of C_6-C_3 phenolic compounds and lignins in higher plants, *Rec Adv Phytochem* **8**: 53-79.

Stafford, H. A., 1981, Compartmentation in natural product biosynthesis by multienzyme complexes. In: *The Biochemistry of Plants*, E.E. Conn (Ed.), New York: Academic Press, pp. 117-137.

Stafford, H.A., 1990, *Flavonoid Metabolism*. Boca Raton: CRC Press.

Stafford, H.A., 1991, Flavonoid evolution: an enzymic approach, *Plant Physiol* **96**: 680-685.

Strack, D., Vogt, T., and Schliemann, W., 2003, Recent advances in betalain research, *Phytochem* **62**: 247-269.

Styles, E. D., and Ceska, O., 1977, *Can J Genet Cytol* **19**: 289-302.

Suzuki, H., Sawada, S., Watanabe, K., Nagae, S., Yamaguchi, M. A., Nakayama, T., and Nishino, T., 2004, Identification and characterization of a novel anthocyanin malonyltransferase from scarlet sage (*Salvia splendens*) flowers: an enzyme that is phylogenetically separated from other anthocyanin acyltransferases, *Plant J* **38**: 994-1003.

Tanner, G. J., Francki, K. T., Abrahams, S., Watson, J. M., Larkin, P. J., and Ashton, A. R., 2003, Proanthocyanidin biosynthesis in plants. Purification of legume leucoanthocyanidin reductase and molecular cloning of its cDNA, *J Biol Chem* **278**: 31647-31656.

Turnbull, J. J., Nagle, M. J., Seibel, J. F., Welford, R. W., Grant, G. H., and Schofield, C. J., 2003. The C-4 stereochemistry of leucocyanidin substrates for anthocyanidin synthase affects product selectivity. *Bioorg Med Chem Lett* **13**: 3853-3857.

Turnbull, J. J., Nakajima, J. I., Welford, R. W., Yamazaki, M., Saito, K., and Schofield, C. J., 2004, Mechanistic studies on three 2-oxoglutarate-dependent oxygenases of flavonoid biosynthesis, *J Biol Chem* **279**: 1206-1216.

Veit, M., and Pauli, G. F., 1999, Major flavonoids from *Arabidopsis thaliana* leaves, *J Nat Prod* **62**: 1301-1303.

Vogt, T., 2004, Regiospecificity and kinetic properties of a plant natural product *O*-methyltransferase are determined by its N-terminal domain, *FEBS Lett* **561**: 159-162.

Vom Endt, D., Kijne, J. W., and Memelink, J., 2002, Transcription factors controlling plant secondary metabolism: what regulates the regulators? *Phytochem* **61**: 107-114.

Wagner, G. J., and Hrazdina, G., 1984, Endoplasmic reticulum as a site of phenylpropanoid and flavonoid metabolism in *Hippeastrum*, *Plant Physiol* **74**: 901-906.

Wang, L., Burhenne, K., Kristensen, B. K., and Rasmussen, S. K., 2004, Purification and cloning of a Chinese red radish peroxidase that metabolises pelargonidin and forms a gene family in Brassicaceae, *Gene* **343**: 323-335.

Wellmann, F., Matern, U., and Lukacin, R., 2004, Significance of C-terminal sequence elements for *Petunia* flavanone 3ß-hydroxylase activity, *FEBS Lett* **561**: 149-154.

Wheldale, M., 1909a, Note on the physiological interpretation of the Mendelian factors for colour in plants, *Rep Evol Com Roy Soc Rpt* **5**: 26-31.

Wheldale, M., 1909b, On the nature of anthocyanin, *Proc Phil Soc* **XV**: 137-161.

Wigand, A., 1862, Einige Sätze über die physiologische Bedeutung des Gerbstoffes und der Pflanzenfarbe, *Bot Ztg* **20**: 121-125.

Willstätter, R., and Everest, A. E., 1913, Ueber den Farbstoff der Kornblume, *Liebigs Ann Chem* **CCCCI**: 189-232.

Willstätter, R., and Weil, F. J., 1916, Ueber das Anthocyan des violetten Stiefmütterchens, *Liebigs Ann Chem* **CCCCXII**: 178-194.

Wilmouth, R. C., Turnbull, J.J., Welford, R. W., Clifton, I. J., Prescott, A. G., and Schofield, C. J., 2002, Structure and mechanism of anthocyanidin synthase from *Arabidopsis thaliana*, *Structure* **10**: 93-103.

Winkel-Shirley, B., 2001a, Flavonoid biosynthesis: a colorful model for genetics, biochemistry, cell biology and biotechnology, *Plant Physiol* **126**: 485-493.

Winkel-Shirley, B., 2001b, It takes a garden. How work on diverse plant species has contributed to an understanding of flavonoid metabolism, *Plant Physiol* **127**: 1399-1404.

Xie, D. Y., Sharma, S. B., Paiva, N. L., Ferreira, D., and Dixon, R. A., 2003, Role of anthocyanidin reductase, encoded by *BANYULS* in plant flavonoid biosynthesis, *Science* **299**: 396-369.

Xie, D. Y., Sharma, S. B., and Dixon, R. A., 2004a, Anthocyanidin reductases from *Medicago truncatula* and *Arabidopsis thaliana*, *Arch Biochem Biophys* **422**: 91-102.

Xie, D. Y., Jackson, L. A., Cooper, J. D., Ferreira, D., and Paiva, N. L., 2004b, Molecular and biochemical analysis of two cDNA clones encoding dihydroflavonol-4-reductase from *Medicago truncatula*, *Plant Physiol* **134**: 979-994.

Yamazaki, M., Gong, Z., Fukuchi-Mizutani, M., Fukui, Y., Tanaka, Y., Kusumi, T., and Saito, K., 1999, Molecular cloning and biochemical characterization of a novel anthocyanin 5-*O*-glucosyltransferase by mRNA differential display for plant forms regarding anthocyanin, *J Biol Chem* **274**: 7405-7411.

Yang, H., Ahn, J. H., Ibrahim, R. K., Lee, S., and Lim, Y., 2004, The three-dimensional structure of *Arabidopsis thaliana* O-methyltransferase predicted by homology-based modelling, *J Mol Graph Model* **23**: 77-87.

CHAPTER 4

THE REGULATION OF FLAVONOID BIOSYNTHESIS

FRANCESCA QUATTROCCHIO[1], ANTOINE BAUDRY[2], LOÏC LEPINIEC[2], AND ERICH GROTEWOLD[3]

[1]*Institut of Molecular and Cell Biology, Dept. of Genetics, Vrije Universiteit, De Boelelaan 1087, 1081 HV Amsterdam, The Netherlands; [2]Seed Biology Laboratory, UMR 204 INRA/INAPG, Institut Jean-Pierre Bourgin (IJPB), Route de Saint-Cyr, 78026 Versailles Cedex, France; [3]Dept. of Plant Cellular and Molecular Biology and Plant Biotechnology Center, The Ohio State University, Columbus, OH 43210 USA*

1. INTRODUCTION

The control of flavonoid biosynthesis provides one of the best-described regulatory systems in plants. The availability of a large number of mutants that affect the expression of several pathway biosynthetic genes, and hence result in significant alterations in pigmentation (Figure 4.1), facilitated the cloning of regulators in plants that classically have been used to investigate the flavonoid pathway (e.g., maize, petunia, snapdragon, gerbera) as well as in plants that are emerging as convenient model systems to further understand the regulation of the pathway (e.g., *Arabidopsis*) (Mol et al., 1998; Winkel-Shirley, 2001). The flavonoid pathway is under tight developmental control, and multiple environmental conditions, of them light and hormones being the best investigated, affect the expression of the flavonoid biosynthetic genes (Irani et al., 2003).

As extensively described in other chapters of this book, the flavonoid pathway provides an intricate grid that results in the formation of various distinct groups of flavonoids, which include the anthocyanin, proanthocyanidin (PA, *syn.* condensed tannins), and phlobaphene pigments, and the nonpigmented flavonols, flavones, and isoflavones. Each group of flavonoid compounds serves specific functions to the plant, under particular developmental or biotic/abiotic conditions. Thus, each branch of the pathway is under separate control, ensuring that the appropriate compounds are produced when and where required.

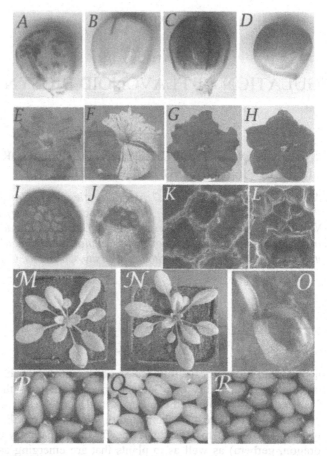

Figure 4.1 *(A–D) Maize kernels showing phlobaphenes and anthocyanins. (A) Accumulation of phlobaphenes in the pericarp under the control of the unstable P1-vv allele and anthocyanins in the aleurone under C1 and the unstable R-m3 allele. (B) P1-vv with phlobaphenes in the pericarp. (C) R C1 aleurones accumulating anthocyanins with phlobaphene sector in the pericarp. (D) Phlobaphene pigments in P-rr pericarp. (E–H) Petunia flowers with different AN and PH alleles. (E) Wild-type (R27 line) petals accumulating cyanidin. (F) Flowers with an1 unstable allele (line W138) in the same background. (G) Wild-type flower (VR line) accumulating mainly malvidin. (H) Same as G, but mutated in the ph2 gene (V26). (I–L) Petunia seed coats showing the effect of the an1 mutation in condensed tannin accumulation, cell size and structure. (I) Seed of wild-type R27 line. (J) Unstable an1 mutant (W138). (K) Scanning electron microsopic (SEM) picture of the seed in I. (L) SEM picture of the seed in J. (M–O) Arabidopsis seedlings accumulating anthocyanins. (M) Wild-type (Col-O ecotype). (N) PAP1 overexpressing line (pap1-D). (O) Three day-old wild-type seedling germinated on norflurazon (chlorophyll synthesis inhibitor). (P–R) Arabidopsis seeds (WS ecotype) accumulating PA or anthocyanins. (P) Wild-type seeds showing PA accumulation. (Q) Pale tt2 seeds. (R) Seeds carrying a mutant ban allele showing anthocyanin accumulation instead of PA. See Color Section for figure in colors.*

The regulation of the pathway is largely at the level of transcription of the regulators and of the corresponding biosynthetic genes. Only in a few exceptional situations, additional posttranscriptional control levels have been described (Damiani and Wessler, 1993; Franken et al., 1991; Pairoba and Walbot, 2003). Thus, this chapter will primarily describe the transcriptional control of the pathway genes and the corresponding regulators and provide a perspective of how the flavonoid regulatory network relates to other important control systems in plants.

2. INDEPENDENT REGULATION OF THE DIFFERENT BRANCHES OF FLAVONOID BIOSYNTHESIS

2.1. Flavonoid Biosynthetic Grid

Flavonoid biosynthesis provides an excellent example of a biosynthetic grid rather than a linear biosynthetic pathway. A detailed description of the flavonoid pathway is presented in Chapter 3. Early studies characterizing the expression of flavonoid structural genes have revealed an important, still intriguing, difference in the regulation of the genes in the pathway, depending on the species of interest. In maize, all the known structural biosynthetic genes are coordinately activated (Irani et al., 2003), whereas in dicotyledonous species, two clusters of coregulated structural genes can be distinguished: early biosynthetic genes (EBG), the induction of which precedes that of the late biosynthetic genes (LBG) (Mol et al., 1998; Nesi et al., 2000; Pelletier et al., 1999). The finding that the maize regulators C1 and R are unable to activate the chalcone synthase CHSA gene of petunia, unlike the putative maize ortholog encoding for CHS (C2), led to the conclusion that dicotyledonous EBG might be controlled by a distinct set of regulators (Quattrocchio et al., 1993, 1998). However, the expression of EBG has not been shown to be affected in any of the single regulatory mutants studied to date, suggesting that these early steps can be redundantly controlled by several regulators. This would add a supplementary level of complexity to the regulation of the flavonoid pathway in dicotyledonous species. The regulation of several of the main flavonoid biosynthetic branches is described below.

2.2. Regulation of 3-Deoxy Flavonoid and Phlobaphene Biosynthesis

Maize, sorghum, and a few other grasses accumulate in floral organs the phlobaphene pigments derived from the likely nonenzymatic condensation of the 3-deoxy flavonoids luteoforol or apiforol. These pigments are likely to contribute to the protective role of the pericarp, similarly as PA function in the testa of numerous dicots (see Section 2.4). The 3-deoxy branch of the pathway is regulated by the P1 (pericarp color1) gene in maize (Styles and Ceska, 1989) and its ortholog, Y1, in sorghum (Chopra et al., 1999). P1 encodes a R2R3-MYB transcription factor (Grotewold et al., 1991), and P1 alleles have different pigmentation patterns in the pericarp and the cob glumes as a consequence of the action of specific cis-regulatory elements in the P1 promoter (Zhang and Peterson, 2005). The R2R3-MYB family is very large in the flowering plants, with around 135 genes in Arabidopsis (Stracke et al., 2001) and over 250 members in maize and its close relatives (Jiang et al., 2004; Rabinowicz et al., 1999).

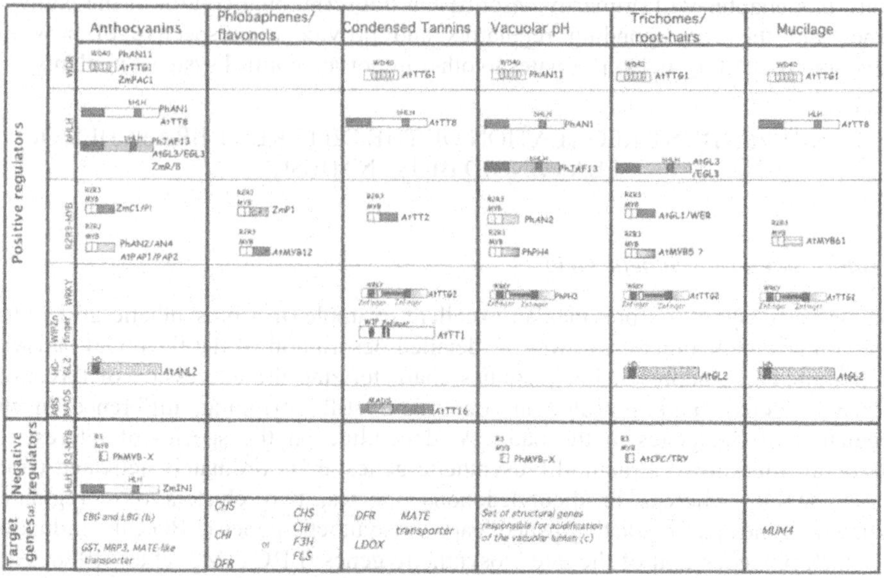

		Anthocyanins	Phlobaphenes/ flavonols	Condensed Tannins	Vacuolar pH	Trichomes/ root-hairs	Mucilage
Positive regulators	WDR	WD40 PhAN11 AtTTG1 ZmPAC1		WD40 AtTTG1	WD40 PhAN11	WD40 AtTTG1	WD40 AtTTG1
	bHLH	bHLH PhAN1 AtTT8 bHLH PhJAF13 AtGL3/EGL3 ZmR/B		bHLH AtTT8	bHLH PhAN1 bHLH PhJAF13	bHLH AtGL3 /EGL3	bHLH AtTT8
	R2R3-MYB	R2R3 MYB ZmC1/P1 R2R3 MYB PhAN2/AN4 A3PAP1/PAP2	R2R3 MYB ZmP1 R2R3 MYB AtMYB12	R2R3 MYB AtTT2	R2R3 MYB PhAN2 R2R3 MYB PhPH4	R2R3 MYB AtGL1/WER R2R3 MYB AtMYB5 ?	R2R3 MYB AtMYB61
	WRKY finger			WRKY AtTTG2	WRKY PhPH3	WRKY AtTTG2	WRKY AtTTG2
	WIP Zn finger			WIP AtTT1			
	HD GL2	AtAN2				AtGL2	AtGL2
	SBP bHLH MADS			AtTT16			
Negative regulators	R3-MYB	R3 MYB PhMYB-X			R3 MYB PhMYB-X	R3 MYB AtCPC/TRY	
	bHLH	bHLH ZmIN1					
Target genes (a)		EBG and LBG (b) GST, MRP3, MATE-like transporter	CHS CHI or DFR CHS CHI F3H FLS	DFR MATE transporter LDOX	Set of structural genes responsible for acidification of the vacuolar lumen (c)		MUM4

Figure 4.2 *Regulatory factors involved in the control of the various branches of flavonoid biosynthesis. (A) Only a few genes have been demonstrated to be the direct targets of the regulators. (B) The set of targets is not the same in all species. For example, in maize, the EBG and LBG are coordinately regulated; in Arabidopsis only the LBG downstream from DFR and in petunia the LBG plus one of the CHS genes. (C) This group of genes, identified as regulated by AN1, PH3, and PH4 have not yet been fully characterized.*

P1 activates a subset of the flavonoid biosynthetic genes (Figure 4.2), including C2 (CHS), CHI1 (CHI, chalcone isomerase), and A1 (DFR, dihydroflavonol 4-reductase) (Grotewold et al., 1994), but not F3H (flavanone 3-hydroxylase) or other genes unique to the anthocyanin branch of the pathway (Grotewold et al., 1998). The transcriptional activity of P1 can be monitored in yeast, in the absence of other plant-specific factors (Hernandez et al., 2004), provided that it is furnished with a promoter containing the high-affinity P1-binding site ([ha]PBS) present in the A1 promoter. Similar to other plant R2R3-MYB factors, P1 binds *in vitro* DNA sequences with the consensus sequence $ACC^{T}/_{A}ACC$ (Grotewold et al., 1994). Recent findings suggest that the regulation of DNA binding by P1 might be under REDOX control (Heine et al., 2004), providing a possible link between the metabolic status of the cells and the activity of the regulators.

2.3. Regulation of Anthocyanin Biosynthesis

A specific subgroup of R2R3-MYB genes, which includes maize C1 (Paz-Ares et al., 1987), petunia AN2 (Quattrocchio et al., 1999), and *Arabidopsis* PAP1/PAP2 (Borevitz et al., 2000), is responsible for the regulation of the anthocyanin biosynthetic genes (Figure 4.2). In contrast to the P1 regulator of 3-deoxy flavonoid biosynthesis, these R2R3-MYB transcription factors cannot function on their own and require the presence of a member of the R/B family of basic helix–loop–helix (bHLH) coactivators for function. Similar to the R2R3-MYB family, the plant bHLH family is very large, with more than 150 members in *Arabidopsis* (Atchley et al., 2000; Buck and Atchley, 2003; Heim et al., 2003; Toledo-Ortiz et al., 2003).

The R2R3-MYB/bHLH physical interaction is mediated by the R3 repeat of the MYB domain and the N-terminal region of the bHLH factor, as was first demonstrated for the maize C1 and B regulators (Goff et al., 1992). Although C1 binds to similar DNA sequences as P1 does, C1 has an intrinsically low DNA-binding affinity (Sainz et al., 1997). It remains to be established whether this low DNA-binding affinity is a property shared by other R2R3-MYB anthocyanin regulators. It is interesting, however, that a C1 site-directed mutant that displays an *in vitro* DNA-binding affinity comparable to that of P1 and which activates transcription in yeast from the [ha]PBS remains dependent on the R/B bHLH partner for transcriptional activity in maize cells (Hernandez et al., 2004), suggesting that the low affinity of C1 for DNA is not sufficient to explain why the C1 activity is R/B-dependent.

To understand the mechanisms of R/B function, a P1 mutant was generated that can interact with R/B (Grotewold et al., 2000). This P1 mutant, named P1*, behaves like P1 in the absence of R/B (i.e., activates only a subset of the flavonoid biosynthetic genes and displays R-independent activity). In the presence of R/B, P1* displays an enhanced activity (R-enhanced activity) on A1, and in sharp contrast with P1 efficiently activates the entire anthocyanin pathway (Hernandez et al., 2004). P1* thus permitted the uncovering of two functions for R/B. On one hand, R/B plays an essential function in the activity of C1. This activity could be associated with the release by R of a C1 inhibitor or with an increase by R of the stability of C1. On the other hand, the R-enhanced activity was shown to require the anthocyanin regulatory element (ARE), a *cis*-regulatory element present in various anthocyanin biosynthetic genes (Lesnick and Chandler, 1998). These studies permitted the establishment of a model for the cooperation between the R2R3-MYB and bHLH regulators of anthocyanin biosynthesis (Figure 4.3).

The regulation of anthocyanins in petunia presents some interesting variations that are likely to reflect aspects still ignored in maize and other plants. For example, in this dicot, there are two bHLH proteins, AN1 and JAF13 (Spelt et al., 2000), that interact with AN2, which is one of the two R2R3-MYBs that regulate the LBG (Quattrocchio et al., 1999). JAF13 is phylogenetically more related to R/B, while AN1 is more related to TT8, a regulator of PA accumulation (see Section 2.4).

AN1 requires the physical interaction with AN2 or AN4 to activate biosynthetic genes such as DFR (Spelt et al., 2000). The two petunia bHLH proteins seem to participate together in the formation of an active transcription complex. In yeast two-hybrid assays, both AN1 and JAF13 can form homo- as well as heterodimers. Transient expression assays show that each of them, in combination with AN2, is sufficient to induce activity of the DFR promoter. However, the two proteins are not completely functionally equivalent as AN1 can, in the same assay, boost the activity of P1, while JAF13 cannot (Spelt et al., 2000). Given that P1 functions independently of any known bHLH factor, the significance of this remains unknown.

While all other anthocyanin regulators in petunia have been cloned from pigmentation mutants, *JAF13* has been cloned by homology with the maize *Lc* gene (a member of the *R/B* family). To better define the function of this gene, an allele containing a transposon insertion upstream of the bHLH encoding region of *JAF13* has been isolated by the screening of a petunia population. The homozygous mutant plants show a reduction in the levels of the *DFR* transcript and about 50% reduction in anthocyanins in petals, sufficient to result in revertant spots on the pale mutated background (Urbanus and Quattrocchio, unpublished results). The phenotype of this mutant can be explained in one of two ways: either the *JAF13* function is partially redundant, or the presence of JAF13 in the protein complex is not necessary, but has an "intensifying" effect on the activity of the complex itself.

In addition to the R2R3-MYB and bHLH regulators, a specific sub-group of WDR (WD-repeat) proteins participates in the regulation of anthocyanin genes. First identified in petunia, *AN11* encodes a small protein with 5-6 conserved WDR (de Vetten et al., 1997). Mutations in *an11* lack flower pigmentation, which can be partially complemented by the over-expression of the AN2 or AN1 regulators in petals. Together with the finding that *an11* mutant lines have normal expression of the known regulators, this prompted the suggestion that AN11 modulates post-transcriptionally the activity of the R2R3-MYB or bHLH factors (de Vetten et al., 1997). The *TTG1* gene in *Arabidopsis* encodes a highly related protein that affects trichome and root hair formation and mucilage deposition, in addition to pigment accumulation (Walker et al., 1999). In maize, *PAC1* (*pale aleurone color1*) encodes the AN11/TTG1 ortholog (Selinger and Chandler, 1999). PAC1 complements *Arabidopsis ttg1* mutants (Carey et al., 2004).

2.4. Regulation of PA Biosynthesis

PA polymers serve various functions in plants, including protection against diverse biotic and abiotic stresses and strengthening of seed dormancy and longevity (Debeaujon et al., 2000; Dixon et al., 2005; Marles et al., 2003; Winkel-Shirley, 1998). They occur in fruits, vegetables, flowers, and seeds. In *Arabidopsis*, PAs accumulate specifically in the seed coat (integuments), giving the mature seed its brown colour after oxidation (Debeaujon et al., 2003). Since the seed coat is a tissue of maternal origin, the phenotype of PA mutants is determined by the genotype of the mother plant. Consequently, the transparent testa (*tt*) phenotype, characteristic of *Arabidopsis* flavonoid mutants (Figure 4.1), can be observed only in the progeny of a plant homozygous for the mutation. Flavonols are also largely represented in *Arabidopsis* seeds, but these compounds accumulate both in the embryo and in the integuments. Along with PA, they contribute to the seed quality of many important crops.

To date, more than 20 loci involved in PA metabolism have been identified and according to the abnormal pigmentation of mature mutant seeds named *TRANSPARENT TESTA1* (*TT1*) to *TT19, TTG1, TTG2,* and *BANYULS* (*BAN*). Sixteen mutants already have led to the cloning of the respective genes, among which 12 could be placed to the flavonoid pathway (Debeaujon et al., 2003; Dixon et al., 2005). In addition, six complementation groups defective in tannins biosynthesis (*tds1-6*) have been described (Abrahams et al., 2002). With the exception of *TDS4/TT18*, which encodes a LDOX (Abrahams et al., 2003; Shikazono et al., 2003), all the other groups represent new loci or correspond to new *tt* alleles.

Among all these loci, six (*TT1, TT2, TT8, TT16, TTG1,* and *TTG2*) appear to have a regulatory function. *TT1* encodes a zinc-finger protein of the new WIP family (Sagasser et al., 2002), *TT2* encodes a R2R3-MYB domain protein with high similarity to the maize C1 regulator (Nesi et al., 2001), and *TT16* corresponds to the ARABIDOPSIS BSISTER (ABS) MADS-domain protein (Nesi et al., 2002). These three proteins are specifically expressed in ovules. In contrast, *TT8* is expressed in seeds and in young seedlings (Baudry et al., unpublished) and encodes an R/B-like bHLH transcription factor (Nesi et al., 2000). *TTG2* encodes a ubiquitously expressed WRKY transcription factor that, similar to TTG1, is involved in the control of trichome organogenesis and mucilage deposition (Johnson et al., 2002).

It was shown recently that anthocyanidin reductase (ANR), a core enzyme in PA biosynthesis that converts anthocyanidins to their corresponding 2,3-*cis*-flavan-3-ols, is encoded by *BAN* (Devic et al., 1999; Xie et al., 2003). The activity of the *BAN* promoter is restricted to PA-accumulating cells, the inner integument and pigment strand (chalaza) cells (Debeaujon et al., 2003). Functional analyses of the *BAN* promoter showed that an 86-bp DNA fragment functions as a PA-accumulating cells-specific enhancer. This enhancer contains *cis*-regulatory elements similar to the [ha]PBS (Debeaujon et al., 2003; Grotewold et al., 1994). Mutations in the *TT2, TT8,* and *TTG1* regulatory loci abolished *BAN* promoter activity, while mutants in *TT1* and *TT16* modified the spatial expression of *BAN*. Similarly, *TT2, TT8,* and *TTG1* mutants impaired the expression of *TT3* (encoding DFR, a LBG), suggesting that these loci coordinately affect the expression of LBG (Nesi et al., 2000).

TT16/ABS is necessary for *BAN* expression and PA accumulation in the endothelium of seed coats, with the exception of the chalazal-micropylar area. The phenotype of the *TT16* mutant, together with the results of the ectopic expression of TT16, suggested that this gene is involved in the specification of endothelial cells, but it is not required for proper ovule function (Debeaujon et al., 2003; Nesi et al., 2002). *TTG2* is likely to participate in the regulation of the pathway after the leucoanthocyanidin branch point (Johnson et al., 2002). However, the direct targets of TT16, TTG2, and TT1 remain to be identified.

Recent progress in the functional analysis of TT2, TT8, and TTG1 has demonstrated that these three proteins act at the same level and cooperate to control directly the expression of a subset of LBG of the PA pathway (Baudry et al., 2004; Nesi et al., 2001). Using an approach that already had been successful in the characterization of the target genes of other flavonoid regulators (Spelt et al., 2000), fusion proteins with the glucocorticoid receptor hormone-binding domain (GR), revealed that *BAN* is a direct target of TT2, TT8, and TTG1 (Baudry et al., 2004). Investigation of the physical interactions between these factors and the *BAN* promoter in yeast revealed a mechanism similar to that described for the regulation of anthocyanins (Hernandez et al., 2004), involving the formation of a R2R3-MYB/bHLH complex (in the case of PA between TT2 and TT8) directly binding to the *BAN* promoter. In addition, TTG1 is likely to be an integral part of this transcription factor complex, thanks to the ability of TTG1 to interact simultaneously with TT2 and TT8 and to boost the transcriptional activity of these proteins in yeast (Baudry et al., 2004). In these experiments, the finding that TT2 and TTG1 directly interact was at first perceived as unexpected. However, a similar protein–protein interaction in yeast was found for AN4 and AN11 but not between AN2 and AN11. These results suggest some conservation and selectivity in the formation of R2R3-MYB/bHLH/WDR modules. The significance of these interactions in the functionality of the complex remains to be established.

The identification of several R2R3-MYB/bHLH/WDR regulatory modules with very similar properties (Figure 4.2) poses the question of how is regulatory specificity achieved. Using the *BAN* promoter as a reporter gene, several combinations of the closest *Arabidopsis* relatives of the TT2 and TT8 proteins were tested in transient expression experiments in protoplasts (Baudry et al., 2004). Based on the phenotype of the *tt8* mutant, it could be assumed that TT8 is the only *Arabidopsis* bHLH necessary for PA regulation *in planta*. However, TT8 could be replaced in these experiments by other TTG1-dependent bHLH proteins, including GL3 and EGL3 (Zhang et al., 2003). Conversely, TT2 was the only *Arabidopsis* R2R3-MYB able to activate *BAN* expression, indicating that in the TT2/TT8 complex the R2R3-MYB is responsible for the specific activation of PA biosynthesis. Potentially, the bHLH factor might participate in the recognition of the target genes via a mechanism conserved among the TTG1-dependent bHLH proteins (even from other species), since a maize bHLH (Sn) also could replace TT8 (Baudry et al., 2004).

2.5. Regulation of Flavonol Biosynthesis

Flavonols play important functions in plants, serving as co-pigments (Forkmann, 1991), as shields from UV radiation (Kootstra, 1994; Ryan et al., 2001) and contributing to male fertility by facilitating pollen germination (Mo et al., 1992; Taylor and Grotewold, 2005; Ylstra et al., 1992). Flavonol formation requires the expression of *CHS, CHI, F3H,* and *FLS,* this last gene encoding flavonol synthase, which converts dihydrokaempferol (DHK) or dihydroquercetin (DHQ) to the corresponding flavonols, kaempferol (K) and quercetin (Q), respectively. Little continues to be known on how this branch of the flavonoid pathway is regulated. The regulators of anthocyanin biosynthesis can induce the accumulation of flavonols, as recently shown by the comprehensive analysis of the metabolic changes associated with the expression of the R2R3-MYB transcription factor *PAP1* gene in *Arabidopsis* (Tohge et al., 2005). Interestingly, however, while the over-expression of *PAP1* resulted in a significant increase in the expression of many flavonoid biosynthetic genes, *FLS* was not among the up-regulated genes, suggesting that the increase in flavonol accumulation is likely a consequence of the presence of a constitutively expressed *FLS*, together with an increased flux through the flavonoid pathway. On this line, preliminary molecular and biochemical results indicate that flavonol biosynthesis is affected in the seeds of *tt2* and *ttg1* mutants (Baudry, Routaboul, and Lepiniec, unpublished results).

A similar situation was found when the maize R and C1 regulators of anthocyanins were expressed from a fruit-specific promoter in the tomato flesh (Bovy et al., 2002). In this case, the expression of *DFR*, *ANS*, and *FLS* was significantly enhanced, but the inability of the tomato DFR to utilize either DHQ or DHK as substrate resulted in a very significant increase in the accumulation of Q derivatives in the fruit flesh. While these results suggest that the anthocyanin regulators have the potential to control the accumulation of flavonols, it is evident from the analysis of maize and Petunia mutants that the flavonols are independently regulated from the anthocyanins, particularly in male floral organs, since all the regulatory mutants in the anthocyanin pathway are male fertile.

AtMYB12, encoding a R2R3-MYB transcription factor with high identity to the maize P1 regulator of the phlobaphene pigments (Dias et al., 2003), has recently been identified as a regulator of flavonol biosynthesis in *Arabidopsis* (Mehrtens et al., 2005). *AtMYB12* regulates the expression of *CHS*, *CHI*, *F3H* and *FLS*, but not of the anthocyanin biosynthetic genes. A mutation in *AtMYB12* results in a significant decrease in flavonol accumulation, suggesting that other regulators may contribute to the control of this branch of the pathway. These study also showed that P1 can control the expression of the flavonol biosynthetic genes in *Arabidopsis* (Mehrtens et al., 2005), suggesting that P1 may regulate flavonol biosynthesis in maize. However, in maize, P1 is unable to activate *F3H*, and no flavonols were detected among the several flavonoids induced by P1 expression (Grotewold et al., 1998). Similar to P1, AtMYB12 functions independently of the known bHLH co-activators (Mehrtens et al., 2005). The studies on *AtMYB12* make this gene the only *bona fide* regulator of the flavonol pathway that has so far been described.

3. REGULATORY LINKS BETWEEN THE BIOSYNTHESIS AND THE TRANSPORT/STORAGE OF FLAVONOIDS

3.1. Relationship between the Regulation and the Transport of Flavonoids

The last step of anthocyanin accumulation is their sequestration into the vacuole of epidermal or subepidermal cells (See Chapter 5 for a full description of mechanisms of transport and sequestration of flavonoids). The maize BZ2 and the petunia AN9 genes encode glutathione S-transferases required for pigment accumulation (Alfenito et al., 1998). Possibly, the activity of these genes is required to tag the anthocyanin molecules for transport to the vacuole. The sequestration of these molecules into the vacuolar lumen, evidenced by the phenotype of the mutants (Figure 4.1), is essential for proper pigmentation. Careful observation of revertant spots in unstable mutant flowers shows that AN1, AN2, and AN11 control pigmentation in a cell-autonomous manner in the epidermal cells of the petals. The same holds for R and C1 in the aleurone tissue of maize kernels (Goodman et al., 2004). On the contrary, structural genes, including AN9 and BZ2, yield non-cell-autonomous revertant spots (spots with not well-defined borders). This observation suggests that the regulators of anthocyanin biosynthesis also control the expression of genes that participate in the sequestration of the anthocyanins into the vacuole. This is supported further by the finding that mutations in the *tt2* gene affect the expression of TT12, encoding a transporter of the MATE family (Nesi et al., 2001).

3.2. Control of Vacuolar pH

Anthocyanins are stored in the vacuole and their color not only depends on the substitution pattern of the molecule but also on the conditions in the vacuolar lumen, such as the pH, the concentration of metal ions and (flavonoid) copigments, and the way anthocyanins are packed (Grotewold, 2006). Thus, anthocyanins can be used as "natural" indicators of the conditions inside the vacuole. The vacuolar lumen is always more acidic than the surrounding cytoplasm and the pH gradient across the vacuolar membrane is actively built and maintained by ATP- or pyrophosphate (PPi)-powered proton pumps residing in the tonoplast (Maeshima, 2001). In petunia flowers, the acidification of the vacuole results in a red color of the flower, and mutations affecting vacuolar pH regulation can be recognized because of the bathochromic shift (toward blue) of the flower color (Figure 4.1G, H). A similar phenomenon is observed in flowers of *Ipomea*, where development of the normal blue color during petal maturation requires the alkalinization of the vacuole, a process in which the PURPLE (PR) protein, encoding a putative Na^+/H^+ pump (Fukada-Tanaka et al., 2000), participates. PR was proposed to transport Na^+ into and H^+ out of the vacuole, resulting in higher vacuolar pH and blue color. In petals of petunia flowers, epidermal cells have large vacuoles that under the microscope look like "big bags" filled with red pigment. These vacuoles undergo an acidification step around the time of flower opening. Later, when the flower is about to wilt, the pH increases further (Quattrocchio et al., 2005).

3.2.1. Identification of pH Mutants and Cloning of the Corresponding Genes

By screening petunia populations for flower color mutants, several genes were identified that control acidification of the vacuole. This includes seven *PH* genes, which, when mutated, make the flower more blue and increase the pH of the petal homogenates from ~5 to ~6 (Spelt et al., 2002). This indicates that in wild type petunia flowers, *PH* genes are involved in the acidification of the vacuole. Mutations in the anthocyanin regulatory genes (*AN1*, *AN2*, *AN11*, and *JAF13*) affect the pH of the flower homogenate in a similar way. This was initially unnoticed because the lack of pigmentation in these mutants masks the effect of the pH on flower color. Mutants in biosynthetic genes (e.g., *AN3* encoding F3H) have normal pH values, indicating that it is not the absence of the anthocyanins that results in the decrease of pH present in the regulatory gene mutants. Thus, the regulators of anthocyanin biosynthesis are likely to also control a parallel pathway that results in the establishment of the vacuolar pH.

3.2.2. Alleles of AN1 Uncover a Dual Function for Some bHLH Regulators

The cloning of the *PH6* gene (Chuck et al., 1993) and, independently, the anthocyanin regulator *AN1* (Spelt et al., 2000, 2002) by transposon tagging revealed that *ph6* is an allele of *AN1*. The *ph6* allele (and *ph6*–like alleles) conditions bluish flowers that contain normal quantities of anthocyanin pigments but have a higher pH. These alleles express truncated versions of the *AN1* mRNA (Spelt et al., 2002). The best-analyzed allele (*an1-G621*) results in the accumulation of high amounts (~ 30-fold higher than wild-type *AN1*) of a truncated AN1 protein that lacks the C-terminal half including the bHLH domain. At first glance, this suggests that the C-terminal half of AN1 is required for vacuolar acidification but not for anthocyanin synthesis. However, analysis of additional *an1* alleles showed that the situation is more complex. First, an allele that expresses the AN1 protein truncated at roughly the same position as those derived from the *ph6* alleles and that accumulates at wild-type levels conditioned strongly reduced anthocyanin synthesis and high pH. Second, transient assays in which the wild-type *AN1* and *ph6* alleles are over-expressed showed that *ph6* is impaired in its ability to efficiently activate *DFR*. Finally, an *AN1* allelic series, generated by independent excisions of a transposon insertion, showed that insertions of one, two, or three amino acids into the first helix of the bHLH domain progressively reduced the capacity of AN1 to induce both pigment biosynthesis and vacuolar pH (Spelt et al., 2002). Together, these results show that the C-terminal half of AN1 is essential for vacuolar acidification. This region of AN1 also participates in the activation of anthocyanin synthesis, but with a less important role, given that the removal of this region of the protein, as in the *ph6* alleles, can be compensated for by overexpression of the protein.

In addition to the anthocyanin and pH phenotypes, *an1* mutants have abnormal seed coats displaying peculiar epidermal cell expansion and/or division after fertilization during the development of the seed (Spelt et al., 2002). Because this phenotype is not observed in *an2* mutants (Quattrocchio et al., 1993) and it is very weak in *ph4* mutants (see below), it is possible that it results as a consequence of the interaction of AN1 with PH4 and/or with another as yet unidentified R2R3-MYB protein.

3.2.3. The PH Genes

Besides *AN1/PH6*, three more petunia *PH* genes have been isolated from transposon-tagged mutants. All these *ph* mutants increase vacuolar pH to a similar extent as *an1* and *an11* mutants, but do not affect anthocyanin biosynthesis.

PH4 encodes a R2R3-MYB factor that interacts with the conserved amino terminal domain of AN1 (Quattrocchio et al., 2005). The AN1–PH4 complex is dependent on the presence of AN11 for the control of vacuolar pH, in a fashion similar to how the anthocyanin regulatory complexes (AN1–AN2 and AN1–AN4) are dependent on AN11 (Koes et al., 2005).

PH3 encodes a WRKY transcription factor with high similarity to TTG2 (Verweij et al., submitted for publication). Expression studies show that *PH3*, like *PH4*, is expressed only in petals and ovaries and appears to be controlled, at least in part, by *AN1*, *PH4*, and *AN11*. Constitutive overexpression of AN1 and PH4 is sufficient to activate PH3 expression in leaves, indicating that PH3 acts downstream of AN1 and PH4 (Verweij et al., submitted for publication).

PH2 encodes a protein kinase related to STE20 from yeast (Spelt, Quattrocchio, Koes, unpublished results). STE20 is a serine/threonine kinase that represents the connecting point in the signal transduction pathway between trimeric G proteins and the MAP kinases cascade (Dan et al., 2001). Substrates for the PH2 kinase have not yet been identified.

All the *PH* loci so far isolated encode regulatory proteins that reveal little about the molecular mechanisms associated with vacuolar acidification. To identify genes directly involved in the acidification of the vacuole, mRNA profiling experiments were conducted by screening for genes strongly downregulated in *an1*, *ph3*, and *ph4* mutants, by cDNA AFLP and microarray hybridization. From this screen, 20 gene fragments have been identified and cloned. Similar to *PH3*, the expression of all these genes is enhanced in petunia plants ectopically over-expressing *AN1* and *PH4*. Furthermore, the analysis of plants transgenic for 35S::AN1-GR (Spelt et al., 2000) confirmed that several of these genes are direct target genes of AN1 (Spelt, Quattrocchio, and Koes, unpublished results). Interestingly, none of these newly identified genes is dependent on PH2, suggesting that *PH2* operates in a separate acidification pathway or that it is involved in post-transcriptional modification of one (or more) structural genes of the same pathway.

4. FLAVONOID REGULATORY NETWORKS

4.1. Models for the Cooperation between R2R3-MYB, bHLH and WDR Proteins

Extensive work on flavonoid regulation has demonstrated that the specific activation of each pathway relies on the participation of the appropriate R2R3-MYB factors. The participating R2R3-MYB proteins can be divided into two classes, depending on fundamental differences in their mode of action (Figure 4.2). A simpler mechanism concerns flavonol and phlobaphene biosynthesis and involves R2R3-MYB acting without known bHLH "cofactors" (Figure 4.3A). However, R2R3-MYB proteins controlling anthocyanin and PA biosynthesis require both bHLH and WDR cofactors acting in a complex interplay (Figure 4.3B). This model suggests that one role of the bHLH cofactor is to allow the R2R3-MYB factors to activate transcription (Grotewold et al., 2000). This function of the bHLH factor, which does not appear to require a recruitment of the bHLH to DNA, is unlikely to involve solely an increase in the DNA-binding affinity of the R2R3-MYB factor, since a mutant of C1 that binds DNA with high affinity continues to be dependent on R (Hernandez et al., 2004). In addition, the bHLH factors can function by enhancing the regulatory activity of the R2R3-MYB factor, presumably by being recruited to DNA. Although a direct binding of the bHLH of R or R-related proteins to DNA has not yet been demonstrated, the need for the presence of the conserved ARE, present in many flavonoid genes, for R-enhanced activation (Elomaa et al., 2003; Lesnick, 1997; Pooma et al., 2002) strongly suggests that the bHLH factor binds DNA or is recruited to DNA by another protein.

The role of the WDR proteins remains poorly understood. They can physically interact with bHLH factors and with some R2R3-MYB protein (e.g., TT2), supporting the idea that they may form part of the regulatory complex (Baudry et al., 2004; Kroon, 2004; Payne et al., 2000; Zhang et al., 2003). Previous findings indicating that AN11 is mainly localized in the cytosol led to the conclusion that the WDR is part of a cytoplasmic-signaling pathway (de Vetten et al., 1997). However, recent results indicate that the Perilla ortholog, PfWD (Sompornpailin et al., 2002), can be mobilized to the nucleus when coexpressed with the bHLH regulator of the pathway. In addition, three-hybrid experiments conducted in yeast indicated that TTG1 can simultaneously interact in the nucleus with the TT2/TT8 complex and the BAN promoter (Baudry et al., 2004). However, in this case, TTG1 was not required for DNA recognition by the complex and the role of the WDR protein remains poorly understood. One attractive hypothesis is that the WDR proteins are necessary to prevent the effect of a negative regulator.

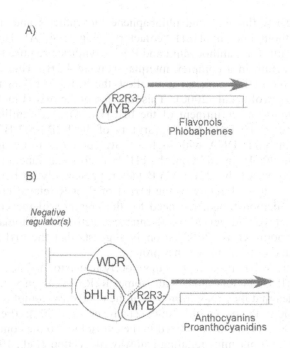

Figure 4.3 *Models for the participation of WDR, MYB, and bHLH in the regulation of flavonoid pathways.*

4.2. Negative Regulators of Flavonoid Gene Activation

From what has been described so far in this chapter, it appears that the expression of the flavonoid genes and largely a result of the action of transcription activators. Nevertheless, substantial experimental evidence also involves negative regulators in the control of transcription of flavonoid biosynthetic genes.

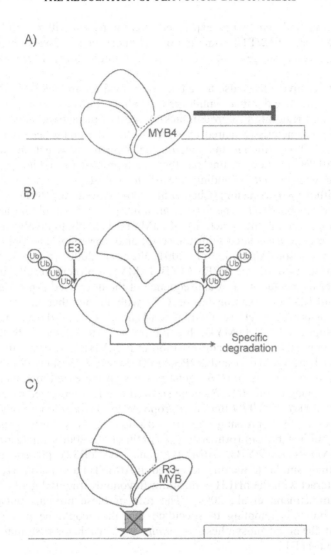

Figure 4.4 *Model for the negative regulation of flavonoid biosynthesis.*

The maize *IN1* (for *Intensifier1*) factor was defined by recessive mutations that increase pigmentation of the kernel (Burr et al., 1996). The full-length cDNA encodes a protein similar to AN1 and TT8. However, 99.8% of the mRNAs are incorrectly spliced and encode truncated proteins. It is possible that some of these truncated bHLH proteins act as dominant negative factors, for example, by titrating out active C1 protein. However, in transient expression experiments in *Arabidopsis* protoplasts, a close IN1 relative (TT8) was not able to activate the expression of an

Arabidopsis structural gene, when expressed in combination with C1, unlike similar C1/Sn, TT2/Sn, and TT2/TT8 combinations (Baudry et al., 2004) (Figure 4.4B). This intriguing result suggests that TT8 and IN1 might form inactive complexes with C1.

Dominant negative alleles also have been described for an R2R3-MYB regulator, the maize *C1* gene. The dominant inhibitor *C1-I* allele (Paz-Ares et al., 1990) carries a frameshift in C1 resulting in a protein lacking most of the activation domain (Sainz et al., 1997), a characteristic shared by several other dominant negative *C1* alleles (Singer et al., 1998). Although the specific mechanisms by which these dominant negative C1 alleles function are unclear, they were proposed to inhibit pigmentation by competing with the DNA-binding site of the wild-type C1 protein or for the interaction with the R/B cofactors (Chen et al., 2004; Goff et al., 1991).

The strawberry *FaMYB1* gene acts as an inhibitor of flavonol and anthocyanin gene expression. The inhibitory activity of FaMYB1 is likely to involve a conserved amino acid sequence associated with transcriptional repression and present also in the AtMYB4 repressor (Aharoni et al., 2001; Jin et al., 2000) (Figure 4.4A). Over-expression in tobacco of the R2R3-MYB FaMYB1 protein results in strongly reduced flower pigmentation, as a consequence of the decreased expression of ANS and a flavonoid-UDP-glucose transferase (GT), with all the other biosynthetic genes remaining at comparable levels as in wild type (Aharoni et al., 2001). Yeast two-hybrid experiments suggested that FaMYB1 has the potential to interact with the petunia AN1 protein, consistent with the conservation in FaMYB1 of residues implicated in the interaction between bHLH and R2R3-MYB domains (Grotewold et al., 2000). *Arabidopsis* mutants for the *AtMYB4* gene possess an increased tolerance to UV-B light, and the cloning of the *AtMYB4* gene showed that it is repressed by exposure to UV (Jin et al., 2000). *AtMYB4* thus was proposed to play a role in modulating UV screen production by repressing the transcription of very early genes of the phenylpropanoid metabolism (primarily *C4H*) (Jin et al., 2000). Interestingly, three other R2R3-MYBs (AtMYB3, AtMYB7, and AtMYB32) present a putative inhibitory domain similar to the one described in AtMYB4 and FaMYB1, and could potentially interact with the bHLH controlling flavonoid structural genes (Stracke et al., 2001; Zimmermann et al., 2004). The role of these proteins remains to be investigated, but it is tempting to speculate that they could be involved in the modulation of flavonoid biosynthesis *in planta* by a mechanism similar to that of AtMYB4 or FaMYB1.

The *Arabidopsis ICX1* locus (for increased chalcone synthase expression1) is a negative regulator of pathways inducing flavonoid biosynthetic genes (Wade et al., 2003) (Figure 4.4C). Mutant analyses suggested that ICX1 negatively modulates gene expression in response to light (mediated by UV-B, *CRY1*, and *PHYA*), environmental stimuli (cold), metabolism, and hormones. The action of this locus, however, is not confined to flavonoid biosynthetic genes and, similarly to the pleiotropic effect of the *ttg1* mutation, *icx1* shows alterations in other aspects of epidermal cells development (e.g., cell division and expansion, trichome number, root elongation, and root-hair initiation). *ICX1* could encode a general regulator of epidermal cell development and interfere with several TTG1-dependent pathways.

The *UPL3/KAKTUS* gene is another negative regulator involved in the control of a TTG1-dependent pathway (Downes et al., 2003; El Refy et al., 2003). It encodes a protein with a HECT domain, a classical signature of E3 ubiquitin ligases, and it represses excess branching and endoreduplication of *Arabidopsis* trichomes. A possible target for UPL3/KAKTUS is GL3 (Downes et al., 2003), suggesting that a posttranslational control of the stability of this bHLH might be important during trichome organogenesis. This hypothesis fits well with the proposition that TTG1 could act in transient stabilization of its bHLH partner (Payne et al., 2000). This model is also consistent with the quantitative effect of TTG1 on TT8 activity (Baudry et al., 2004).

The MYBX protein of petunia was identified as a partner of AN1 in yeast two-hybrid experiments (Kroon, 2004). MYBX is a small R3-MYB protein that lacks a transcription activator domain, thus resembling structurally the *Arabidopsis* CPC and TRY proteins (Kroon, 2004; Schellmann et al., 2002; Wada et al., 1997). Over-expression of this factor in transgenic petunia plants results in flowers with reduced expression of all *AN1* target genes, loss of pigmentation, and vacuolar pH alkalinization. The observation that *MYBX* expression is affected in *an1* flowers and that the endogenous *MYBX* gene is downregulated in plants overexpressing *MYBX* suggests that this small R3-MYB is part of a regulatory loop that modulates *AN1* activity. Transgenic plants overexpressing *AN1* do not show any kind of phenotypic alteration, which is probably a consequence of the negative feedback loop. In addition, MYBX might function by titrating out an R2R3-MYB factor to restrict more efficiently the activity of R2R3-MYB/bHLH/WDR module(s) to specific cellular domains, as proposed recently for CPC and TRY in *Arabidopsis* (Esch et al., 2003; Hulskamp, 2004) (Figure 4.4C).

4.3. Similar R2R3-MYB/bHLH/WDR Regulatory Complexes Control Other Processes

The development of leaf hairs in *Arabidopsis* is a process extensively studied as a model for cell differentiation. On leaves and stems, some epidermal cells develop into trichomes. Cells that adopt a trichome fate do not appear in regular (fixed) positions, but do show more or less regular spacing (i.e., only one of two neighboring cells normally develops into a trichome). In the root, certain files of cells develop into root hairs, depending on their position relative to the underlying cortical cells.

The definition of the trichome fate of a leaf epidermal cell is dependent on TTG1, the redundant bHLH factors GL3 and EGL3, and the MYB factor GL1 (Ramsay and Glover, 2005). In addition, the single MYB repeat protein TRY acts as an inhibitor of trichome development and may play a role in determining trichome spacing, since *try* mutants show an increased number of trichomes that often occur in clusters (Schellmann et al., 2002). Underground, the default pathway of epidermal is root hair formation, and a mechanism very similar to that determining trichome fate establishes the fate of a root non-hair (atrichoblast) cell. In the root, *GL3* and *EGL3* (Zhang et al., 2003), together with the *GL1* paralogue *WER* (Lee and Schiefelbein, 1999) and *TTG1*, are necessary for non-hair cell specification, and a paralogue of *TRY*, *CPC* (Wada et al., 2002), functions as an inhibitor of the process. TRY and CPC are able to interact with the bHLH proteins such as R from maize and AN1 and JAF13 from petunia (Spelt, Kroon, Quattrocchio, and Koes, in preparation).

Interestingly, the various bHLH proteins are interchangeable across the different pathways. The overexpression of *EGL3* and/or *GL3* partially complement the defect in PA of the *tt8* mutant (Zhang et al., 2003). Ectopic coexpression of EGL3 or GL3 together with GL1 results in supernumerary trichomes, while the ectopic expression of EGL3 or GL3 together with PAP1 induces anthocyanins production (Zhang et al., 2003). These results indicate that the specificity for the set of target genes is provided by the MYB protein, while the bHLH factor is likely to be shared among multiple different processes. Curiously, this principle holds even across species. Although mutations in the anthocyanin regulators of maize and petunia have no effect on trichome development, the expression of R or the WDR PAC1 readily restores the defects in root hair development of *Arabidopsis ttg1* mutants (Carey et al., 2004; Lloyd et al., 1994).

Once the pattern of trichome differentiation (distribution of trichomes on the leaf surface) is determined and the differentiation program is induced, the specification of the trichome structure (expansion, elongation, branching, orientation of the branches) requires additional regulators that are controlled by the WDR/bHLH/R2R3-MYB network (Ramsay and Glover, 2005). One of these "secondary" regulators is the WRKY factor TTG2 (Johnson et al., 2002), which is involved in branching and proper development of the trichome structure. Like the closely related petunia PH3 protein, TTG2 is expressed under the control of the WDR/bHLH/R2R3-MYB network and can complement the *ph3* when expressed in petunia (Verweij, Spelt, Koes, and Quattrocchio, in preparation). A second downstream regulator is the homeodomain-leucine zipper protein encoded by GL2 (Di Cristina et al., 1996; Masucci et al., 1996). *GL2* mutants have small trichomes with reduced branching and abnormal expansion. Interestingly, a protein of the same family (ANL2) is involved in anthocyanins accumulation in subepidermal cells of *Arabidopsis* seedlings and mature plant tissue (Kubo et al., 1999). The emerging picture is one of a repetition of similar, successful networks of interacting factors to which small variations are applied to control distinct cellular processes.

4.4. Regulating the Regulators

The control of the regulators described in the previous sections is likely to be the primary step in the determination of the precise pattern of flavonoid accumulation in plants. While R2R3-MYB/bHLH/WDR interactions are well documented, essential data are still lacking concerning the activation of the expression of the corresponding genes. For instance, there is an obvious correlation between the activities of *BAN* and *TT2* promoters in PA-accumulating cells (Debeaujon et al., 2003). Additionally, recent analysis has revealed that the activity of the *TTG1* promoter is tightly regulated and restricted to specific cellular domains, suggesting that the expression of the WDR is also an important control point in the specificity of flavonoid accumulation in plants (Baudry *et al.*, 2004).

One example of the analysis of the mechanisms controlling the expression of a flavonoid regulator is the study of the *C1* promoter (Kao et al., 1996). At least three *cis*-acting regulatory elements are of particular importance in the C1 promoter, including an RY motif responsible for VP1-mediated activation and two elements responding to abscisic acid or light. Further evidence for the direct binding of VP1 to the RY motif have highlighted the role of this transcription factor in the direct activation of the expression of *C1* in maize seeds (Suzuki et al., 1997). Factors homologous to VP1 have been identified in *Arabidopsis*. These include the ABI3, LEC2, and FUS3 proteins, which constitute the B3-domain transcription factor family (Gazzarrini et al., 2004). By contrast, the *Arabidopsis* factors appear to be negative regulators of flavonoid biosynthesis, as anthocyanin accumulation is ectopically induced in *lec2* and *fus3* mutant embryos. Consistently, FUS3 prevents *TTG1* expression in precise cellular domains of the embryo (Tsuchiya et al., 2004). Thus, unlike the strong similarities observed in flavonoid R2R3-MYB/bHLH/WDR regulatory modules, mechanisms of control of these regulators can exhibit strong divergence between species. In this regard, an interesting interplay has been revealed between R2R3-MYB and the expression of their bHLH cofactors in dicotyledonous species. In petunia, the expression of *AN1* (bHLH) is activated by AN2 and AN4, two R2R3-MYB controlling anthocyanin production in flowers (Spelt et al., 2000). Similarly, the ectopic expression of *TT2* can induce the production of *TT8* in roots (Nesi et al., 2001). In these species, the control of the expression of the bHLH genes by other flavonoid regulators might be an important feature of the regulatory network, but no such mechanism so far has been revealed in maize (Carey et al., 2004).

4.5. Evolutionary Considerations

The anthocyanin biosynthetic pathway shows a high level of conservation of both the structural and regulatory genes, in spite of the very different pigmentation patterns and pathway end products. To explain regulatory differences between the anthocyanin pathways from maize and petunia, a model has been postulated for the evolution of the pathway (a full description of this and other models can be found in Chapter 7). This model assumed that the regulators are older than the structural genes of the anthocyanin pathway itself and that they have (or had) distinct, more ancient functions (Koes et al., 1994). The coupling of different gene sets (by acquisition of the appropriate *cis*-regulatory elements) to the regulators, combined with the acquisition of tissue-specific expression patterns by the regulators themselves, gave rise to the large variety of flavonoid accumulation patterns that we can observe nowadays.

The finding that the same set of WDR, bHLH, and R2R3-MYB regulators control various other aspects of epidermal cell differentiation (like vacuolar acidification and hair specification) seems to ask for an extension of the model. Moreover, new sets of target genes appear to be acquired by preexisting regulators. However, in these examples, the addition of a specific *cis*-regulatory element to the new target genes is not sufficient and may have been accompanied by the recruitment of new (MYB-type) protein partners to give a new specificity to the protein complex. Models to explain how small variations in R2R3-MYB may result in novel regulatory functions, particularly from the perspective of their ability to regulate metabolic pathways, recently has been described (Grotewold, 2005).

Curiously, the ability of the anthocyanin regulators to control pigment formation is largely conserved throughout the angiosperms, although there are (small) differences in the way they regulate EBG. In contrast, their involvement in hair development or vacuolar acidification seems to be more limited (Carey et al., 2004; Lloyd et al., 1992, 1994). Thus, alterations of the WDR and bHLH regulators do not seem to determine whether they control hair development or not, but changes in the downstream genes and/or the MYB partners do. The role of the anthocyanin regulators in vacuolar acidification is only seen in petunia, but it cannot be excluded that the same mechanisms exists in other flowers, where no mutants yet have been isolated.

One way to explain why the role in hair development or vacuolar acidification is limited to relatively few plant species is to assume that the anthocyanin regulators have acquired these functions relatively recently. However, an exciting alternative is that hair development and vacuolar acidification depend on the same or very similar (conserved) downstream processes, for example, the assembly or function of endomembrane systems. The comparative analysis of down stream structural genes involved in vacuolar acidification and hair development will be essential to shed light on the question of how apparently dissimilar processes are regulated by similar sets of proteins.

5. CONCLUSIONS, REMARKS, AND FUTURE DIRECTIONS

The regulation of flavonoid biosynthesis continues to provide a powerful system to investigate basic mechanisms of the regulation of plant gene expression. The interaction between transcription factors from various different families makes it an ideal case to understand combinatorial control of transcription. Given that proteins similar to those controlling flavonoid biosynthesis are used in the regulation of a number of other cellular processes indicates a more general mechanism of cooperation between these groups of regulatory proteins. Uncovering the biochemical bases of the cooperation is one area in which our efforts are currently focusing.

The finding that distinct branches of the flavonoid pathway are independently regulated opens significant avenues for the manipulation of specific groups of phytochemicals by metabolic engineering. The use of transcription factors for the manipulation of plant metabolic pathways is gaining momentum (Braun et al., 2001; Broun, 2004; Memelink et al., 2000). The possibility that transcription factors not only control the expression of biosynthetic enzymes, but that they also contribute to the establishment of the proper conditions (e.g., pH) and compartments (e.g., vacuoles) for the accumulation of the corresponding phytochemicals is attractive and deserves to be investigated in more detail.

6. ACKNOWLEDGMENTS

We appreciate the comments of Rebecca Lamb on this manuscript. Research on regulators of anthocyanin biosynthesis and on the mechanisms of pH regulation in petunia in the Quattrocchio laboratory is supported by grants of the Netherlands Organization for Life Science (ALW) and the Netherlands Technology Foundation (STW) with financial aid from the Netherlands Organization for the Advancement of Research (NWO). A. Baudry and part of the research carried out in the Lepiniec laboratory are supported by a grant from GABI-Genoplante. Research on the regulation of flavonoid biosynthesis in the Grotewold laboratory is funded by grants from the National Science Foundation (MCB 0210413) and the National Research Initiative of the USDA Cooperative State Research, Education and Extension Service (grant number 2003-35318-13689).

7. REFERENCES

Abrahams, S., Lee, E., Walker, A., Tanner, G., Larkin, T., and Ashton, A., 2003, The *Arabidopsis tds4* gene encodes leucoanthocyanidin dioxygenase (ldox) and is essential for proanthocyanidin synthesis and vacuole development, *Plant J* **35**: 624-636.

Abrahams, S., Tanner, G. J., Larkin, P. J., and Ashton, A. R., 2002, Identification and biochemical characterization of mutants in the proanthocyanidin pathway in *Arabidopsis*, *Plant Physiol* **130**: 561-576.

Aharoni, A., De Vos, C. H., Wein, M., Sun, Z., Greco, R., Kroon, A., Mol, J. N., and O'Connell, A. P., 2001, The strawberry *FaMYB1* transcription factor suppresses anthocyanin and flavonol accumulation in transgenic tobacco, *Plant J* **28**: 319-332.

Alfenito, M. R., Souer, E., Goodman, C. D., Buell, R., Mol, J., Koes, R., and Walbot, V., 1998, Functional complementation of anthocyanin sequestration in the vacuole by widely divergent glutathione *S*-transferases, *Plant Cell* **10**: 1135-1149.

Atchley, W. R., Wollenberg, K. R., Fitch, W. M., Terhalle, W., and Dress, A. W., 2000, Correlations among amino acid sites in bHLH protein domains: an information theoretic analysis, *Mol Biol Evol* **17**: 164-178.

Baudry, A., Heim, M., Dubreucq, B., Caboche, M., Weisshaar, B., and Lepiniec, L., 2004, TT2, TT8 and TTG1 synergistically specify the expression of *BANYULS* and proanthocyanidin biosynthesis in *Arabidopsis thaliana*, *Plant J* **39**: 366.

Borevitz, J. O., Xia, Y., Blount, J., Dixon, R. A., and Lamb, C., 2000, Activation tagging identifies a conserved MYB regulator of phenylpropanoid, *Plant Cell* **12**: 2383-2394.

Bovy, A., de Vos, R., Kemper, M., Schijlen, E., Almenar Pertejo, M., Muir, S., Collins, G., Robinson, S., Verhoeyen, M., Hughes, S., Santos-Buelga, C., and van Tunen, A., 2002, High-flavonol tomatoes resulting from the heterologous expression of the maize transcription factor genes *Lc* and *C1*, *Plant Cell* **14**: 2509-2526.

Braun, E. L., Matulnik, T., Dias, A., and Grotewold, E., 2001, Transcription factors and metabolic engineering: Novel applications for ancient tools, *Recent Adv Phytochem* **35**: 79-109.

Broun, P., 2004, Transcription factors as tools for metabolic engineering in plants, *Curr Opin Plant Biol* **7**: 202-209.

Buck, M. J., and Atchley, W. R., 2003, Phylogenetic analysis of plant basic helix-loop-helix proteins, *J Mol Evol* **56**: 742-750.

Burr, F. A., Burr, B., Scheffler, B. E., Blewitt, M., Wienand, U., and Matz, E. C., 1996, The maize repressor-like gene *intensifier1* shares homology with the *r1/b1* multigene family of transcription factors and exhibits missplicing, *Plant Cell* **8**: 1249-1259.

Carey, C. C., Strahle, J. T., Selinger, D. A., and Chandler, V. L., 2004, Mutations in the *pale aleurone color1* regulatory gene of the *Zea mays* anthocyanin pathway have distinct phenotypes relative to the functionally similar TRANSPARENT TESTA GLABRA1 gene in *Arabidopsis thaliana, Plant Cell* **16**: 450-464.

Chen, B., Wang, X., Hu, Y., Wang, Y., and Lin, Z., 2004, Ectopic expression of a *c1-I* allele from maize inhibits pigment formation in the flower of transgenic tobacco, *Mol Biotechnol* **26**: 187-192.

Chopra, S., Brendel, V., Zhang, J., Axtell, J. D., and Peterson, T., 1999, Molecular characterization of a mutable pigmentation phenotype and isolation of the first active transposable element from *Sorghum bicolor, Proc Natl Acad Sci U S A* **96**: 15330-15335.

Chuck, G., Robbins, T., Nijjar, C., Ralston, E., Courtney-Gutterson, N., and Dooner, H. K., 1993, Tagging and cloning of a petunia flower color gene with the maize transposable element *Activator, Plant Cell* **5**: 371-378.

Damiani, R. D., and Wessler, S. R., 1993, An upstream open reading frame represses expression of Lc, a member of the R/B family of maize transcriptional activators, *Proc Natl Acad Sci USA* **90**: 8244-8248.

Dan, I., Watanabe, N. M., and Kusumi, A., 2001, The Ste20 group kinases as regulators of MAP kinase cascades, *Trends Cell Biol* **11**: 220-230.

de Vetten, N., Quattrocchio, F., Mol, J., and Koes, R., 1997, The *an11* locus controlling flower pigmentation in petunia encodes a novel WD-repeat protein conserved in yeast, plants, and animals, *Genes Dev* **11**: 1422-1434.

Debeaujon, I., Leon-Kloosterzeil, K. M., and Koornneef, M., 2000, Influence of the testa on seed dormancy, germination, and longevity in *Arabidopsis, Plant Physiol* **122**: 403-413.

Debeaujon, I., Nesi, N., Perez, P., Devic, M., Grandjean, O., Caboche, M., and Lepiniec, L., 2003, Proanthocyanidin-accumulating cells in *Arabidopsis* testa: regulation of differentiation and role in seed development, *Plant Cell* **15**: 2514-2531.

Devic, M., Guilleminot, J., Debeaujon, I., Bechtold, N., Bensaude, E., Koornneef, M., Pelletier, G., and Delseny, M., 1999, The BANYULS gene encodes a DFR-like protein and is a marker of early seed coat development., *Plant J* **19**: 387-398.

Di Cristina, M., Sessa, G., Dolan, L., Linstead, P., Baima, S., Ruberti, I., and Morelli, G., 1996, The *Arabidopsis* Athb-10 (GLABRA2) is an HD-Zip protein required for regulation of root hair development, *Plant J* **10**: 393-402.

Dias, A. P., Braun, E. L., McMullen, M. D., and Grotewold, E., 2003, Recently duplicated maize R2R3 Myb genes provide evidence for distinct mechanisms of evolutionary divergence after duplication, *Plant Physiol* **131**: 610-620.

Dixon, R. A., Xie, D. Y., and Sharma, S. B., 2005, Proanthocyanidins—a final frontier in flavonoid research? *New Phytol* **165**: 9-28.

Downes, B. P., Stupar, R. M., Gingerich, D. J., and Vierstra, R. D., 2003, The HECT ubiquitin-protein ligase (UPL) family in *Arabidopsis*: UPL3 has a specific role in trichome development, *Plant J* **35**: 729-742.

El Refy, A., Perazza, D., Zekraoui, L., Valay, J. G., Bechtold, N., Brown, S., Hulskamp, M., Herzog, M., and Bonneville, J. M., 2003, The *Arabidopsis KAKTUS* gene encodes a HECT protein and controls the number of endoreduplication cycles, *Mol Genet Genomics* **270**: 403-414.

Elomaa, P., Uimari, A., Metho, M., Albert, V., Laitinen, R. A. E., and Teeri, T. H., 2003, Activation of anthocyanin biosynthesis in *Gerbera hybrida* (Asteraceae) suggests conserved protein-protein and protein-promoter interactions between the anciently diverged monocots and eudicots, *Plant Physiol* **133**: 1831-1842.

Esch, J. J., Chen, M., Sanders, M., Hillestad, M., Ndkium, S., Idelkope, B., Neizer, J., and Marks, M. D., 2003, A contradictory *GLABRA3* allele helps define gene interactions controlling trichome development in *Arabidopsis, Development* **130**: 5885-5894.

Forkmann, G., 1991, Flavonoids as flower pigments: The formation of the natural spectrum and its extension by genetic engineering, *Plant Breeding* **106**: 1-26.

Franken, P., Niesbach-Klosgen, U., Weydemann, U., Marechal-Drouard, L., Saedler, H., and Wienand, U., 1991, The duplicated chalcone synthase genes *C2* and *Whp* (white pollen) of *Zea mays* are independently regulated; evidence for translational control of Whp expression by the anthocyanin intensifying gene *In*, *EMBO J* **10**: 2605-2612.

Fukada-Tanaka, S., Inagaki, Y., Yamaguchi, T., Saito, N., and Iida, S., 2000, Colour-enhancing protein in blue petals., *Nature* **407**: 581.

Gazzarrini, S., Tsuchiya, Y., Lumba, S., Okamoto, M., and McCourt, P., 2004, The transcription factor FUSCA3 controls developmental timing in *Arabidopsis* through the hormones gibberellin and abscisic acid, *Dev Cell* **7**: 373-385.

Goff, S. A., Cone, K. C., and Chandler, V. L., 1992, Functional analysis of the transcriptional activator encoded by the maize *B* gene: evidence for a direct functional interaction between two classes of regulatory proteins, *Genes Dev* **6**: 864-875.

Goff, S. A., Cone, K. C., and Fromm, M. E., 1991, Identification of functional domains in the maize transcriptional activator C1: comparison of wild-type and dominant inhibitor proteins, *Genes Dev* **5**: 298-309.

Goodman, C. D., Casati, P., and Walbot, V., 2004, A multidrug resistance–associated protein involved in anthocyanin transport in *Zea mays*, *Plant Cell* **16**: 1812-1826.

Grotewold, E., 2005, Plant metabolic diversity: A regulatory perspective, *Trends Plant Sci* **10**: 57-62.

Grotewold, E., 2006, The genetics and biochemistry of floral pigments, *Annu Rev Plant Biol* in press.

Grotewold, E., Athma, P., and Peterson, T., 1991, Alternatively spliced products of the maize P gene encode proteins with homology to the DNA-binding domain of myb-like transcription factors, *Proc Natl Acad Sci USA* **88**: 4587-4591.

Grotewold, E., Chamberlin, M., Snook, M., Siame, B., Butler, L., Swenson, J., Maddock, S., Clair, G. S., and Bowen, B., 1998, Engineering secondary metabolism in maize cells by ectopic expression of transcription factors, *Plant Cell* **10**: 721-740.

Grotewold, E., Drummond, B. J., Bowen, B., and Peterson, T., 1994, The myb-homologous P gene controls phlobaphene pigmentation in maize floral organs by directly activating a flavonoid biosynthetic gene subset, *Cell* **76**: 543-553.

Grotewold, E., Sainz, M. B., Tagliani, L., Hernandez, J. M., Bowen, B., and Chandler, V. L., 2000, Identification of the residues in the Myb domain of maize C1 that specify the interaction with the bHLH cofactor R, *Proc Natl Acad Sci U S A* **97**: 13579-13584.

Heim, M. A., Jakoby, M., Werber, M., Martin, C., Weisshaar, B., and Bailey, P. C., 2003, The basic helix-loop-helix transcription factor family in plants: a genome-wide study of protein structure and functional diversity, *Mol Biol Evol* **20**: 735-747.

Heine, G. F., Hernandez, M. J., and Grotewold, E., 2004, Two cysteines in plant R2R3 MYB domains participate in REDOX-dependent DNA binding, *J Biol Chem* **279**: 37878-37885.

Hernandez, J., Heine, G., Irani, N. G., Feller, A., Kim, M.-G., Matulnik, T., Chandler, V. L., and Grotewold, E., 2004, Different mechanisms participate in the R-dependent activity of the R2R3 MYB transcription factor C1, *J Biol Chem* **279**: 48205-48213.

Hulskamp, M., 2004, Plant trichomes: a model for cell differentiation, *Nat Rev Mol Cell Biol* **5**: 471-480.

Irani, N. G., Hernandez, J. M., and Grotewold, E., 2003, Regulation of anthocyanin pigmentation, *Rec Adv Phytochem* **38**: 59-78.

Jiang, C., Gu, J., Chopra, S., Gu, X., and Peterson, T., 2004, Ordered origin of the typical two- and three-repeat *Myb* genes, *Gene* **326**: 13-22.

Jin, H., Cominelli, E., Bailey, P., Parr, A., Mehrtens, F., Jones, J., Tonelli, C., Weisshaar, B., and Martin, C., 2000, Transcriptional repression by AtMYB4 controls production of UV-protecting sunscreens in *Arabidopsis.*, *EMBO J* **19**: 6150-6161.

Johnson, C. S., Kolevski, B., and Smyth, D. R., 2002, TRANSPARENT TESTA GLABRA2, a trichome and seed coat development gene of *Arabidopsis*, encodes a WRKY transcription factor, *Plant Cell* **14**: 1359-1375.

Kao, C-Y., Cocciolone, S. M., Vasil, I. K., and McCarty, D. R., 1996, Localization and interaction of the *cis*-acting elements for abscisic acid, VIVIPAROUS1, and light activation of the C1 gene of maize, *Plant Cell* **8**: 1171-1179.

Koes, R., Verweij, W., and Quattrocchio, F., 2005, Flavonoids: a colorful model for the regulation and evolution of biochemical pathways, *Trends Plant Sci* **10**: 236-242.

Koes, R. E., Quattrocchio, F., and Mol, J. N. M., 1994, The flavonoid biosynthetic pathway in plants: function and evolution, *BioEssays* **16**: 123-132.

Kootstra, A., 1994, Protection from UV-B-induced DNA damage by flavonoids, *Plant Mol Biol* **26**: 771-774.

Kroon, A. R., 2004, Thesis: Transcriptional regulation of the anthocyanin pathway in *Petunia hybrida*, Vrije Universiteit, Amsterdam.

Kubo, H., Peeters, A. J., Aarts, M. G., Pereira, A., and Koornneef, M., 1999, ANTHOCYANINLESS2, a homeobox gene affecting anthocyanin distribution and root development in *Arabidopsis*, *Plant Cell* **11**: 1217-1226.

Lee, M. M., and Schiefelbein, J., 1999, WEREWOLF, a MYB-related protein in *Arabidopsis*, is a position-dependent regulator of epidermal cell patterning, *Cell* **99**: 473-483.

Lesnick, M. L., 1997, Thesis: Analysis of the *cis*-acting sequences required for C1/B activation of the maize anthocyanin biosynthetic pathway, University of Oregon, Eugene.

Lesnick, M. L., and Chandler, V. L., 1998, Activation of the maize anthocyanin gene a2 is mediated by an element conserved in many anthocyanin promoters, *Plant Physiol* **117**: 437-445.

Lloyd, A. M., Schena, M., Walbot, V., and Davis, R. W., 1994, Epidermal cell fate determination in *Arabidopsis*: Patterns defined by a steroid-inducible regulator, *Science* **266**: 436-439.

Lloyd, A. M., Walbot, V., and Davis, R. W., 1992, *Arabidopsis* and *Nicotiana* anthocyanin production activated by maize regulators *R* and *C1*, *Science* **258**: 1773-1775.

Maeshima, M., 2001, Tonoplast transporters: organization and function, *Annu Rev Plant Physiol Plant Mol Biol* **52**: 469-497.

Marles, M. A. S., Ray, H., and Gruber, M. Y., 2003, New perspectives on proanthocyanidin biochemistry and molecular regulation, *Phytochem* **64**: 367-383.

Masucci, J. D., Rerie, W. G., Foreman, D. R., Zhang, M., Galway, M. E., Marks, M. D., and Schiefelbein, J. W., 1996, The homeobox gene GLABRA2 is required for position-dependent cell differentiation in the root epidermis of *Arabidopsis thaliana*, *Development* **122**: 1253-1260.

Mehrtens, F., Kranz, H., Bednarek, P., and Weissaar, B., 2005, The *Arabidopsis thaliana* transcription factor MYB12 is a flavonol-specific regulator of phenylpropanoid biosynthesis, *Plant Physiol* in press.

Memelink, J., Menke, F. L. W., van der Fits, L., and Kijne, J. W., 2000, Transcriptional regulators to modify secondary metabolism. In: *Metabolic Engineering of Plant Secondary Metabolism*, R. Verpoorte, and A. W. Alfermann (Eds.), Dordrecht, Kluwer Academic Publishers, pp. 111-125.

Mo, Y., Nagel, C., and Taylor, L. P., 1992, Biochemical complementation of chalcone synthase mutants defines a role for flavonols in functional pollen, *Proc Natl Acad Sci USA* **89**: 7213-7217.

Mol, J., Grotewold, E., and Koes, R., 1998, How genes paint flowers and seeds, *Trends Plant Sci* **3**: 212-217.

Nesi, N., Debeaujon, I., Jond, C., Pelletier, G., Caboche, M., and Lepiniec, L., 2000, The TT8 gene encodes a basic helix-loop-helix domain protein required for expression of DFR and BAN genes in *Arabidopsis* siliques, *Plant Cell* **12**: 1863-1878.

Nesi, N., Debeaujon, I., Jond, C., Stewart, A. J., Jenkins, G. I., Caboche, M., and Lepiniec, L., 2002, The TRANSPARENT TESTA16 locus encodes the ARABIDOPSIS BSISTER MADS domain protein and is required for proper development and pigmentation of the seed coat, *Plant Cell* **14**: 2463-2479.

Nesi, N., Jond, C., Debeaujon, I., Caboche, M., and Lepiniec, L., 2001, The *Arabidopsis* TT2 gene encodes an R2R3 MYB domain protein that acts as a key determinant for proanthocyanidin accumulation in developing seed, *Plant Cell* **13**: 2099-2114.

Pairoba, C. F., and Walbot, V., 2003, Post-transcriptional regulation of expression of the *Bronze2* gene of *Zea mays* L., *Plant Mol Biol* **53**: 75-86.

Payne, C., Zhang, F., and Lloyd, A., 2000, *GL3* encodes a bHLH protein that regulate trichome development in *Arabidopsis* through interaction with GL1 and TTG1, *Genetics* **156**: 1349-1362.

Paz-Ares, J., Ghosal, D., and Saedler, H., 1990, Molecular analysis of the C1-I allele from *Zea mays*: a dominant mutant of the regulatory C1 locus, *EMBO J* **9**: 315-321.

Paz-Ares, J., Ghosal, D., Weinland, U., Peterson, P. A., and Saedler, H., 1987, The regulatory *c1* locus of *Zea mays* encodes a protein with homology to *myb* proto-oncogene products and with structural similarities to transcriptional activators, *EMBO J* **6**: 3553-3558.

Pelletier, M. K., Burbulis, I. E., and Winkel-Shirley, B., 1999, Disruption of specific flavonoid genes enhances the accumulation of flavonoid enzymes and end-products in *Arabidopsis* seedlings, *Plant Mol Biol* **40**: 45-54.

Pooma, W., Gersos, C., and Grotewold, E., 2002, Transposon insertions in the promoter of the *Zea mays a1* gene differentially affect transcription by the Myb factors P and C1, *Genetics* **161**: 793-801.

Quattrocchio, F., Verweij, C. W., Kroon, A., Spelt, C., Mol, J., and Koes, R., 2005, The *PH4* gene of petunia encodes a MYB domain protein that interacts with transcription factors of the anthocyanins pathway, submitted for publication.

Quattrocchio, F., Wing, J., van der Woude, K., Mol, J., and Koes, R., 1998, Analysis of bHLH and MYB domain proteins: Species-specific regulatory differences are caused by divergent evolution of target anthocyanin genes, *Plant J* **13**: 475-488.

Quattrocchio, F., Wing, J., van der Woude, K., Souer, E., de Vetten, N., Mol, J., and Koes, R., 1999, Molecular analysis of the *anthocyanin2* gene of petunia and its role in the evolution of flower color, *Plant Cell* **11**: 1433-1444.

Quattrocchio, F., Wing, J. F., Leppen, H. T. C., Mol, J. N. M., and Koes, R. E., 1993, Regulatory genes controlling anthocyanin pigmentation are functionally conserved among plant species and have distinct sets of target genes, *Plant Cell* **5**: 1497-1512.

Rabinowicz, P. D., Braun, E. L., Wolfe, A. D., Bowen, B., and Grotewold, E., 1999, Maize *R2R3 Myb* genes: Sequence analysis reveals amplification in higher plants, *Genetics* **153**: 427-444.

Ramsay, N. A., and Glover, B. J., 2005, MYB-bHLH-WD40 protein complex and the evolution of cellular diversity, *Trends Plant Sci* **10**: 63-70.

Ryan, K. G., Swinny, E. E., Winefield, C., and Markham, K. R., 2001, Flavonoids and UV photoprotection in *Arabidopsis* mutants, *Z Naturforsch* **56**: 745-754.

Sagasser, M., Lu, G. H., Hahlbrock, K., and Weisshaar, B., 2002, *A. thaliana* TRANSPARENT TESTA 1 is involved in seed coat development and defines the WIP subfamily of plant zinc finger proteins, *Genes Dev* **16**: 138-149.

Sainz, M. B., Grotewold, E., and Chandler, V. L., 1997, Evidence for direct activation of an anthocyanin promoter by the maize C1 protein and comparison of DNA binding by related Myb domain proteins, *Plant Cell* **9**: 611-625.

Sainz, M. B., Goff, S. A., and Chandler, V. L., 1997, Extensive mutagenesis of a transcriptional activation domain identifies single hydrophobic and acidic amino acids important for activation *in vivo*, *Mol Cell Biol* **17**: 115-122.

Schellmann, S., Schnittger, A., Kirik, V., Wada, T., Okada, K., Beermann, A., Thumfahrt, J., Jurgens, G., and Hulskamp, M., 2002, *TRIPTYCHON* and *CAPRICE* mediate lateral inhibition during trichome and root hair patterning in *Arabidopsis*, *EMBO J* **21**: 5036-5046.

Selinger, D., and Chandler, V., 1999, A mutation in the *pale aleurone color1* gene identifies a novel regulator of the maize anthocyanin pathway, *Plant Cell* **11**: 5-14.

Shikazono, N., Yokota, Y., Kitamura, S., Suzuki, C., Watanabe, H., Tano, S., and Tanaka, A., 2003, Mutation rate and novel *tt* mutants of *Arabidopsis thaliana* induced by carbon ions, *Genetics* **163**: 1449-1455.

Singer, T., Gierl, A., and Peterson, P. A., 1998, Three new dominant *C1* suppressor alleles in *Zea mays*, *Genet Res* **71**: 127-132.

Sompornpailin, K., Makita, Y., Yamazaki, M., and Saito, K., 2002, A WD-repeat-containing putative regulatory protein in anthocyanin biosynthesis in *Perilla frutescens*, *Plant Mol Biol* **50**: 485-495.

Spelt, C., Quattrocchio, F., Mol, J., and Koes, R., 2002, ANTHOCYANIN1 of petunia controls pigment synthesis, vacuolar pH, and seed coat development by genetically distinct mechanisms, *Plant Cell* **14**: 2121-2135.

Spelt, C., Quattrocchio, F., Mol, J. N., and Koes, R., 2000, Anthocyanin1 of petunia encodes a basic helix-loop-helix protein that directly activates transcription of structural anthocyanin genes, *Plant Cell* **12**: 1619-1632.

Stracke, R., Werber, M., and Weisshaar, B., 2001, The R2R3 MYB gene family in *Arabidopsis thaliana*, *Curr Opin Plant Biol* **4**: 447-456.

Styles, E. D., and Ceska, O., 1989, Pericarp flavonoids in genetic strains of *Zea mays*, *Maydica* **34**: 227-237.

Suzuki, M., Kao, C. Y., and McCarty, D. R., 1997, The conserved B3 domain of VIVIPAROUS1 has a cooperative DNA binding activity, *Plant Cell* **9**: 799-807.

Taylor, L. P., and Grotewold, E., 2005, Flavonoids as developmental regulators, *Curr Op Plant Biol* **8**: 317-323.

Tohge, T., Nishiyama, Y., Hirai, M. Y., Yano, M., Nakajima, J., Awazuhara, M., Inoue, E., Takahashi, H., Goodenowe, D. B., Kitayama, M., et al., 2005, Functional genomics by integrated analysis of metabolome and transcriptome of *Arabidopsis* plants over-expressing an MYB transcription factor, *Plant J* **42**: 218-235.

Toledo-Ortiz, G., Huq, E., and Quail, P. H., 2003, The *Arabidopsis* basic/helix-loop-helix transcription factor family, *Plant Cell* **15**: 1749-1770.

Tsuchiya, Y., Nambara, E., Naito, S., and McCourt, P., 2004, The FUS3 transcription factor functions through the epidermal regulator TTG1 during embryogenesis in *Arabidopsis*, *Plant J* **37**: 73-81.

Wada, T., Kurata, T., Tominaga, R., Koshino-Kimura, Y., Tachibana, T., Goto, K., Marks, M. D., Shimura, Y., and Okada, K., 2002, Role of a positive regulator of root hair development, *CAPRICE*, in *Arabidopsis* root epidermal cell differentiation, *Development* **129**: 5409-5419.

Wada, T., Tachibana, T., Shimura, Y., and Okada, K., 1997, Epidermal cell differentiation in *Arabidopsis* determined by a *Myb* homolog, CPC, *Science* **277**: 1113-1116.

Wade, H. K., Sohal, A. K., and Jenkins, G. I., 2003, *Arabidopsis ICX1* is a negative regulator of several pathways regulating flavonoid biosynthesis genes, *Plant Physiol* **131**: 707-715.

Walker, A. R., Davison, P. A., Bolognesi-Winfield, A. C., James, C. M., Srinivasan, N., Blundell, T. L., Esch, J. J., Marks, M. D., and Gray, J. C., 1999, The TRANSPARENT TESTA GLABRA1 locus, which regulates trichome differentiation and anthocyanin biosynthesis in *Arabidopsis*, encodes a WD40 repeat protein., *Plant Cell* **11**: 1337-1349.

Winkel-Shirley, B., 1998, Flavonoids in seeds and grains: physiological function, agronomic importance and the genetics of biosynthesis, *Seed Sci Res* **8**: 415-422.

Winkel-Shirley, B., 2001, It takes a garden. How work on diverse plant species has contributed to an understanding of flavonoid metabolism, *Plant Physiol* **127**: 1399-1404.

Xie, D., Sharma, S., Paiva, N., Ferreira, D., and Dixon, R., 2003, Role of anthocyanidin reductase, encoded by *BANYULS* in plant flavonoid biosynthesis, *Science* **299**: 396-399.

Ylstra, B., Touraev, A., Moreno, R. M. B., Stoger, E., van Tunen, A. J., Vicentie, O., Mol, N. N. M., and Heberle-Bors, E., 1992, Flavonols stimulate development, germination, and tube growth of tobacco pollen, *Plant Physiol* **100**: 902-907.

Zhang, F., Gonzalez, A., Zhao, M., Payne, C. T., and Lloyd, A., 2003, A network of redundant bHLH proteins functions in all TTG1-dependent pathways of *Arabidopsis*, *Development* **130**: 4859-4869.

Zhang, F., and Peterson, T., 2005, Comparisons of maize *pericarp color1* alleles reveal paralogous gene recombination and an organ-specific enhancer region, *Plant Cell* **17**: 903-914.

Zimmermann, I. M., Heim, M. A., Weisshaar, B., and Uhrig, J. F., 2004, Comprehensive identification of *Arabidopsis thaliana* MYB transcription factors interacting with R/B-like BHLH proteins, *Plant J* **40**: 22-34.

CHAPTER 5

TRANSPORT OF FLAVONOIDS

From Cytosolic Synthesis to Vacuolar Accumulation

SATOSHI KITAMURA

Japan Atomic Energy Research Institute (JAERI), Gunma, Japan

1. INTRODUCTION

Flavonoid compounds are one of the most analyzed groups of secondary metabolites in higher plants. The main reason for the interest in flavonoids is that they are major constituents of plant pigments. Anthocyanins, a flavonoid subclass, have been of special interest because of their ability to confer red, orange, blue, and purple coloration to leaves, flowers, and fruits (Mol et al., 1998). As pigments, flavonoids have facilitated the testing of hypotheses related to Mendel's law and transposable elements. Flavonoids have been the focus of attempts to modify flower color by genetic engineering (Tanaka et al., 1998). There also is interest in using them as drugs or dietary supplements because of their strong antioxidant activities (Harborne and Williams, 2000; Bartel and Matsuda, 2003). In plants, flavonoids have several functions including attracting insects for pollination and dispersal of seeds, acting in defense systems (e.g., as UV-B protectants and phytoalexins), signaling between plants and microbes, and regulating auxin transport (Winkel-Shirley, 2001). Many of these functions cannot occur unless flavonoids are properly localized within the cells.

Anthocyanins, proanthocyanidins (also called condensed tannins), and flavonols are three major subclasses of flavonoid compounds. Anthocyanins do not show their brilliant colors until they are accumulated in the acidic vacuoles. Proanthocyanidins are polymers of the catechins and/or epicatechins monomeric precursors (Xie et al., 2003). In the seed coat, proanthocyanidin precursors accumulate in the vacuole, followed by polymerization and oxidation to proanthocyanidin within this organelle. The oxidation of proanthocyanidins hardens the seed coat, which induces moderate dormancy in the seeds and limits the detrimental effects of physical and biological attacks (Shirley, 1998; Debeaujon et al., 2000). Thus, the proper subcellular localization of these flavonoids is crucial for fulfilling their functions in plant cells. The subcellular localization of flavonols is still a matter of debate. Some flavonols have a protective role as UV-B filters, and they also could function as copigments for anthocyanins in specific tissues. To fulfil these functions, such flavonols are assumed to accumulate in the vacuole (Koes et al., 1994; Mol et al., 1998). In contrast, other flavonols appear to be present in the cytoplasm, as has been found in *Arabidopsis* root tissue (Buer and Muday, 2004). The cytosolic flavonols appear to regulate auxin movement from cell to cell.

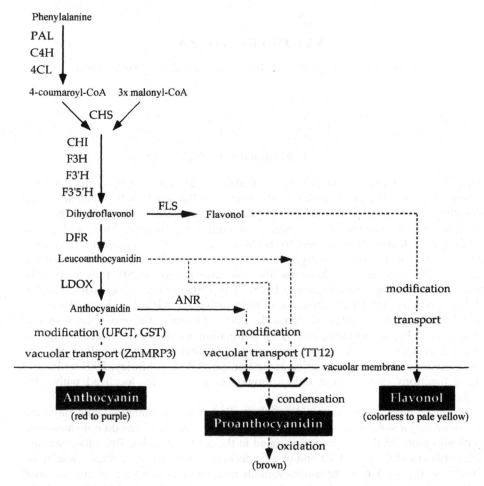

Figure 5.1 Flavonoid biosynthetic and accumulation pathways.

The results of many experiments indicate that the sites at which flavonoids accumulate are different from the sites at which they are synthesized. It thus is believed that plant cells have flavonoid transport mechanisms. Although the biosynthesis of these compounds has been described in detail (Figure 5.1), their transport and accumulation mechanisms are not clear. This chapter focuses on flavonoids that accumulate in the vacuole, which are mainly anthocyanins and proanthocyanidins. I briefly describe here their subcellular biosynthetic sites, introduce the constituents of subcellular flavonoid transport, and discuss possible transport mechanisms.

2. FLAVONOID BIOSYNTHETIC SITES

Flavonoids are believed to be synthesized in the cytosol. Flavonoid biosynthesis requires elaborate metabolic mechanisms, for example, for preventing reactive and potentially toxic intermediates from diffusing into the cytoplasm. Accordingly, Stafford (1974) proposed that flavonoid biosynthetic enzymes form a multienzyme complex in the cytosol. The enzymes involved in flavonoid synthesis are shown in Figure 5.1. In flower protoplasts of *Hippeastrum* and *Tulipa*, the highest activities of chalcone synthase (CHS), chalcone isomerase (CHI), and glucosyltransferase were found in the cytosolic fraction, whereas low but significant activities also were detected in the microsomal fraction (Hrazdina et al., 1978). These results suggested that these enzymes are localized in close proximity to endomembranes. In *Hippeastrum* petals, cinnamate 4-hydroxylase (C4H) cofractionated with the endoplasmic reticulum (ER) (Wagner and Hrazdina, 1984). Cofractionation of C4H with the ER or microsomal membrane has been observed in several plant species (Fritsch and Grisebach, 1975; Czichi and Kindl, 1977; Hrazdina and Wagner, 1985), whereas many of the other flavonoid enzymes appeared to be mainly in the cytosol (Fujiwara et al., 1998; Devic et al., 1999; Jones and Vogt, 2001). C4H and flavonoid 3'-hydroxylase (F3'H) belong to the cytochrome P450 enzyme family and have an ability to bind to membranes. Immunoelectron microscopy confirmed that CHS has a loose association with the ER membrane in buckwheat hypocotyls (Hrazdina et al., 1987). These results have led to a modification of Stafford's (1974) proposal in which flavonoid biosynthetic enzymes are organized as a multienzyme complex loosely associated with the cytoplasmic face of the ER and are anchored to the ER membrane through C4H and F3'H (Hrazdina, 1992; Winkel-Shirley, 1999). Recent biochemical, immunohistochemical, and molecular cytological investigations have supported convincingly this concept. Burbulis and Winkel-Shirley (1999) demonstrated protein–protein interactions between CHS, CHI, and dihydroflavonol reductase (DFR) by yeast two-hybrid assays, affinity chromatography, and immunoprecipitation, and suggested that these proteins interacted both *in vitro* and *in vivo*. These results are consistent with the hypothesis that the flavonoid enzymes assemble as a multienzyme complex. Furthermore, CHS and CHI were found to co-localize at the ER and tonoplast in *Arabidopsis* roots (Saslowsky and Winkel-Shirley, 2001). Other components in the flavonoid enzyme pathway have been suggested to colocalize close to the ER membrane. These include 2-hydroxyisoflavanone synthase and isoflavone O-methyltransferase in alfalfa (Liu and Dixon, 2001), and phenylalanine ammonia-lyase (PAL) and C4H in tobacco (Achnine et al., 2004). Although the association of flavonoid enzymes with the ER membrane has not been completely established (Saslowsky and Winkel-Shirley, 2001), determination of the subcellular site of flavonoid biosynthesis could provide important clues to understand the flavonoid transport mechanisms in plant cells. This topic is further discussed in Chapter 3.

3. ENTRANCE OF FLAVONOIDS INTO MEMBRANE-BOUND COMPARTMENTS

Anthocyanins and proanthocyanidins are synthesized in the cytosol and finally accumulated in the vacuoles. To accomplish this, vacuolar membrane-bound permeases and/or transporters are required to take up cytosolic-synthesized flavonoids into the vacuole. As described in the following sections, several types of transporters participate in the vacuolar uptake of flavonoids.

3.1. Proton-Dependent Transporters

Vacuoles have various physiological functions in plant development and differentiation. One of their functions is to act as a storage site for various secondary metabolites including flavonoids. In plants, the acidic conditions of the vacuolar lumen are established mainly by two tonoplast-localized proton pumps: the v-ATPase and the v-PPase (Sze et al., 1999; Maeshima, 2001). The acidic pH makes anthocyanins colored by the formation of flavylium ions within the vacuole (Mol et al., 1998; Nakajima et al., 2001). The pH gradient between the cytosol and the vacuole provides the potential energy needed to take up substrates into the vacuole. Hopp and Seitz (1987) investigated the uptake efficiency of radiolabeled anthocyanins into isolated carrot vacuoles, and suggested that vacuolar uptake of flavonoids was dependent on the energy provided by the proton gradient. Flavone glucosides are endogenously synthesized in barley. Klein et al. (1996) suggested the existence of a proton antiporter in barley for the uptake of radiolabeled flavone glucosides in vitro. Many other secondary metabolites, such as coumaric acid glucosides, have been suggested to be imported into isolated vacuoles in a proton-dependent manner (Martinoia et al., 2000). However, proton-dependent transporter genes that import flavonoids into the vacuole have not been isolated from these species.

3.2. ABC-Type Transporters

Some secondary metabolites and their intermediates are considered to be highly reactive and potentially toxic in the cytoplasm (Marrs, 1996). To avoid toxicity, flavonoids are transported or sequestered to vacuoles or to the exterior of the cell.

Plants have developed detoxification systems to defend themselves against exogenous phytotoxic chemicals, also called xenobiotics. The main detoxification pathway for xenobiotics in plants is composed of three phases: (I) an activation phase, which usually involves hydrolysis or oxidation, resulting in xenobiotics becoming more reactive; (II) a conjugation phase in which compounds metabolized in phase I are conjugated with hydrophilic molecules such as glucose, malonate, or glutathione; and (III) a compartmentalization phase in which the conjugates are separated from the cytosol by membrane-associated transport proteins (Coleman et al., 1997). The major reaction in phase I is catalyzed by cytochrome P450 proteins, and some P450s, including C4H and F3'H, participate in the flavonoid biosynthetic pathway. Many flavonoids have glucosyl and malonyl moieties, and genes encoding the corresponding transferases have been identified in some species (Fedoroff et al., 1984; Suzuki et al., 2001). Glutathione S-transferase (GST) proteins participate in anthocyanin pigmentation in maize (Marrs et al., 1995) and petunia (Alfenito et al., 1998). These results suggest that the flavonoid pathway, like the detoxification system, has activation and conjugation phases and raise the possibility that it also has a compartmentalization phase mediated by a membrane-associated transporter protein that is similar to that of the detoxification system. It should be noted that the existence in plant cells of flavonoid–glutathione conjugates has not yet been conclusively demonstrated.

In the detoxification pathway, exogenous compounds are conjugated with glutathione by GST, which is a crucial reaction in the following compartmentalization step. The glutathione conjugates are recognized and sequestered to the vacuoles or exported to the cell wall by ATP-dependent, proton-gradient-independent transporters, named ATP-binding cassette (ABC)-type transporters (Martinoia et al., 1993). No other types of transporters are known to mediate compartmentalization of the glutathione conjugates. The involvement of an ABC-type transporter in flavonoid accumulation was first suggested by Marrs et al. (1995). They cloned a maize GST gene, *Bronze2* (*BZ2*), from the anthocyanin-deficient *bz2* mutant and showed that BZ2 was required for normal accumulation of anthocyanins. They further found that treating cultured maize cells with vanadate, an inhibitor of ABC-type transporters, conferred a flavonoidless phenotype, which mimicked the *bz2* mutant. These results strongly suggested that an ABC-type transporter was required for vacuolar uptake of anthocyanins in maize via the presumptive formation of anthocyanin–glutathione conjugates by *BZ2* (Marrs et al., 1995). In mung bean, the uptake of the isoflavonoid medicarpin by isolated vacuolar membrane vesicles was found to be accelerated by Mg-ATP and to be strongly inhibited by vanadate (Li et al., 1997). Flavone glucuronides of rye also were taken up by the vacuolar membrane *in vitro* through an ABC-type transporter (Klein et al., 2000). These results suggest that some flavonoids are taken up into the vacuoles via ABC-type transporters.

A subclass of ABC proteins, the multidrug resistance-associated proteins (MRPs), facilitate transport of flavonoid–glutathione conjugates in plants. Two

Arabidopsis MRP-type ABC transporters (AtMRP1 and AtMRP2) mediate the vacuolar uptake of anthocyanin–glutathione conjugates in a heterologous host (yeast) (Lu et al., 1997, 1998). The *Arabidopsis* genome encodes for 15 putative *MRP* genes, some of which appear to be able to transport glutathione conjugates in yeast (Rea et al., 1998; Kolukisaoglu et al., 2002). Although knockout mutants for some *AtMRPs* have been identified, none of them have resulted in flavonoidless phenotypes (Gaedeke et al., 2001; Klein et al., 2004). As the substrate specificity for the respective AtMRPs is relatively broad (Lu et al., 1997, 1998; Liu et al., 2001), the vacuolar uptake of anthocyanins may be handled in a redundant fashion by some MRPs in *Arabidopsis*.

A putative ABC-type transporter gene, *ZmMRP3*, recently was found to be involved in anthocyanin accumulation in maize (Goodman et al., 2004). The expression pattern of *ZmMRP3* correlated with the anthocyanin accumulation pattern and with the expression patterns of other flavonoid structural genes. An antisense mutant of *ZmMRP3* had lower anthocyanin levels than did the wild type. The antisense mutant resembled the *bz2* mutant in having a brownlike coloration and similar HPLC profile. These similarities may be due to the nonuptake of anthocyanins into vacuoles in both the antisense mutant and the *bz2* mutant. ZmMRP3-green fluorescent protein (GFP) localized at the tonoplast in etiolated maize protoplasts. Thus, *ZmMRP3* appears to play a role in the vacuolar accumulation of anthocyanins. ZmMRP3 most likely functions *in planta* as a membrane-bound transporter that is required for vacuolar uptake of anthocyanins in maize (Goodman et al., 2004).

3.3. MATE-Type Transporters

Genetic approaches have suggested that a novel type of transporter is involved in flavonoid transport in plants. A number of *Arabidopsis* mutants deficient in flavonoid biosynthesis have been isolated. Many of these mutants, called *transparent testa* (*tt*) mutants, have altered seed color (Koornneef, 1990; Winkel-Shirley, 2001). One of the *tt* mutants, *tt12*, is deficient in flavonoid pigments in the seed coat but not in vegetative organs (Debeaujon et al., 2001). Analyses of double mutants of *tt12* with other *tt* mutants and histochemical analyses showed that *TT12* was involved in vacuolar accumulation of proanthocyanidins in the seed coat. Cloning and sequencing of *TT12* revealed that it encoded a 507-amino acid protein with 12 transmembrane segments. These results suggest that TT12 is a transporter that accumulates proanthocyanidin precursors into vacuoles (Debeaujon et al., 2001). The predicted amino acid sequence of TT12 is similar to the amino acid sequences of ERC1 from *Saccharomyces cerevisiae* (Shiomi et al., 1991), NorM from *Vibrio parahaemolyticus* (Morita et al., 1998), and DinF from *Escherichia coli* (Kenyon and Walker, 1980; Krueger et al., 1983). Based on their sequence characteristics, these proteins, including TT12, are classified as multidrug efflux transporters, but are distinct from other members of this group that have been characterized to date. For example, these proteins lack a diagnostic motif of ABC-type transporters, namely a nucleotide-binding fold (NBF) motif, which contains a Walker A box, a Walker B box, and an ABC signature (Rea, 1999). For this reason, they have been placed in a new family called the multidrug and toxic compound extrusion (MATE) family (Brown et al., 1999). The MATE family is characterized by the presence of 12 putative transmembrane segments and by the absence of signature sequences specific to the other multidrug transporter superfamilies. The molecular function of the MATE-type transporters is less well understood, although NorM was suggested to act as a Na^+/drug antiporter (Morita et al., 2000).

Another putative MATE-type transporter participating in flavonoid transport was identified in tomato. By screening activation-tagged mutants, Mathews et al. (2003) isolated one line in which the whole plant body had a deep purple pigmentation. They cloned and sequenced the activation-tagged gene, *ANT1*, and showed that it encoded a MYB transcriptional factor (see Chapter 4 for a full description of the regulators of the flavonoid pathway). Activation and/or overexpression of *ANT1* resulted in the coordinated up-regulation of the expression of some flavonoid genes including structural (*CHS, CHI, DFR*), modification (glucosyltransferase genes), and a transporterlike (defined as *MTP77* in Mathews et al., 2003) genes. *MTP77* encodes an amino acid sequence similar to that of *Arabidopsis TT12*. The similarity of MTP77 to TT12 and its coordinated up-regulation by *ANT1* activation make MTP77 the most likely candidate for an anthocyanin transporter in tomato. A similar strategy has been used to comprehensively identify genes related to flavonoid biosynthesis and accumulation in *Arabidopsis*. An activation-tagged mutant, *pap1-D*, accumulated large amounts of secondary metabolites including anthocyanins. This phenotype resulted from activation of a MYB transcriptional factor, *PAP1* (Borevitz et al., 2000). Tohge et al. (2005) investigated the gene expression profile as well as metabolic profiles in plants overexpressing *PAP1* and identified many flavonoid structural genes and two transporterlike genes that were induced. The two transporterlike genes were classified as "sugar transporterlike" proteins, rather than MATE-type transporters. Sugar transporter proteins belong to the major facilitator superfamily (MFS) of multidrug transporters and possess a common motif of 12 transmembrane segments (Marger and Saier, 1993), shared by the MATE and MFS transporters. Although it is now difficult to conclude whether anthocyanin accumulation involves MATE-type transporters in dicots, elucidation of the function of these candidates is important for understanding the diversity of flavonoid transporter proteins and the evolutionary processes in flavonoid transport mechanisms.

Little is known about the functional mechanisms (e.g., the energization modes and substrate specificity) of MATE-type transporters. MATE-type transporters are abundant in bacteria and plants, but have not been identified in animals. This may mean that MATE-type transporters have unique functions in these organisms. The *Arabidopsis* genome has more than 50 genes encoding putative MATE-type proteins, and a few mutants with defective MATE-type genes have been isolated. MATE-type proteins are thought to be involved in the secretion of exogenous toxins (Diener et al., 2001) and in signaling pathways for disease resistance (Nawrath et al., 2002). Further biochemical and cell biological analyses are needed to clarify the functions of these newly categorized transporter proteins.

4. OTHER FACTORS INVOLVED IN FLAVONOID TRANSPORT

The preceding sections described the site of flavonoid biosynthesis and possible flavonoid transporters. This raises the question of how flavonoids that are synthesized near the ER make their way to membrane-bound transporter proteins within the cytosol. Does flavonoid transport within the cytosol occur by simple diffusion or is it mediated by other factor(s)? There is some evidence that other factors are involved in flavonoid transport, as described below.

4.1. GSTs

As mentioned above, GST proteins are required for anthocyanin accumulation. In maize, a GST appears to operate after glycosylation of anthocyanin precursors by UDP-glucose:flavonoid 3-O-glucosyltransferase (UFGT), a final enzyme necessary for producing stable anthocyanin molecules in plant cells (Reddy and Coe, 1962). A GST encoded by $BZ2$ is necessary for taking up anthocyanin precursors into the vacuole in maize (Marrs et al., 1995). This was corroborated by the finding of the same phenotype in loss-of-function mutants of $ZmMRP3$, an anthocyanin transporter gene, and $BZ2$ in maize (Goodman et al., 2004). GSTs are multifunctional enzymes and one of their most interesting functions is detoxification of exogenous substrates, such as herbicides, by conjugation with glutathione (Edwards et al., 2000; Dixon et al., 2002). Because the detoxification pathway for xenobiotics was presumed to be parallel to the pathway for flavonoids, the flavonoid-specific GSTs were previously believed to form flavonoid–glutathione conjugates before their vacuolar compartmentalization (Marrs et al., 1995). This putative function was supported by the facts that (1) BZ2 and Anthocyanin9 (AN9, the counterpart of BZ2 in petunia) had glutathionation activity against a model substrate, 1-chloro 2,4-dinitrobenzene (CDNB) *in vitro* (Marrs et al., 1995; Alfenito et al., 1998), and (2) radiolabeled glutathione was colocalized with anthocyanin pigments on two-dimensional thin layer chromatography in cultured maize cells (Marrs et al., 1995). However, anthocyanin–glutathione conjugates have not been identified in plant extracts. Furthermore, glutathionation activity of BZ2 and AN9 proteins against flavonoids was not detected *in vitro* or *in vivo* (Mueller et al., 2000; Walbot et al., 2000). AN9 was found to bind to flavonoids rather than to cause them to form glutathione conjugates (Mueller et al., 2000). These findings raise another possibility, i.e., that flavonoid-specific GSTs, such as AN9, bind to flavonoids in order to sequester the flavonoids into vacuoles, acting as cytoplasmic escort or carrier proteins rather than glutathionation agents (Mueller et al., 2000). As further evidence that glutathionation activity is not required for normal accumulation of anthocyanins, site-directed mutagenized BZ2 proteins, in which the amino acid that is important for glutathionation activity in animal and insect GSTs is mutated, did complement the null-mutant phenotype of *bz2* (Mueller and Walbot, 2001). As flavonoid-specific carrier proteins, GSTs have been proposed to prevent oxidation of flavonoids within the cytosol and/or deliver "flavonoid cargo" directly to the membrane-bound transporters (Mueller and Walbot, 2001).

GST and GST-like genes other than *BZ2* and *AN9* have been shown to be involved in anthocyanin accumulation in *Arabidopsis* (*TT19*), soybean (*GmGST26A*), and carnation (*Flavonoid3*, *FL3*) (Table 5.1). The anthocyaninless phenotypes of the respective mutants were complemented by other GSTs shown in Table 5.1. For example, the lack of anthocyanins in *bz2*, *tt19*, and *fl3* was complemented by the expression of *AN9*, and the lack of anthocyanins in *an9* was complemented by the expression of *BZ2*. Thus, the GSTs shown in Table 5.1 appear to have the same function in anthocyanin transport in these species. Plant GSTs are divided on the basis of sequence identity into five types (Phi, Zeta, Theta, Tau, and Lambda) (Dixon et al., 2002). It is interesting that a mutation in a Tau-type GST that prevented anthocyanin accumulation was complemented by a Phi-type GST (Alfenito et al., 1998). Among the GST genes, *Arabidopsis TT19* appears to have unique and intriguing functions. Loss-of-function mutants of *TT19* result in a significant reduction in anthocyanin pigments in the vegetative parts, as well as a large reduction of brown pigments in the seed coat, where proanthocyanidin-derived pigments accumulate. Overexpression of the petunia *AN9* gene in the *tt19* mutants restored the anthocyanin pigments in the vegetative parts, but did not restore the brown pigments in the seed coat (Figure 5.2). In the immature seed coat, the proanthocyanidin precursors accumulated within the large central vacuoles in the wild type, whereas they accumulated within small vacuolelike structures in the *tt19* mutant. These results indicate that *TT19* is required not only for vacuolar uptake of anthocyanins but also for normal sequestration of proanthocyanidin precursors into vacuoles (Kitamura et al., 2004). GSTs were previously thought not to be involved in the proanthocyanidin pathway in *Arabidopsis*, because the putative proanthocyanidin transporter TT12 is not a member of the ABC transporter family, which is the only transporter that is involved in the transport of GST-related compounds (Debeaujon et al., 2001). In other species such as petunia and maize, different GST proteins might be used for vacuolar accumulation of distinct flavonoid subclasses.

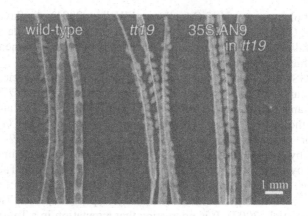

Figure 5.2 *Seed color at the ripening stage in Arabidopsis. See Color Section for figure in colors.*

GSTs generally work as dimers and each of the subunits has two potential binding sites for glutathione (Reinemer et al., 1996). It is unclear whether glutathione molecules are necessary for proper flavonoid accumulation. In rye, *in vitro* vacuolar uptake of flavone glucuronides was stimulated by glutathione, without the formation of glutathione conjugates (Klein et al., 2000). This result may indicate that the function of flavonoid-specific GSTs is to efficiently take up flavonoids using glutathione as a cofactor. On the other hand, the finding that site-directed mutagenesis of BZ2, which generates mutantproteins that presumably lack glutathionation activity, could restore the mutant phenotype of *bz2* (Mueller and Walbot, 2001) suggests that glutathione molecules are not important for flavonoid transport in maize. Our biochemical analyses of *Arabidopsis* TT19 suggested that glutathione molecules are not required for flavonoid transport (unpublished data). However, we cannot rule out the possibility that the point-mutagenized BZ2 protein described above is still able to form a heterodimer with another class of GST, and that the glutathione held by the partner GST in the heterodimer enables the flavonoid-specific GST to fulfil its function in flavonoid transport *in planta*. Further studies are needed to determine whether glutathione is involved in flavonoid transport.

Table 5.1. GST genes involved in flavonoid accumulation in plants

Gene name	Species	Gene isolated?	Type of GST	Mutant isolated?	Reference
BZ2	maize	yes	Tau	yes	Marrs et al. (1995)
AN9	petunia	yes	Phi	yes	Alfenito et al. (1998)
TT19	Arabidopsis	yes	Phi	yes	Kitamura et al. (2004)
FL3	carnation	no	—	yes	Larsen et al. (2003)
GmGST26A[1]	soybean	yes	Tau	no	Alfenito et al. (1998)

[1]Involvement of anthocyanin accumulation was suggested by molecular complementation tests in a bz2 background

4.2. Acylation of Flavonoids

In vitro experiments with parsley showed that flavone glucosides needed to be malonylated to be efficiently taken up by isolated vacuoles (Matern et al., 1986). Similarly, the uptake of anthocyanins by isolated vacuoles of carrot suspension cells was not efficient, unless the anthocyanins were modified with sinapic acid (Hopp and Seitz, 1987). These *in vitro* observations suggest that acylation with malonic acid or sinapic acid is crucial for efficient flavonoid accumulation. Acylation might be a prerequisite molecular tag for efficient vacuolar uptake of flavonoids in these species. Acylation of flavonoids may serve to trap flavonoids within vacuoles (Matern et al., 1986; Hopp and Seitz, 1987).

5. CYTOLOGICAL ASPECTS OF FLAVONOID TRANSPORT

In addition to the biochemical and molecular studies described above, cytological approaches have provided further clues to flavonoid transport mechanisms. Anthocyanin-producing cells often have unique subcellular structures called "cyanoplasts" and "anthocyanoplasts" with high concentrations of anthocyanin pigments. Over 70 species in 33 families of angiosperms possess spherical anthocyanoplasts (Pecket and Small, 1980). Anthocyanoplasts are observed as numerous small red vesicle-like structures at the early stage of development, while they are observed as a smaller number of large anthocyanoplasts at a later stage (Pecket and Small, 1980; Nozue and Yasuda, 1985). These observations suggest that anthocyanoplasts are formed as a result of the progressive coalescence of smaller vesicles. Although flavonoid biosynthesis was first thought to occur in anthocyanoplasts, there is growing evidence that it occurs at the surface of the ER. This has led to the speculation that anthocyanoplasts are involved in transporting anthocyanins from the ER to the vacuole (Nozzolillo and Ishikura, 1988; Nozue et al., 1993). Similar red spherical bodies that are present within vacuoles are called anthocyanic vacuolar inclusions (AVIs). They do not appear to be encircled by membranes (Nozue et al., 1993; Markham et al., 2000). It remains to be resolved whether AVIs are derived from the digestion of membranes of anthocyanoplasts after they enter the vacuole or from the aggregation of pigments within the vacuole. Anthocyanoplasts and AVIs might have different origins. Nozue et al. (1997) isolated a 24-kDa protein (VP24) from cultured sweet potato cells. VP24 is localized mainly in anthocyanoplasts and vacuoles. The mature VP24 protein sequence is similar to the sequences of carboxypeptidases (Xu et al., 2001). Carboxypeptidases have been proposed to be involved in the hydrolysis of the glutathione moieties from the glutathione conjugates in the plant vacuole (Wolf et al., 1996). Thus, VP24 may be involved in the processing of flavonoid–glutathione conjugates (Xu et al., 2001). Alternatively, VP24 might function as a membrane-bound protein that is involved in the transport or accumulation of anthocyanins. This is possible, because the precursor form of VP24 possesses a highly hydrophobic C-terminus containing eight putative transmembrane domains (Xu et al., 2001). However, this function would last only until the cleavage of the C-terminal propeptide.

Anthocyanin-accumulating vesicles were found to form in the cytoplasm of suspension cells of black Mexican sweet (BMS) maize that had been transformed with maize transcriptional factors (*C1* and *R*) (Grotewold et al., 1998). These vesicles appeared to eventually coalesce into a single one. These characteristics are quite similar to those of anthocyanoplasts mentioned above. Interestingly, when BMS cells were transformed with another regulator, *P1*, instead of the anthocyanoplasts, two types of autofluorescent bodies, designated yellow fluorescent bodies (YFBs) and green fluorescent bodies (GFBs), were found (Lin et al., 2003). Epifluorescence microscopy demonstrated that YFBs and GFBs first occurred as small spherical vesicles in the cytoplasm and finally accumulated in the vacuole and the cell wall, respectively (Grotewold et al., 1998; Lin et al., 2003). An ultrastructural analysis of autofluorescing cells revealed electron-dense spherical structures that were in the expanded ER at the early stage of *P1* expression and in the vacuole and cell wall at a later stage. *P1* is required for the production of 3-deoxy flavonoids in the floral organs of maize (Grotewold et al., 1994). In fact, BMS cells expressing *P1* were shown to produce flavan-4-ols, one of the 3-deoxy flavonoids, as well as some phenylpropanoids (Grotewold et al., 1998). These results raise the possibility that YFBs and GFBs are involved in subcellular trafficking of secondary metabolites from the sites of synthesis to two major accumulating sites, the vacuole and the cell wall.

All of the flavonoid-specific GST genes characterized to date are involved in the accumulation of anthocyanins, whereas the *Arabidopsis TT19* is the only one that is involved also in the accumulation of proanthocyanidins. In *Arabidopsis*, synthesis of the proanthocyanidin precursors is limited to the innermost layer of the seed coat (Chapple et al., 1994) and begins after fertilization (Debeaujon et al., 2003). The precursors can be visualized by histochemical staining with vanillin or DMACA (Debeaujon et al., 2000; Abrahams et al., 2002). At one day after flowering (DAF), the red color that results from the reaction of the precursors with vanillin is visible as dots in both the wild type and the *tt19* mutant. However, the patterns of proanthocyanidin accumulation were different between the wild type and *tt19* at 2~3

DAF. At 5 DAF, the precursors were located in a single expanded vacuole within each constitutive cell in the wild type, whereas they were restricted to a few smaller vacuoles in *tt19* (Figure 5.3). Thus, in the mutant of the flavonoid-specific GST in *Arabidopsis*, the proanthocyanidin precursors are accumulated in membranelike structures. This is surprising because in mutants of the flavonoid-specific GSTs in other species, flavonoids such as anthocyanins are not taken up into the membrane compartments and remain dispersed in the cytosol (e.g., Marrs et al., 1995). These unusual membranelike structures in the *tt19* mutant might reflect disruption of the normal transport mechanism for proanthocyanidin precursors in the mutant seed coat (Kitamura et al., 2004). These findings raise the possibility that the function of *TT19* in the proanthocyanidin pathway is different from its function in the anthocyanin pathway.

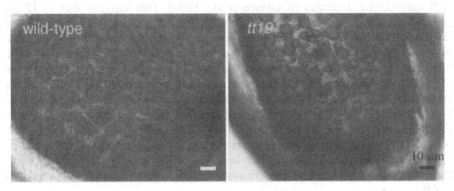

Figure 5.3 Depositional patterns of proanthocyanidin precursors in immature seeds of Arabidopsis. See Color Section for figure in colors.

6. PUTATIVE FLAVONOID TRANSPORT MODELS

The mechanisms of flavonoid transport in different plant species are difficult to reconcile, and therefore it is hypothesized that different species utilize different transport mechanisms (Mueller and Walbot, 2001). Nevertheless, two characteristics of flavonoid transport appear to be shared by the different mechanisms. First, analyses of mutants of several species have clearly shown that GSTs are involved in flavonoid transport (Table 5.1). In addition to the genes shown in Table 5.1, a GST-like gene (*MTP4*) was found to be up-regulated in *ANT1*-overexpressing tomato (Mathews et al., 2003), and a GST-like gene was found to be down-regulated in mutants for flavonoid-specific transcriptional factors in morning glory (Y. Morita, A. Hoshino, S. Iida, personal communication). These results suggest that the function of the GST and GST-like proteins is relatively conserved among different species. Second, membrane-bound transporter proteins seem to be required for normal accumulation of flavonoids into vacuoles, although a consensus type of transporter for flavonoid accumulation has not been established. In the maize anthocyanin pathway and in the *Arabidopsis* proanthocyanidin pathway, the genes for both GST and the transporter have been isolated. In view of its flavonoid-binding ability, the GST encoded by *BZ2* appears to act as a flavonoid carrier protein before the uptake of flavonoids into the membrane compartments. GST or GST-glutathione-bound flavonoids might be recognized for uptake into membrane compartments by ZmMRP3, which is a putative ABC-type transporter. The similar lack of anthocyanin pigmentation in *zmmrp3* and *bz2* mutants also could be explained by a failure at the same biological step: the uptake of flavonoids into the membrane-bound compartments. On the other hand, it is rather difficult to explain the *Arabidopsis* proanthocyanidin pathway. In a mutant with a defective proanthocyanidin transporter gene (*tt12*), proanthocyanidin precursors are distributed throughout the cytoplasm (Debeaujon et al., 2001), whereas in a mutant with a defective GST gene (*tt19*), proanthocyanidin precursors are concentrated in small vacuolelike structures (Kitamura et al., 2004). This suggests that TT12 and TT19 function at different points in the proanthocyanidin pathway. Because TT12 is classified as a MATE-type transporter, tagging of substrates with GST or GST-glutathione should not be required for their transport via TT12 (Debeaujon et al., 2001). Therefore, it is likely that knocking out *TT19* would not completely prevent the uptake of the precursors into the membrane compartments. This idea was supported by the finding of a precursor-vanillin reaction at the maturation stage of the *tt19* seed (*ca.* 10~18 DAF), which suggests that the precursors are not degraded or oxidized by the cellular machinery in *tt19* before this stage (Kitamura et al., 2004). If TT19 acts as a flavonoid carrier protein as suggested above, how then are the small vacuoles formed in *tt19*? Flavonoid transport models are discussed in the following two sections. These models assume that flavonoids are synthesized near the ER.

6.1. Transport of "Naked" Flavonoids without Membrane Structures

In plants, the ER is a large network of continuous tubules and sheets that is spread throughout the cytosol (Staehelin, 1997). Therefore it is possible that the cytoplasmic face of the ER where flavonoids are synthesized is physically close to the vacuole where flavonoids are finally accumulated. This suggests one model in which flavonoids, with the help of GST, are transported in a "naked" form (i.e., not in transport vesicleslike structures) directly to the vacuolar membrane-bound transporter (Figure 5.4, pathway I). In this case, ZmMRP3 and TT12 are localized on the membrane of the vacuoles. The finding that ZmMRP3 is localized at the tonoplast (Goodman et al., 2004) is consistent with this model. This model also is consistent with the hypothesis that flavonoid-specific GST acts as a flavonoid carrier protein to prevent oxidation of flavonoids and/or to deliver flavonoids directly to the tonoplast-bound transporters. In *Arabidopsis*, the finding of smaller vacuolelike structures filled with proanthocyanidin precursors in the *tt19* mutant (Kitamura et al., 2004) can be attributed to reduced interactions between flavonoids and tonoplast-bound TT12 proteins due to loss of the carrier proteins. Since flavonoid-specific GSTs, including BZ2 and TT19, do not possess vacuolar transport signals, other proteins possessing such signal peptides may interact with the GSTs to deliver the flavonoids to the vacuoles.

6.2. Flavonoid Transport by Vesicles or Small Vacuoles

The second flavonoid transport model is like a vesicle-trafficking model (Figure 5.4, pathway II). This model already has been proposed for proanthocyanidin accumulation. In the early stage of proanthocyanidin accumulation in Douglas fir and loblolly pine, proanthocyanidins are found as small osmiophilic deposits in minute vesicles or the ER, while in later stages of accumulation they occurred as larger deposits in the vacuoles, probably as the result of fusion of smaller vesicles (Parham and Kaustinen, 1977). One hypothesis, based on studies of enzymes involved in flavonoid biosynthesis, is that proanthocyanidin precursors synthesized on the ER membranes enter into the ER lumen and then ER-derived vesicles containing the precursors are transported to the vacuole, fuse with the vacuolar membrane, and empty their contents into the vacuole (Stafford, 1989).

Transport of some macromolecules (such as proteins and polysaccharides) between organelles is mediated by small vesicles in eukaryotic cells (Rothman and Wieland, 1996; Robinson et al., 1998; Ueda and Nakano, 2002). The transport vesicles are released from ER foci that concentrate the cargo and are transported and fused to other membranous organelles (Golgi apparatus, endosomes, and vacuoles) or the plasma membrane, where they empty their cargo. Vesicle trafficking of some proteins from the ER directly to the vacuole is consistent with the putative pathway for flavonoids. The vesicle-trafficking model provides plausible mechanisms for avoiding the effects of potentially toxic secondary metabolites moving within the cytosol and for accumulating large quantities of flavonoids in the vacuoles. The vesicle-trafficking mechanism may transport not only proteins but also other molecules such as flavonoids to the vacuole. However, there is no direct evidence that flavonoids are transported by vesicle trafficking. Genes encoding small GTPases and SNAREs that are required for budding, targeting, and docking of the vesicular membranes in the secretory pathway (Sanderfoot and Raikhel, 1999; Molendijk et al., 2004) have not yet been implicated in the accumulation of flavonoids.

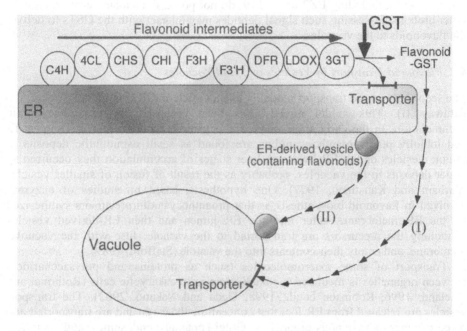

Figure 5.4 *Proposed flavonoid transport mechanisms. The flavonoid multienzyme complex is shown as a linear array for convenience. See Color Section for figure in colors.*

The mechanisms by which flavonoids are taken up into the ER in the vesicle-trafficking model are thought to be similar or identical to the mechanisms by which flavonoids are taken up into vacuolar lumens in pathway I, i.e., mechanisms that involve a GST and a transporter (Figure 5.4) (Grotewold, 2001). If this is correct, then in the vesicle-trafficking model the flavonoid-specific transporter proteins could be localized mainly in the ER membranes and possibly in vesicular membranes and the tonoplast. However, there is not yet any evidence that TT12 and ZmMRP3 are found on the ER or vesicular membranes (Debeaujon et al., 2001; Goodman et al., 2004). In the vesicle-trafficking model, the flavonoid-specific GSTs would localize very close to the ER so that they can act as flavonoid carrier proteins. This is because the flavonoid biosynthesis site and the flavonoid import site are both on the ER and are thought to be almost adjacent to each other (Figure 5.4, pathway II). Biochemical studies suggested that BZ2 is loosely associated with microsomal membranes (Pairoba and Walbot, 2003), although its exact localization is unknown. On the other hand, the vesicle-trafficking model can not easily explain the *Arabidopsis tt19* phenotype, in which endothelial cells of the seed coat have a few small vacuoles filled with the proanthocyanidin precursors. If flavonoids are taken up into the ER lumen in local areas, then the loss of the ability to escort flavonoids would have a negligible effect in this model. The *tt19* phenotype, in which proanthocyanidin precursors are restricted to a few small vacuoles, may indicate that TT19 has other unknown functions. One explanation is that TT19 also acts after proanthocyanidin precursors are taken up into the ER by TT12, for example, at the stage of intracellular vesicle trafficking of flavonoids. To test this hypothesis, further studies are needed to determine the subcellular localizations of TT19 and TT12, to isolate other factors that interact with TT19, and to biochemically characterize TT19.

It is not clear whether the autofluorescent bodies in maize and the anthocyanoplasts in some species are compatible with the vesicle-trafficking model. The sizes of the autofluorescent bodies (0.3~3 μm in diameter) and anthocyanoplasts (3~10 μm) are considerably larger than those of the vesicles used in the secretory pathway (e.g., COPII vesicle = ~0.1 μm). Because anthocyanoplasts are red, the lumen of anthocyanoplasts is thought to be acidic. It therefore is conceivable that anthocyanoplasts are intermediates of the flavonoid-accumulating vacuoles. The flavonoid transport models may need to be modified to explain the cytological observations. Flavonoids are first accumulated into a number of small vacuoles, rather than directly transported to the preexisting large vacuoles. Transport of flavonoids between the membrane compartments (such as between vesicle–vesicle and/or vesicle–vacuole) may depend on (1) fusion of mutual membranes, such as occurs in the vesicle trafficking of macromolecules, (2) loading of vesicles into other larger compartments, as is suggested by the detection of auto-fluorescent bodies in the vacuole of maize, or (3) unknown novel mechanisms (see Grotewold, 2004).

7. PERSPECTIVES

Flavonoid transport mechanisms previously have been analyzed mainly in terms of transport of flavonoids across membranes. Some mechanisms have been suggested to take up flavonoids across membranes *in vitro*, such as a proton-dependent mechanism and a proton-independent and ATP-dependent mechanism. Genetic approaches have been used to isolate both the ABC-type transporter gene for anthocyanins in maize (*ZmMRP3*) (Goodman et al., 2004) and the MATE-type transporter gene for proanthocyanidins in *Arabidopsis* (*TT12*) (Debeaujon et al., 2001). Activation tagging analyses have shown that the anthocyanin pathway involves a MATE-type protein in tomato (Mathews et al., 2003) and sugar-transporter-like proteins in *Arabidopsis* (Tohge et al., 2005). These results may indicate that different species utilize different transport mechanisms or that individual species possess different transport mechanisms for different flavonoid end products (Winkel-Shirley, 2002). Plant cells have more than one kind of vacuole (Marty, 1999). It is unclear what kinds of vacuoles accumulate flavonoids and whether the same type of vacuole accumulates both anthocyanins and proanthocyanidins (Gruber et al., 1999). If species-specific or end product-specific transport mechanisms exist, they might be due to the existence of specialized vacuoles that accumulate specific flavonoids via the corresponding transporter proteins. Alternatively, they might be due to the temporal, spatial, or developmental expression of the transporter genes. To understand the complex flavonoid transport mechanisms, further studies are needed to identify the flavonoid-accumulating vacuoles and to determine the distribution patterns of the respective transporter proteins in different tissues at different developmental stages.

At present, a GST is the only player that links flavonoid synthesis and its vacuolar accumulation. Therefore, GSTs have a crucial role in the subcellular transport of flavonoids. Although the function of the GSTs in flavonoid transport is unclear, their most likely function is to act as flavonoid carrier proteins (Mueller et al., 2000). To obtain further insights into the function of the GSTs, we now are investigating the subcellular localization of TT19. Our preliminary results (unpublished data) suggest that the GSTs are occasionally localized at certain cytoplasmic foci within the cells. Determining the distribution patterns of flavonoids as well as endomembrane compartments and comparing them with the distribution pattern of GSTs will clarify the biological significance of the subcellular localization of GSTs. On the other hand, we recently obtained another *tt19* mutant in which only one amino acid in TT19 is changed (Shikazono et al., 2005). This mutant is phenotypically indistinguishable from other *tt19* null mutants, which suggests that the point mutation nearly completely disrupts TT19 function. We are presently characterizing the point-mutated protein, with the hope that it will elucidate the function of flavonoid-specific GSTs.

In order to understand flavonoid transport, we also should determine the state of the endomembrane system, not only in vacuoles but also in other membranous organelles in the cells being analyzed. Cultured cells, such as maize BMS cells, are good materials for studying flavonoid transport because the interaction between flavonoid accumulation and development of endomembranes in these cells can be revealed easily by a number of reagents. For example, the endomembrane system can be visualized by a fluorescent dye and vesicle trafficking in the secretory pathway can be examined with various inhibitors. Although the biogenesis and development of vacuoles are not well understood, the vacuolar membrane has been suggested to be derived from the ER (Bethke and Jones, 2000; Surpin and Raikhel, 2004). Plant cells have several kinds of membrane compartments, and most of them [protein bodies, ER bodies, prevacuolar compartments (PVCs), and precursor-accumulating (PAC) vesicles] are derived from the ER (Hara-Nishimura et al., 2004). In the endothelial cells of the *Arabidopsis* seed coat, pigment accumulation is preceded by the formation of a central vacuole at the early stage of seed development (Beeckman et al., 2000). On the other hand, loss-of-function mutants of *LDOX/TT18/Tannin Deficient Seed 4* (*TDS4*) gene resulted in the appearance of multiple small vacuoles in the *Arabidopsis* seed coat, suggesting that proanthocyanidin accumulation is a signal for vacuolar maturation (Abrahams et al., 2003). Interestingly, in the *Arabidopsis* mutant *atvsr1*, abnormal trafficking of vacuolar storage proteins by PAC vesicles resulted in the formation of multiple small vacuoles (Shimada et al., 2003). Other *tds* mutants isolated by Abrahams et al. (2002) may include mutants for vacuolar biogenesis or development. Such mutants would help to identify novel genes that participate in proanthocyanidin transport of *Arabidopsis*. In addition, many *tt* mutants remain to be analyzed (Winkel-Shirley,

2001). Surprisingly, a novel *tt* mutant was recently attributed to the loss-of-function of a putative plasma membrane-type H^+-ATPase gene, which clearly indicates that this gene is specifically involved in proanthocyanidin transport or accumulation (Baxter et al., 2005). Further studies using a combination of forward and reverse genetics, cell biological approaches, and molecular mutagenesis (including RNAi) should help to understand the molecular mechanisms of flavonoid transport in plant cells as well as to manipulate flavonoid contents in plant breeding programs.

8. ACKNOWLEDGMENTS

The author wishes to thank Atsushi Tanaka and Naoya Shikazono in JAERI for their helpful discussions and critical reviews of this manuscript. I am also grateful to Kazuki Saito (Chiba University, Chiba, Japan), Shigeru Iida and Atsushi Hoshino (National Institute for Basic Biology, Okazaki, Japan), and Erich Grotewold (The Ohio State University, Columbus, USA) for sharing unpublished results or data in preparation for publication.

9. REFERENCES

Abrahams, S., Lee, E., Walker, A. R., Tanner, G. J., Larkin, P. J., and Ashton, A. R., 2003, The *Arabidopsis TDS4* gene encodes leucoanthocyanidin dioxygenase (LDOX) and is essential for proanthocyanidin synthesis and vacuole development, *Plant J* **35**: 624-636.

Abrahams, S., Tanner, G. J., Larkin, P. J., and Ashton, A. R., 2002, Identification and biochemical characterization of mutants in the proanthocyanidin pathway in Arabidopsis, *Plant Physiol* **130**: 561-576.

Achnine, L., Blancaflor, E. B., Rasmussen, S., and Dixon, R. A., 2004, Colocalization of L-phenylalanine ammonia-lyase and cinnamate 4-hydroxylase for metabolic channeling in phenylpropanoid biosynthesis, *Plant Cell* **16**: 3098-3109.

Alfenito, M. R., Souer, E., Goodman, C. D., Buell, R., Mol, J., Koes, R., and Walbot, V., 1998, Functional complementation of anthocyanin sequestration in the vacuole by widely divergent glutathione *S*-transferases, *Plant Cell* **10**: 1135-1149.

Bartel, B., and Matsuda, S. P. T., 2003, Seeing red, *Science* **299**: 352-353.

Baxter, I. R., Young, J. C., Armstrong, G., Foster, N., Bogenschutz, N., Cordova, T., Peer, W. A., Hazen, S. P., Murphy, A. S., and Harper, J. F., 2005, A plasma membrane H^+-ATPase is required for the formation of proanthocyanidins in the seed coat endothelium of *Arabidopsis thaliana*, *Proc Natl Acad Sci USA* **102**: 2649-2654.

Beeckman, T., De Rycke, R., Viane, R., and Inzé, D., 2000, Histological study of seed coat development in *Arabidopsis thaliana*, *J Plant Res* **113**: 139-148.

Bethke, P. C., and Jones, R. L., 2000, Vacuoles and prevacuolar compartments, *Curr Opin Plant Biol* **3**: 469-475.

Borevitz, J. O., Xia, Y., Blount, J., Dixon, R. A., and Lamb, C., 2000, Activation tagging identifies a conserved MYB regulator of phenylpropanoid biosynthesis, *Plant Cell* **12**: 2383-2393.

Brown, M. H., Paulsen, I. T., and Skurray, R. A., 1999, The multidrug efflux protein NorM is a prototype of a new family of transporters, *Mol Microbiol* **31**: 394-395.

Buer, C. S., and Muday, G. K., 2004, The *transparent testa4* mutation prevents flavonoid synthesis and alters auxin transport and the response of *Arabidopsis* roots to gravity and light, *Plant Cell* **16**: 1191-1205.

Burbulis, I. E., and Winkel-Shirley, B., 1999, Interactions among enzymes of the *Arabidopsis* flavonoid biosynthetic pathway, *Proc Natl Acad Sci USA* **96**: 12929-12934.

Chapple, C. C. S., Shirley, B. W., Zook, M., Hammerschmidt, R., and Somerville, S. C., 1994, Secondary metabolism in *Arabidopsis*, in: *Arabidopsis*, E. M. Meyerowitz, and C. R. Somerville (Eds), New York: Cold Spring Harbor Laboratory Press, pp. 989-1030.

Coleman, J. O. D., Blake-Kalff, M. M. A., and Davies, T. G. E., 1997, Detoxification of xenobiotics by plants: chemical modification and vacuolar compartmentation, *Trends Plant Sci* **2**: 144-151.

Czichi, U., and Kindl, H., 1977, Phenylalanine ammonia lyase and cinnamic acid hydroxylases as assembled consecutive enzymes on microsomal membranes of cucumber cotyledons: cooperation and subcellular distribution, *Planta* **134**: 133-143.

Debeaujon, I., Léon-Kloosterziel, K. M., and Koornneef, M., 2000, Influence of the testa on seed dormancy, germination, and longevity in *Arabidopsis*, *Plant Physiol* **122**: 403-413.

Debeaujon, I., Nesi, N., Perez, P., Devic, M., Grandjean, O., Caboche, M., and Lepiniec, L., 2003, Proanthocyanidin-accumulating cells in *Arabidopsis* testa: regulation of differentiation and role in seed development, *Plant Cell* **15**: 2514-2531.

Debeaujon, I., Peeters, A. J. M., Léon-Kloosterziel, K. M., and Koornneef, M., 2001, The *TRANSPARENT TESTA12* gene of *Arabidopsis* encodes a multidrug secondary transporter-like protein required for flavonoid sequestration in vacuoles of the seed coat endothelium, *Plant Cell* **13**: 853-871.

Devic, M., Guilleminot, J., Debeaujon, I., Bechtold, N., Bensaude, E., Koornneef, M., Pelletier, G., and Delseny, M., 1999, The *BANYULS* gene encodes a DFR-like protein and is a marker of early seed coat development, *Plant J* **19**: 387-398.

Diener, A. C., Gaxiola, R. A., and Fink, G. R., 2001, *Arabidopsis ALF5*, a multidrug efflux transporter gene family member, confers resistance to toxins, *Plant Cell* **13**: 1625-1637.

Dixon, D. P., Lapthorn, A., and Edwards, R., 2002, Plant glutathione transferases, *Genome Biol* **3**: 3004.1-3004.10.

Edwards, R., Dixon, D. P., and Walbot, V., 2000, Plant glutathione S-transferases: enzymes with multiple functions in sickness and in health, *Trends Plant Sci* **5**: 193-198.

Fedoroff, N. V., Furtek, D. B., and Nelson, O. E., Jr., 1984, Cloning of the *bronze* locus in maize by a simple and generalizable procedure using the transposable controlling element *Activator* (*Ac*), *Proc Natl Acad Sci USA* **81**: 3825-3829.

Fritsch, H., and Grisebach, H., 1975, Biosynthesis of cyanidin in cell cultures of *Haplopappus gracilis*, *Phytochem* **14**: 2437-2442.

Fujiwara, H., Tanaka, Y., Yonekura-Sakakibara, K., Fukuchi-Mizutani, M., Nakao, M., Fukui, Y., Yamaguchi, M., Ashikari, T., and Kusumi, T., 1998, cDNA cloning, gene expression and subcellular localization of anthocyanin 5-aromatic acyltransferase from *Gentiana triflora*, *Plant J* **16**: 421-431.

Gaedeke, N., Klein, M., Kolukisaoglu, U., Forestier, C., Müller, A., Ansorge, M., Becker, D., Mamnun, Y., Kuchler, K., Schulz, B., Mueller-Roeber, B., and Martinoia, E., 2001, The *Arabidopsis thaliana* ABC transporter *At*MRP5 controls root development and stomata movement, *EMBO J* **20**, 1875-1887.

Goodman, C. D., Casati, P., and Walbot, V., 2004, A multidrug resistance-associated protein involved in anthocyanin transport in *Zea mays*, *Plant Cell* **16**: 1812-1826.

Grotewold, E., 2001, Subcellular trafficking of phytochemicals, *Recent Res Devel Plant Physiol* **2**: 31-48.

Grotewold, E., 2004, The challenges of moving chemicals within and out of cells: insights into the transport of plant natural products, *Planta* **219**: 906-909.

Grotewold, E., Chamberlin, M., Snook, M., Siame, B., Butler, L., Swenson, J., Maddock, S., St. Clair, G., and Bowen, B., 1998, Engineering secondary metabolism in maize cells by ectopic expression of transcription factors, *Plant Cell* **10**: 721-740.

Grotewold, E., Drummond, B. J., Bowen, B., and Peterson, T., 1994, The *myb*-homologous *P* gene controls phlobaphene pigmentation in maize floral organs by directly activating a flavonoid biosynthetic gene subset, *Cell* **76**: 543-553.

Gruber, M. Y., Ray, H., Auser, P., Skadhauge, B., Falk, J., Thomsen, K. K., Stougaard, J., Muir, A., Lees, G., Coulman, B., McKersie, B., Bowley, S., and von Wettstein, D., 1999, Genetic systems for condensed tannin biotechnology, in: *Plant Polyphenols 2: Chemistry and Biology*, G. G. Gross, R. W. Hemingway, and T. Yoshida (Eds), New York: Kluwer Academic/Plenum Publishers, pp. 315-341.

Hara-Nishimura, I., Matsushima, R., Shimada, T., and Nishimura, M., 2004, Diversity and formation of endoplasmic reticulum-derived compartments in plants. Are these compartments specific to plant cells? *Plant Physiol* **136**: 3435-3439.

Harborne, J. B., and Williams, C. A., 2000, Advances in flavonoid research since 1992, *Phytochem* **55**: 481-504.

Hopp, W., and Seitz, H. U., 1987, The uptake of acylated anthocyanin into isolated vacuoles from a cell suspension culture of *Daucus carota*, *Planta* **170**: 74-85.

Hrazdina, G., 1992, Compartmentation in aromatic metabolism, in: *Phenolic Metabolism in Plants*, H. A. Stafford and R. K. Ibrahim (Eds.), New York: Plenum Press, pp. 1-23.

Hrazdina, G., and Wagner, G. J., 1985, Compartmentation of plant phenolic compounds; sites of synthesis and accumulation, *Annu Proc Phytochem Soc Eur* **25**: 119-133.

Hrazdina, G., Wagner, G. J., and Siegelman, H. W., 1978, Subcellular localization of enzymes of anthocyanin biosynthesis in protoplasts, *Phytochem* **17**: 53-56.

Hrazdina, G., Zobel, A. M., and Hoch, H. C., 1987, Biochemical, immunological, and immunocytochemical evidence for the association of chalcone synthase with endoplasmic reticulum membranes, *Proc Natl Acad Sci USA* **84**: 8966-8970.

Jones, P., and Vogt, T., 2001, Glycosyltransferases in secondary plant metabolism: tranquilizers and stimulant controllers, *Planta* **213**: 164-174.

Kenyon, C. J., and Walker, G. C., 1980, DNA-damaging agents stimulate gene expression at specific loci in *Escherichia coli*, *Proc Natl Acad Sci USA* **77**: 2819-2823.

Kitamura, S., Shikazono, N., and Tanaka, A., 2004, *TRANSPARENT TESTA 19* is involved in the accumulation of both anthocyanins and proanthocyanidins in *Arabidopsis*, *Plant J* **37**: 104-114.

Klein, M., Geisler, M., Suh, S. J., Kolukisaoglu, H. U., Azevedo, L., Plaza, S., Curtis, M. D., Richter, A., Weder, B., Schulz, B., and Martinoia, E., 2004, Disruption of *AtMRP4*, a guard cell plasma membrane ABCC-type ABC transporter, leads to deregulation of stomatal opening and increased drought susceptibility, *Plant J* **39**: 219-236.

Klein, M., Martinoia, E., Hoffmann-Thoma, G., and Weissenböck, G., 2000, A membrane-potential dependent ABC-like transporter mediates the vacuolar uptake of rye flavone glucuronides: regulation of glucuronide uptake by glutathione and its conjugates, *Plant J* **21**: 289-304.

Klein, M., Weissenböck, G., Dufaud, A., Gaillard, C., Kreuz, K., and Martinoia, E., 1996, Different energization mechanisms drive the vacuolar uptake of a flavonoid glucoside and a herbicide glucoside, *J Biol Chem* **271**: 29666-29671.

Koes, R. E., Quattrocchio, F., and Mol, J. N. M., 1994, The flavonoid biosynthetic pathway in plants: function and evolution, *BioEssays* **16**: 123-132.

Kolukisaoglu, H. U., Bovet, L., Klein, M., Eggmann, T., Geisler, M., Wanke, D., Martinoia, E., and Schulz, B., 2002, Family business: the multidrug-resistance related protein (MRP) ABC transporter genes in *Arabidopsis thaliana*, *Planta* **216**: 107-119.

Koornneef, M., 1990, Mutations affecting the testa colour in *Arabidopsis*, *Arabidopsis Inf Serv* **27**: 1-4.

Krueger, J. H., Elledge, S. J., and Walker, G. C., 1983, Isolation and characterization of *Tn5* insertion mutations in the *lexA* gene of *Escherichia coli*, *J Bacteriol* **153**: 1368-1378.

Larsen, E. S., Alfenito, M. R., Briggs, W. R., and Walbot, V., 2003, A carnation anthocyanin mutant is complemented by the glutathione *S*-transferases encoded by maize *Bz2* and petunia *An9*, *Plant Cell Rep* **21**: 900-904.

Li, Z. -S., Alfenito, M., Rea, P. A., Walbot, V., and Dixon, R. A., 1997, Vacuolar uptake of the phytoalexin medicarpin by the glutathione conjugate pump, *Phytochemistry* **45**: 689-693.

Lin, Y., Irani, N. G., and Grotewold, E., 2003, Subcellular trafficking of phytochemicals explored using auto-fluorescent compounds in maize cells, *BMC Plant Biol* **3**: 10.

Liu, C-J., and Dixon, R. A., 2001, Elicitor-induced association of isoflavone *O*-methyltransferase with endomembranes prevents the formation and 7-*O*-methylation of daidzein during isoflavonoid phytoalexin biosynthesis, *Plant Cell* **13**: 2643-2658.

Liu, G., Sanchez-Fernandez, R., Li, Z.-S., and Rea, P. A., 2001, Enhanced multispecificity of *Arabidopsis* vacuolar multidrug resistance-associated protein-type ATP-binding cassette transporter, AtMRP2, *J Biol Chem* **276**: 8648-8656.

Lu, Y-P., Li, Z. -S., Drozdowicz, Y. M., Hörtensteiner, S., Martinoia, E., and Rea, P. A., 1998, AtMRP2, an *Arabidopsis* ATP binding cassette transporter able to transport glutathione *S*-conjugates and chlorophyll catabolites: functional comparisons with AtMRP1, *Plant Cell* **10**: 267-282.

Lu, Y-P., Li, Z. -S., and Rea, P. A., 1997, *AtMRP1* gene of *Arabidopsis* encodes a glutathione *S*-conjugate pump: Isolation and functional definition of a plant ATP-binding cassette transporter gene, *Proc Natl Acad Sci USA* **94**: 8243-8248.

Maeshima, M., 2001, Tonoplast transporters: organization and function, *Annu Rev Plant Physiol Plant Mol Biol* **52**: 469-497.

Marger, M. D., and Saier, M. H., Jr., 1993, A major superfamily of transmembrane facilitators that catalyse uniport, symport and antiport, *Trends Biochem Sci* **18**: 13-20.

Markham, K. R., Gould, K. S., Winefield, C. S., Mitchell, K. A., Bloor, S. J., and Boase, M. R., 2000, Anthocyanic vacuolar inclusions—their nature and significance in flower colouration, *Phytochem* **55**: 327-336.

Marrs, K. A., 1996, The functions and regulation of glutathione *S*-transferases in plants, *Annu Rev Plant Physiol Plant Mol Biol* **47**: 127-158.

Marrs, K. A., Alfenito, M. R., Lloyd, A. M., and Walbot, V., 1995, A glutathione *S*-transferase involved in vacuolar transfer encoded by the maize gene *Bronze-2*, *Nature* **375**: 397-400.

Martinoia, E., Grill, E., Tommasini, R., Kreuz, K., and Amrhein, N., 1993, ATP-dependent glutathione *S*-conjugate "export" pump in the vacuolar membrane of plants, *Nature* **364**: 247-249.

Martinoia, E., Massonneau, A., and Frangne, N., 2000, Transport processes of solutes across the vacuolar membrane of higher plants, *Plant Cell Physiol* **41**: 1175-1186.

Marty, F., 1999, Plant vacuoles, *Plant Cell* **11**: 587-599.

Matern, U., Reichenbach, C., and Heller, W., 1986, Efficient uptake of flavonoids into parsley (*Petroselinum hortense*) vacuoles requires acylated glycosides, *Planta* **167**: 183-189.

Mathews, H., Clendennen, S. K., Caldwell, C. G., Liu, X. L., Connors, K., Matheis, N., Schuster, D. K., Menasco, D. J., Wagoner, W., Lightner J., and Wagner, D. R., 2003, Activation tagging in tomato identifies a transcriptional regulator of anthocyanin biosynthesis, modification, and transport, *Plant Cell* **15**: 1689-1703.

Mol, J., Grotewold, E., and Koes, R., 1998, How genes paint flowers and seeds, *Trends Plant Sci* **3**: 212-217.

Molendijk, A. J., Ruperti, B., and Palme, K., 2004, Small GTPases in vesicle trafficking, *Curr Opin Plant Biol* **7**: 694-700.

Morita, Y., Kataoka, A., Shiota, S., Mizushima, T., and Tsuchiya, T., 2000, NorM of *Vibrio parahaemolyticus* is an Na⁺-driven multidrug efflux pump, *J Bacteriol* **182**: 6694-6697.

Morita, Y., Kodama, K., Shiota, S., Mine, T., Kataoka, A., Mizushima, T., and Tsuchiya, T., 1998, NorM, a putative multidrug efflux protein, of *Vibrio parahaemolyticus* and its homolog in *Escherichia coli*, *Antimicrob Agents Chemother* **42**: 1778-1782.

Mueller, L. A., Goodman, C. D., Silady, R. A., and Walbot, V., 2000, AN9, a petunia glutathione *S*-transferase required for anthocyanin sequestration, is a flavonoid-binding protein, *Plant Physiol* **123**: 1561-1570.

Mueller, L. A., and Walbot, V., 2001, Models for vacuolar sequestration of anthocyanins, *Recent Adv Phytochem* **35**: 297-312.

Nakajima, J., Tanaka, Y., Yamazaki, M., and Saito, K., 2001, Reaction mechanism from leucoanthocyanidin to anthocyanidin 3-glucoside, a key reaction for coloring in anthocyanin biosynthesis, *J Biol Chem* **276**: 25797-25803.

Nawrath, C., Heck, S., Parinthawong, N., and Metraux, J. -P., 2002, EDS5, an essential component of salicylic acid-dependent signaling for disease resistance in Arabidopsis, is a member of the MATE transporter family, *Plant Cell* **14**: 275-286.

Nozue, M., Kubo, H., Nishimura, M., Katou, A., Hattori, C., Usuda, N., Nagata, T., and Yasuda, H., 1993, Characterization of intravacuolar pigmented structures in anthocyanin-containing cells of sweet potato suspension cultures, *Plant Cell Physiol* **34**: 803-808.

Nozue, M., Yamada, K., Nakamura, T., Kubo, H., Kondo, M., and Nishimura, M., 1997, Expression of a vacuolar protein (VP24) in anthocyanin-producing cells of sweet potato in suspension culture, *Plant Physiol* **115**: 1065-1072.

Nozue, M., and Yasuda, H., 1985, Occurrence of anthocyanoplasts in cell suspension cultures of sweet potato, *Plant Cell Rep* **4**: 252-255.

Nozzolillo, C., and Ishikura, N., 1988, An investigation of the intracellular site of anthocyanoplasts using isolated protoplasts and vacuoles, *Plant Cell Rep* **7**: 389-392.

Pairoba, C. F., and Walbot, V., 2003, Post-transcriptional regulation of expression of the *Bronze2* gene of *Zea mays* L., *Plant Mol Biol* **53**: 75-86.

Parham, R. A., and Kaustinen, H. M., 1977, On the site of tannin synthesis in plant cells, *Bot Gaz* **138**: 465-467.

Pecket, R. C., and Small, C. J., 1980, Occurrence, location and development of anthocyanoplasts, *Phytochem* **19**: 2571-2576.

Rea, P. A., 1999, MRP subfamily ABC transporters from plants and yeast, *J Exp Bot* **50**: 895-913.

Rea, P. A., Li, Z-S., Lu, Y-P., Drozdowicz, Y. M., and Martinoia, E., 1998, From vacuolar GS-X pumps to multispecific ABC transporters, *Annu Rev Plant Physiol Plant Mol Biol* **49**: 727-760.

Reddy, G. M., and Coe, E. H., Jr., 1962, Inter-tissue complementation: a simple technique for direct analysis of gene-action sequence, *Science* **138**: 149-150.

Reinemer, P., Prade, L., Hof, P., Neuefeind, T., Huber, R., Zettl, R., Palme, K., Schell, J., Koelln, I., Bartunik, H. D., and Bieseler, B., 1996, Three-dimensional structure of glutathione *S*-transferase from *Arabidopsis thaliana* at 2.2 Å resolution: structural characterization of herbicide-conjugating plant glutathione *S*-transferases and a novel active site architecture, *J Mol Biol* **255**: 289-309.

Robinson, D. G., Hinz, G., and Holstein, S. E. H., 1998, The molecular characterization of transport vesicles, *Plant Mol Biol* **38**: 49-76.

Rothman, J. E., and Wieland, F. T., 1996, Protein sorting by transport vesicles, *Science* **272**: 227-234.

Sanderfoot, A. A., and Raikhel, N. V., 1999, The specificity of vesicle trafficking: coat proteins and SNAREs, *Plant Cell* **11**: 629-641.

Saslowsky, D., and Winkel-Shirley, B., 2001, Localization of flavonoid enzymes in *Arabidopsis* roots, *Plant J.* **27**: 37-48.

Shikazono, N., Suzuki, C., Kitamura, S., Watanabe, H., Tano, S., and Tanaka, A., 2005, Analysis of mutations induced by carbon ions in *Arabidopsis thaliana*, *J Exp Bot* **56**: 587-596.

Shimada, T., Fuji, K., Tamura, K., Kondo, M., Nishimura, M., and Hara-Nishimura, I., 2003, Vacuolar sorting receptor for seed storage proteins in *Arabidopsis thaliana*, *Proc Natl Acad Sci USA* **100**: 16095-16100.

Shiomi, N., Fukuda, H., Fukuda, Y., Murata, K., and Kimura, A., 1991, Nucleotide sequence and characterization of a gene conferring resistance to ethionine in yeast *Saccharomyces cerevisiae*, *J Ferment Bioeng* **4**: 211-215.

Shirley, B. W., 1998, Flavonoids in seeds and grains: physiological function, agronomic importance and the genetics of biosynthesis, *Seed Sci Res* **8**: 415-422.

Staehelin, L. A., 1997, The plant ER: a dynamic organelle composed of a large number of discrete functional domains, *Plant J* **11**: 1151-1165.

Stafford, H. A., 1974, Possible multienzyme complexes regulating the formation of C_6-C_3 phenolic compounds and lignins in higher plants, *Recent Adv Phytochem* **8**: 53-79.

Stafford, H. A., 1989, The enzymology of proanthocyanidin biosynthesis, in: *Chemistry and Significance of Condensed Tannins*, R. W. Hemingway and J. J. Karchesy (Eds.), Plenum Press, New York, pp. 47-70.

Surpin, M., and Raikhel, N., 2004, Traffic jams affect plant development and signal transduction, *Nature Rev Mol Cell Biol* **5**: 100-109.

Suzuki, H., Nakayama, T., Yonekura-Sakakibara, K., Fukui, Y., Nakamura, N., Nakao, M., Tanaka, Y., Yamaguchi, M., Kusumi, T., and Nishio, T., 2001, Malonyl-CoA:anthocyanin 5-*O*-glucoside-6'''-*O*-malonyltransferase from scarlet sage (*Salvia splendens*) flowers, *J Biol Chem* **276**: 49013-49019.

Sze, H., Li, X., and Palmgren, M. G., 1999, Energization of plant cell membranes by H^+-pumping ATPases: regulation and biosynthesis, *Plant Cell* **11**: 677-689.

Tanaka, Y., Tsuda, S., and Kusumi, T., 1998, Metabolic engineering to modify flower color, *Plant Cell Physiol* **39**: 1119-1126.

Tohge, T., Nishiyama, Y., Hirai, M. Y., Yano, M., Nakajima, J., Awazuhara, M., Inoue, E., Takahashi, H., Goodenowe, D. B., Kitayama, M., Noji, M., Yamazaki, M., and Saito, K., 2005, Functional genomics by integrated analysis of metabolome and transcriptome of *Arabidopsis* plants over-expressing an MYB transcription factor, *Plant J* **42**: 218-235.

Ueda, T., and Nakano, A., 2002, Vesicular traffic: an integral part of plant life, *Curr Opin Plant Biol* **5**: 513-517.

Wagner, G. J., and Hrazdina, G., 1984, Endoplasmic reticulum as a site of phenylpropanoid and flavonoid metabolism in *Hippeastrum*, *Plant Physiol* **74**: 901-906.

Walbot, V., Mueller, L. A., Silady, R. A., and Goodman, C. D., 2000, Do glutathione S-transferases act as enzymes or carrier proteins for their natural substrates? in: *Sulfur Nutrition and Sulfur Assimilation in Higher Plants*, C. Brunold, H. Rennenberg, L. J. Kok, I. Stulen, and J-C. Davidian (Eds.), Paul Haupt Publishers, Bern, pp. 155-165.

Winkel-Shirley, B., 1999, Evidence for enzyme complexes in the phenylpropanoid and flavonoid pathways, *Physiol Plant* **107**: 142-149.

Winkel-Shirley, B., 2001, Flavonoid biosynthesis. A colorful model for genetics, biochemistry, cell biology, and biotechnology, *Plant Physiol* **126**: 485-493.

Winkel-Shirley, B., 2002, A mutational approach to dissection of flavonoid biosynthesis in *Arabidopsis*, *Recent Adv Phytochem* **36**: 95-110.

Wolf, A. E., Dietz, K-J., and Schröder, P., 1996, Degradation of glutathione *S*-conjugates by a carboxypeptidase in the plant vacuole, *FEBS Lett* **384**: 31-34.

Xie, D.-Y., Sharma, S. B., Paiva, N. L., Ferreira, D., and Dixon, R. A., 2003, Role of anthocyanidin reductase, encoded by *BANYULS* in plant flavonoid biosynthesis, *Science* **299**: 396-399.

Xu, W., Shioiri, H., Kojima, M., and Nozue, M., 2001, Primary structure and expression of a 24-kD vacuolar protein (VP24) precursor in anthocyanin-producing cells of sweet potato in suspension culture, *Plant Physiol* **125**: 447-455.

CHAPTER 6

FLAVONOID PIGMENTS AS TOOLS IN MOLECULAR GENETICS

SURINDER CHOPRA[1], ATSUSHI HOSHINO[2], JAYANAND BODDU[1], AND SHIGERU IIDA[2]

[1]*Department of Crop & Soil Sciences, The Pennsylvania State University, University Park, PA 16802 USA Tel: 814-865-1159; Fax: 814-863-7043; E-mail: sic3@psu.edu;* [2]*National Institute for Basic Biology, National Institutes of Natural Sciences, Myodaiji, Okazaki 444-8585, Japan, Tel: +81-564-55-7680; Fax: +81-564-55-7685; E-mail: shigiida@nibb.ac.jp*

1. INTRODUCTION

In the plant kingdom, the flavonoid biosynthetic pathway is ubiquitous and produces a variety of pigmented as well as nonpigmented compounds. Flavonoid compounds have been implicated in several biological processes and some of their functions include the attraction of pollinating agents via pigmentation of floral organs (Huits et al., 1994), pollen tube germination (Mo et al., 1992), protection from UV exposure (Bieza and Lois, 2001), and defense against insects by acting as insecticides (Wiseman et al., 1996) and fungal pathogens by acting as phytoalexins (Nicholson and Hammerschmidt, 1992; Dixon and Steele, 1999). Flavonoid pigments have been used as a convenient visible marker in molecular genetic experiments and to study regulation of gene expression (Styles and Ceska, 1977, 1981, 1989; Dooner et al., 1991; Koes et al., 2005). Flavonoid biosynthesis takes place through the phenylpropanoid pathway, and depending on the genetic constitution of the plant naringenin can have several different fates leading to the formation of flavonoid metabolites that include anthocyanins, flavones, and anthocyanidins (Figure 6.1) (Styles and Ceska, 1989; Dooner et al., 1991; Winkel-Shirley, 2001a, 2001b; Schijlen et al., 2004). Since the isolation of chalcone synthase (CHS), which catalyzes the first committed step of this pathway (Kreuzaler et al., 1983), efforts have been focused on the isolation of mutants and the cloning of structural genes that are required for the different biosynthetic steps (Harborne and Williams, 2000; Winkel-Shirley, 2001a). Progress in our understanding of gene expression and regulation has been enhanced by targeting structural as well as regulatory genes that are required for the biosynthesis of these pigmented flavonoids.

Since various aspects of the gene structure and regulation, genetics, epigenetics, biochemistry, and regulation of flavonoid biosynthesis have been reviewed recently (Springob et al., 2003; Schwinn and Davies, 2004; Tanner, 2004; Koes et al., 2005; Tanaka et al., 2005) and also discussed in other chapters of this book, we will focus on the biosynthesis and accumulation of 3-deoxyflavonoid (phlobaphene) pigments in maize (*Zea mays*) and sorghum (*Sorghum bicolor*) (Figure 6.2), proanthocyanidins (condensed tannins) in *Arabidopsis* (*Arabidopsis thaliana*), and anthocyanins in snapdragon (*Antirrhinum majus*), petunia (*Petunia hybrida*), and three *Ipomoea* species: the Japanese morning glory (*Ipomoea nil*), the common morning glory (*Ipomoea purpurea*), and *Ipomoea tricolor* (Figure 6.3). We will also discuss molecular genetic aspects by emphasizing phenotypes of mutant alleles characterized at the sequence level and describe briefly epigenetic aspects of regulation in the genes for phlobaphenes, proanthocyanidins, and anthocyanins.

2. FLAVONOID PIGMENTS IN MAIZE AND SORGHUM

In maize, purple and red anthocyanins are derived from 3-hydroxyflavonoids while the brick-red phlobaphenes are condensed products of 3-deoxyflavonoids (Figure 6.1) (Styles and Ceska, 1989; Dooner et al., 1991). Anthocyanin pigmentation is the oldest trait employed for studies in genetics: Gregor Mendel, the founder of modern genetics, studied inheritance of flower and seed color in pea, and Nobel Laureate Barbara McClintock, the discoverer of transposable elements and epigenetic gene regulation, used pigmentation patterns of maize kernels as a marker in her research. Over the years, genetics of anthocyanins and phlobaphenes have allowed the precise understanding of molecular mechanisms that govern their accumulation in a tissue-specific and developmentally regulated fashion (Kirby and Styles, 1970; Dooner and Nelson, 1977; Larson et al., 1986; Styles and Ceska, 1989; Dooner et al., 1991; Grotewold et al., 1998).

2.1. Phlobaphene pigments in maize and their regulation

The phlobaphenes or the 3-deoxyflavonoid pigments are brick-red in color and the chemical constitution of this class of secondary metabolites remains largely unknown. Although controversies exist in terms of naming and chemistry of certain condensed tannins and phlobaphenes, we will present the knowledge earned through the study of 3-deoxyflavonoid pigments or phlobaphenes observed in maize and sorghum. The steps in this pathway share the same enzymes required for the early steps of the synthesis of anthocyanins. Three structural genes, namely, *C2, CHI1,* and *A1,* which encode CHS, CHI and DFR, respectively (Figure 6.1), are required for the synthesis of 3-deoxyflavonoids (Styles and Ceska, 1989). The phlobaphene pigments are believed to result from polymerization products of flavan-4-ols (Styles and Ceska, 1981; Foo and Karchesy, 1989), and genetics of the accumulation of these precursors is very well understood in maize through the use of mutants in biosynthetic genes (Dooner et al., 1991; Neuffer et al., 1997). The polymerization of flavan-4-ols to condensed phlobaphenes is likely to be a nonenzymatic step (Styles and Ceska, 1989) and this aspect requires further study both at the genetic and biochemical levels.

The synthesis of the flavan-4-ols and the phlobaphenes requires a functional *P1* (*pericarp color*) locus. The *P1* gene encodes an R2R3-MYB DNA-binding domain protein (Grotewold et al., 1991). Gel retardation and transposon mutagenesis studies have demonstrated that the P1 DNA-binding domain binds to the *cis*-regulatory elements of the *A1* gene promoter and this binding facilitates transcription of the *a1* gene (Grotewold et al., 1994; Pooma et al., 2002) (a detailed description of the mechanisms of regulation of phlobaphenes and anthocyanins is presented in Chapter 4).

2.2. Allelic diversity and tissue-specific expression patterns

The most obvious pigmentation phenotypes of *P1* can be observed in the pericarp (outer layer of the ovary wall) and cob glume (palea and lemma, bracts that subtend the kernel) tissues of mature maize ears. The maize *P1* gene was cloned from an *Ac* (*Activator*) tagged allele called *P1-vv* (variegated pericarp, variegated cob); thereafter, several *Ac* insertion alleles of *P1* have been identified (Peterson, 1990; Athma et al., 1992) and used to study the importance of coding as well as regulatory sequences of the *P1* gene (Sidorenko et al., 2000).

Apart from the existence of *Ac* transposon alleles of the *P1* gene, several naturally existing stable as well as unstable alleles have been documented (http://www.maizegdb.org). Thus, using phlobaphene pigments as markers, alleles at the *P1* locus can be distinguished easily based on their variation in kernel pericarp and cob glume pigmentation (Figure 6.2) (Brink and Styles, 1966). These alleles have been conventionally identified by a suffix indicating their expression in pericarp and cob glumes. For example, a *p1* allele with expression in pericarp and cob glumes has been designated as *P1-rr* as opposed to a nonfunctional allele *p1-ww* (white or colorless pericarp and white glumes).

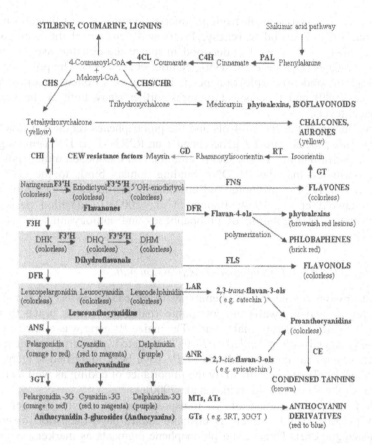

Figure 6.1 *General scheme of the phenylpropanoid and flavonoid pathways. Pathways show flavonoid subbranches of anthocyanins, phlobaphenes, condensed tannins, isoflavonoids, isoflavonoid phytoalexins, 3-deoxyanthocyanidin phytoalexins, CEW (corn ear worm) resistance factors, flavonols, and aurones. Pigment color description of some of the compounds is shown in parenthesis below the respective compound. Dihydrokaempferol (DHK), dihydroquercetin (DHQ), and dihydromyrecetin (DHM). Gene products shown are phenylalanine ammonia lyase (PAL), cinnamate 4-hydroxylase (C4H), 4-coumaroyl: CoA-ligase (4CL), chalcone synthase (CHS), chalcone reductase (CHR), chalcone isomerase (CHI), flavanone 3-hydroxylase (F3H), dihydroflavonol 4-reductase (DFR), flavonoid 3'-hydroxylase (F3'H), flavonoid 3',5'-hydroxylase (F3'5'H), glucosyltransferase (GT), 3-glucosyltransferase (3GT), anthocyanidin synthase (ANS), leucoanthocyanidin reductase (LAR), anthocyanidin reductase (ANR), condensing enzyme (CE), flavone synthase (FNS), flavonol synthase (FLS), rhamnosyltransferase (RT), glucose 4,6-dehydratase (GD), methyltransferases (MTs), acyltransferases (ATs), 3-glucoside rhamnosyltransferase (3RT), and 3-glucoside glucosyltransferase (3GGT).*

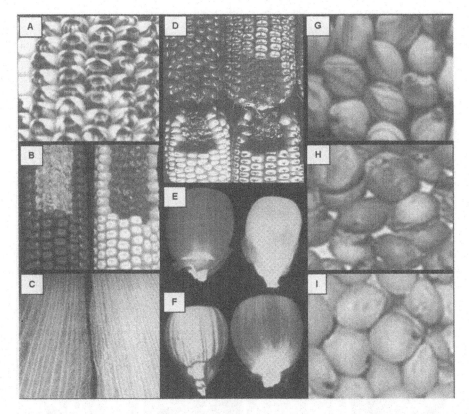

Figure 6.2 Phlobaphene pigmentation phenotypes of maize and sorghum. (A) Variegations in the kernel pericarp of P1-vv allele resulting from the excision of the Ac transposon from the p1 gene of maize. (B) kernel pericarp and cob glume pigmentation and (C) husk pigmentation patterns of two stable alleles P1-rr (left) and P1-wr (right). (D – F) gain of pericarp and cob glume pigmentation phenotypes showing epigenetic variegation of ears and kernels of P1-wr Ufo1. (G – I) sorghum phlobaphenes are regulated by yellow seed1 (Y1) and the alleles of Y1 shown are Y1-cs, Y1-rr, and y1-ww, respectively. See Color Section for figure in colors.

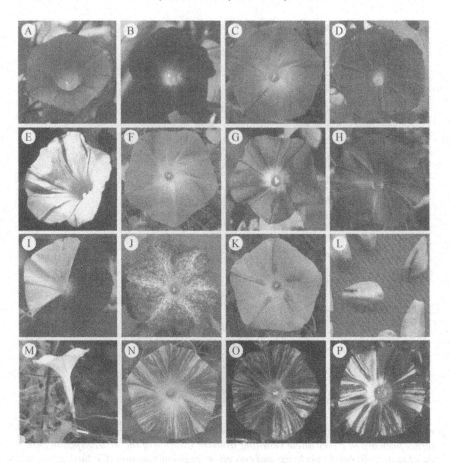

Figure 6.3 *Flower pigmentation phenotypes of morning glories.(A – C) Blue or dark purple flowers of wild-type I. nil, I. purpurea, and I. tricolor, respectively. (D – F) f3'h mutants exhibiting reddish flowers of I. nil with the magenta allele, I. purpurea with the pink and mutable a (flaked) alleles, and I. tricolor with the fuchsia allele, respectively. The a (flaked) allele is the CHS-D gene carrying an autonomous Ac-related transposable element, Tip100.*
(G) Variegated flowers of I. nil with the purple-mutable allele in the vacuolar Na⁺/H⁺ antiporter gene (NHX1) carrying the Tpn4 transposon. (H) Brownish flowers of I. nil with the magenta, purple, and dusky alleles. (I) Flowers with typical chimeric sectors and spots of I. nil with the a3 (flecked) allele, corresponding to the DFR-B gene carrying the Tpn1 transposon. (J) Variegated flowers of I. nil with the mutable speckled allele, which is the CHI gene with Tpn2. (K – L) Flowers and seeds, respectively, of I. tricolor with the mutable ivory seed-variegated allele. (M) White flowers with colored stems of I. nil with the mutant c1, encoding a maize C1-like flower-specific R2R3-MYB transcription factor.(N) variegated flowers of I. tricolor due to transcriptional gene silencing. (O – P) Flowers of I. nil with a recessive and a dominant mutant allele, respectively, displaying nonclonal variegations caused by epigenetic gene silencing. See Color Section for figure in colors.

Other examples of *P1* alleles include *P1-wr* (white pericarp, red cob glumes) and *P1-rw* (red pericarp and white cob glumes). These alleles not only differ in their pigment accumulation in the pericarp and cob glumes but their affects also can be observed in the husk and tassel glume tissues. Specific expression patterns controlled by these individual alleles have been found to be correlated with the transcriptional regulation of *P1* in various tissues (Chopra et al., 1996). Genetic and molecular studies further established that the transcriptional regulation of *P1* alleles is a function of the unique gene structures of each of the alleles. The *P1-rr* allele carries a single coding sequence that is flanked on both 5′ and 3′ ends by a 5.2 kb direct repeat (Lechelt et al., 1989), while the *P1-wr* contains six identical gene copies that are arranged in a tandem head-to-tail fashion (Chopra et al., 1998). The association between the *P1-wr* phenotype and the amplified gene structure was proven by cosegregation analyses using tissue-specific pigmentations of *P1-rr* and *P1-wr* as a marker, and these studies established that *P1-wr* gene copies present in the maize inbred lines with white pericarp and red cob phenotype are tightly linked (Chopra et al., 1998). Apart from the copy number difference, the coding and proximal regulatory sequences of the two alleles are more than 99% similar at the DNA sequence level.

As opposed to the multiple copy complex of the *P1-wr*, the *P1-rw* (red pericarp, white cob) allele is encoded by a singly copy *P1* gene sequence (Zhang and Peterson, 2005). The expression pattern of the *P1-rw* has been shown to be genetically correlated with the sequence modification of a distal regulatory element. Allele-specific expression patterns can be produced by promoter sequence differences, as demonstrated for the alleles of maize regulatory genes *B1* and *R1* (Ludwig and Wessler, 1990; Radicella et al., 1991). As compared to the simple *P1-rw* regulation, which could be explained by the absence of a DNA sequence per se, the regulation of the *P1-wr* allele seems to be more complex (Chopra et al., 1998). Transgenic analysis of *p1* regulatory sequences have demonstrated that other factors, including an epigenetic mode of regulation (discussed in Section 4), may be responsible for the formation of stable and unstable alleles (Cocciolone et al., 2001; Chopra et al., 2003).

The function of *P1* has been diversified further by a duplicated and tightly linked gene, *P2* (Zhang et al., 2000). Although the expression of *P1* can be observed in kernel and silk tissues, the expression of *P2* is restricted to silks (Zhang et al., 2003b). Pigmentation phenotypes produced by the browning of cut silks (silk browning) has led to the identification of the role of the *P2* gene in silks. Both *P1* and *P2* are equivalent in regulating the silk browning phenotype (Cocciolone et al., 2005). It is now clear that silk browning is associated with the presence of *C*-glycosyl flavones, a class of flavonoids that includes maysin (Byrne et al., 1996; Grotewold et al., 1998), which acts as an insecticide to the larvae of corn ear worm (Wiseman et al., 1996).

2.3. Regulation of phlobaphenes in sorghum

In sorghum, phlobaphene pigments have been described and their biosynthesis has been studied to some extent, but the genetic and molecular nature of this pathway is not well elucidated (Kambal and Bate-Smith, 1976). Making use of a visible pericarp phlobaphene phenotype, Chopra et al. (1999) showed that a functional *Y1* (*yellow seed1*) gene is required for the biosynthesis of flavan-4-ols in various sorghum tissues, including the seed pericarp and glumes. Unlike the maize *P1*, functional alleles of *Y1* also regulate phlobaphene accumulation in leaves. The naming of alleles of *Y1* has been based on the maize *P1* nomenclature, such that *Y1-rr* stands for red pericarp, red glumes. Functionality of the *Y1* gene largely has been studied from red revertants obtained from the mutable allele *Y1-cs* (*Y1-candystripe*) (see Section 4.1). As mentioned above, biosynthesis of flavan-4-ols in maize requires a functional allele of the *P1* gene, and now studies have established that *Y1* and *P1* genes are orthologs that encode R2R3-MYB domain proteins (Chopra et al., 1999; Zhang et al., 2000; Boddu et al., 2005a). Their functionality also is conserved in that, similar to the *P1*-regulated pathway, the transcription of the *DFR* gene is more drastically reduced in the mutable *Y1-cs* allele than in the functional revertants of *y1* (Boddu et al., 2005b). These expression data have been obtained from leaves, immature inflorescence, pericarp, and glumes, indicating that the *Y1* gene is transcribed and regulated in sorghum tissues in a developmental fashion. In addition, the characterization of loss of pigmentation mutants designated as *y1-ww* (white pericarp, white glume) has been useful in establishing the genetics and regulation of phlobaphenes biosynthesis and their accumulation in sorghum. Some of these "loss of function" alleles contain large deletions of regulatory as well as coding sequences, and the cause of these deletions is under further investigation (J. Boddu and S. Chopra, unpublished), while others contain partial deletions within the *Y1* coding sequences that have arisen by the imperfect excision of the *Cs1* element (F. Ibraheem and S. Chopra, unpublished). In addition to the accumulation of flavan-4-ols, the *Y1-rr* (red pericarp, red glume) seeds show the presence of compounds called 3-deoxyanthocyanidins (Boddu et al., 2005b). These types of flavonoids have been shown to be induced and act as phytoalexins in sorghum leaves during attempted penetration by *Colletotrichum sublineolum* (Snyder and Nicholson, 1990). The chemical structures of flavan-4-ols and 3-deoxyanthocyanidin phytoalexins in sorghum are very similar, and thus it has been speculated that these compounds may arise through a common pathway (Lo and Nicholson, 1998; Chopra et al., 2002). DNA polymorphisms together with phlobaphene pigmentation were used as tools to perform cosegregation analyses to establish a genetic linkage between the expression of phytoalexins and functional *Y1* alleles (Chopra et al., 2002). Study of the biosynthesis of sorghum phytoalexins and phlobaphene pigments thus have provided clues to the existence of additional subbranches of flavonoid pathways that may involve identification of function of additional structural gene encoding flavonoid 3′-hydroxylase (F3′H) (Boddu et al., 2004).

3. ANTHOCYANINS AND PROANTHOCYANIDINS

In dicotyledonous plants, *Arabidopsis*, snapdragon, petunia, and morning glories, mutations in the biosynthesis of flavonoid pigments were mainly scored as defects in flower coloration or seed pigmentation. Genes for the biosynthesis of anthocyanin pigments in flowers of the ornamental plants, snapdragon, petunia, and *Ipomoea* and those of proanthocyanidins in *Arabidopsis* seeds as well as their mutants are well documented. Complete descriptions of the structural and regulatory genes in the pathway are presented in other chapters of this book (Chapters 3 and 4, respectively).

3.1. Anthocyanins and their biosynthetic genes

Anthocyanins are the most abundant flavonoid pigments that get sequestered in vacuoles and are largely responsible for diverse pigmentation from orange to red, purple and blue in flowers and fruits. Often, these compounds also occur in leaves, stems, seeds, and other tissues. There are six major anthocyanidins acting as central chromophores of anthocyanins: pelargonidin, cyanidin, peonidin, delphinidin, petunidin, and malvidin, which differ only in the extent of B-ring hydroxylation and methylation (Figure 6.4). In general, pigments of pelargonidin and cyanidin derivatives produce red and purple color, respectively, whereas those of delphinidin show purple or blue color (Goto and Kondo, 1991). Genes controlling the B-ring hydroxylation of anthocyanidins, *F3'H* and *F3'5'H* encoding flavonoid 3'-hydroxylase and flavonoid 3',5'-hydroxylase, respectively, are key determinants for the synthesis of cyanidin or delphinidin (Figure 6.1). Indeed, *F3'5'H* transgenes confer violet or mauve coloration to carnation and blue pigmentation to rose flowers (Tanaka et al., 2005; http: //www.suntory.com/news/2004/8826.html). Further modifications of anthocyanidins, mainly by *O*-glycosylations and *O*-acylations, result in a large variety of anthocyanins (Honda and Saito, 2002). Anthocyanidin-3-*O*-rutinosides were found in flowers of petunia and snapdragon, whereas anthocyanidin-3-*O*-sophoroside and sambubioside derivatives were detected in flowers of *Ipomoea* and pigmented leaves of *Arabidopsis*, respectively (Martin and Gerats, 1993a; Brugliera et al., 1994; Kroon et al., 1994; Holton and Cornish, 1995; Tohge et al., 2005). The production of anthocyanidin-3-*O*-rutinosides and anthocyanidin-3-*O*-sophorosides is mediated by UDP-rhamnose:anthocyanidin 3-*O*-glucoside-6″-*O*-rhamnosyltransferase (3RT) and UDP-glucose:anthocyanidin 3-*O*-glucoside-2″-*O*-glucosyltransferase (3GGT), which catalyze the addition of a rhamnose or a glucose molecule to anthocyanidin 3-*O*-glucosides, respectively (Brugliera et al., 1994; Kroon et al., 1994; Morita et al., 2005).

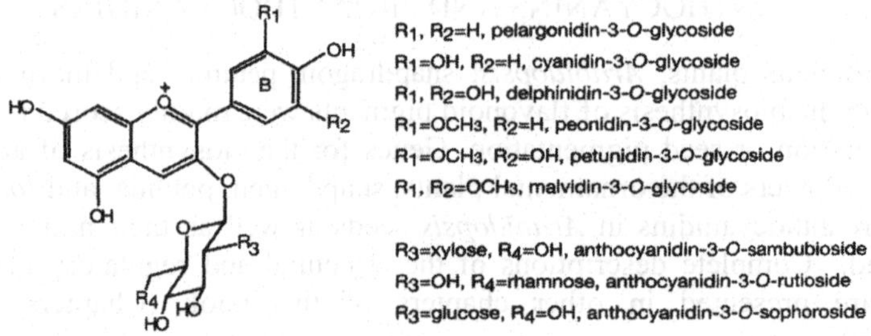

R1, R2=H, pelargonidin-3-O-glycoside
R1=OH, R2=H, cyanidin-3-O-glycoside
R1, R2=OH, delphinidin-3-O-glycoside
R1=OCH3, R2=H, peonidin-3-O-glycoside
R1=OCH3, R2=OH, petunidin-3-O-glycoside
R1, R2=OCH3, malvidin-3-O-glycoside

R3=xylose, R4=OH, anthocyanidin-3-O-sambubioside
R3=OH, R4=rhamnose, anthocyanidin-3-O-rutioside
R3=glucose, R4=OH, anthocyanidin-3-O-sophoroside

Figure 6.4 *Basic structure of anthocyanins. The extent of hydroxylation and methylation of the B-ring, indicated by B, determine the six major anthocyanins.*

Blue and reddish flowers of the morning glories *I. nil* and *I. tricolor* (Figure 6.3) contain peonidin-3-*O*-sophoroside and pelargonidin-3-*O*-sophoroside derivatives with four additional glucose residues and three caffeoyl moieties, which are called Heavenly Blue Anthocyanin (HBA) and Wedding Bell Anthocyanins (WBA), respectively. The latter was found in the *magenta* or *fuchsia* mutants (Table 6.1) (Goto and Kondo, 1991; Hoshino et al., 2003). Not only the structure of HBA but also the vacuolar pH of around 7 or even higher in the open flowers of morning glories plays an important role in blue flower pigmentation (Yoshida et al., 1995; Yamaguchi et al., 2001). The change from reddish-purple buds into blue open flowers correlates with an increase in the vacuolar pH that is mediated by the vacuolar Na^+/H^+ antiporters, InNHX1 and InNHX2 (Fukada-Tanaka et al., 2000; Ohnishi et al., 2005), and the recessive *purple* mutants deficient in the *InNHX1* gene fails to increase vacuolar pH partially during flower opening (Figure 6.3) (Yamaguchi et al., 2001). This contrasts with the petunia flowers, in which the pH values of petal extracts from the wild-type and recessive *ph* mutants are about 5.4 and 6.0, respectively (de Vlaming et al., 1983; van Houwelingen et al., 1998). Some of the petunia *Ph* genes were found to encode transcriptional regulatory proteins controlling both anthocyanin pigmentation and vacuolar pH (Spelt et al., 2002).

3.2. Anthocyanin regulatory genes

Anthocyanin biosynthetic genes can be divided into early and late genes in snapdragon and petunia, whereas those in maize appear to be coordinately controlled as a single module (Martin and Gerats, 1993b; Mol et al., 1998). The regulatory division between early and late genes occurs before and after *F3H* in snapdragon and petunia, respectively (Figure 6.1). Thus, the regulations of the genes in the flavone and flavonol biosynthetic pathways in snapdragon and petunia, respectively, differ from those in their anthocyanin biosynthetic pathways. The similar division after *F3H* also was observed in proanthocyanidin biosynthetic genes in *Arabidopsis* (Nesi et al., 2001; Debeaujon et al., 2003). Although the presence or absence of these regulatory divisions is regarded to stem from the differences between dicots and monocots (Martin and Gerats, 1993b; Mol et al., 1998), available data indicated that the biosynthetic genes in dicotyledonous *Ipomoea* mainly appear to be under coordinate control (Hoshino et al., 1997, 2003; Tiffin et al., 1998; Park et al., 2004; Morita et al., 2005).

Mutations in regulatory genes encoding transcription factors often affect the expression of more than one biosynthetic enzyme genes, since the production of anthocyanins involves the coordinated expression of these genes. These transcriptional regulatory proteins include members of R2R3-MYB, bHLH (basic helix-loop-helix) and WD40 family proteins (Table 6.1), and a combination of the WD40, R2R3-MYB, and bHLH proteins and their molecular interactions determine a set of genes to be expressed (Mol et al., 1998; Winkel-Shirley, 2001b; Koes et al., 2005; Ramsay and Glover, 2005). In *Arabidopsis*, anthocyanin pigmentation can be detected in the hypocotyls of seedlings and in the stems and leaves of aged plants, which can be intensified under various stressed conditions including high light and water stress. Moreover, intense purple coloration can be observed in various vegetative organs by overexpressing the *PAP1* gene encoding an MYB transcription factor (Borevitz et al., 2000). The anthocyanins produced in *Arabidopsis* are cyanidin-3-*O*-sambubioside 5-*O*-glucoside derivatives (Figure 6.4) (Tohge et al., 2005).

In maize the anthocyanins accumulation is regulated by pairs of duplicated transcription factors: R1 and B1, which are bHLH proteins (Ludwig and Wessler, 1990; Goff et al., 1992), while the C1 and PL1 are MYB DNA-binding domain proteins (Cone et al., 1993). Anthocyanin accumulation in the kernel aleurone is controlled by R1 and C1, while B1 and PL1 together are required for anthocyanins in plant parts (Chandler et al., 1989). Recent genetic screens have identified the *pale aleurone color1* (*PAC1*) locus, which functions together with the downstream transcription factors *r1* and *c1* to induce anthocyanins in aleurone and scutellum (Carey et al., 2004). The *pac1* gene encodes a WD40 repeat protein with similarity to AN11 (ANTHOCYANIN11) and TTG1 (TRANSPARENT TESTA GLABRA1) proteins of petunia and *Arabidopsis*, respectively. Comparative studies of *pac1*, *an11*, and *ttg1* mutations and their phenotypes have led to identification of distinct regulatory mechanisms that may control different traits in a species-specific fashion.

Table 6.1 *Flavonoid pigmentation biosynthetic and regulatory genes characterized in different plant species*

Gene products	Gene name	Gene symbols	TE[a] [T-DNA]	References
Maize				
CHS	colorless2	c2	+	(Wienand et al., 1986)
	white pollen	whp		(Franken et al., 1991)
CHI	chalcone isomerase1	chi1		(Grotewold et al., 1994)
F3'H	red aleurone	pr1		(C. Svabek and S. Chopra, unpub)
DFR	anthocyaninless1	a1	+	(O'Reilly et al., 1985)
ANS	anthocyaninless2	a2	+	(Menssen et al., 1990)
UF3GT	bronze1	bz1	+	(Fedoroff et al., 1984)
GST	bronze2	bz2	+	(McLaughlin and Walbot, 1987)
MRP	ZmMRP3			(Goodman et al., 2004)
MYB	colorless1	c1	+	(Cone et al., 1986)
				(Paz-Ares et al., 1986)
	purple plant	pl		(Cone et al., 1993)
	pericarp color	p	+	(Chen et al., 1987)
				(Lechelt et al., 1989)
				(Grotewold et al., 1991)
	yellow seed1 (sorghum)	y1	+	(Chopra et al., 1999)
bHLH	red	r	+	(Dellaporta et al., 1988)
	booster	b		(Chandler et al., 1989)
				(Radicella et al., 1991)
	intensifier	in1	+	(Burr et al., 1996)
WD40	pale aleurone color1	pac1	+	(Carey et al., 2004)
Snapdragon				
CHS	nivea	niv	+	(Wienand et al., 1982)
F3H	incolorata	inc		(Martin et al., 1991)
DFR	pallida	pal	+	(Martin et al., 1985)
ANS	candi	candi		(Martin et al., 1991)
bHLH	dellila	del	+	(Goodrich et al., 1992)
Petunia				
CHS		chsA		(van der Krol et al., 1988)
		chsJ		(Koes et al., 1989)
CHI		po		(van Tunen et al., 1991)
F3H	anthocyanin3	an3	+	(Souer et al., 1995)
F3'H		ht1, 2	+ (ht1)	(Brugliera et al., 1999)
Cyt b5		difF	+	(de Vetten et al., 1999)
F3'5'H		hf1, 2	+ (hf1)	(Holton et al., 1993)
DFR	anthocyanin6	an6	+	(Huits et al., 1994)
UF3GRT		rt	+	(Brugliera et al., 1994)
				(Kroon et al., 1994)
GST	anthocyanin9	an9	+	(Alfenito et al., 1998)
MYB	anthocyanin2	an2	+	(Quattrocchio et al., 1999)
bHLH	anthocyanin1	an1	+	(Spelt et al., 2000)
		jaf13	+	(Quattrocchio et al., 1998)

Table 6.1 (cont.)

Gene products	Gene name		Gene symbols	TE[a] [T-DNA]	References
WD40	*anthocyanin11*		*an11*	+	(de Vetten et al., 1997)
Arabidopsis					
CHS	*TRANSPARENT TESTA4*		*TT4*		(Shirley et al., 1995)
CHI	*TRANSPARENT TESTA5*		*TT5*		(Shirley et al., 1992)
F3H	*TRANSPARENT TESTA6*		*TT6*	+[a]	(Wisman et al., 1998)
F3'H	*TRANSPARENT TESTA7*		*TT7*		(Schoenbohm et al., 2000)
DFR	*TRANSPARENT TESTA3*		*TT3*		(Shirley et al., 1992)
ANS	*TRANSPARENT TESTA18*		*TT18*		(Shikazono et al., 2003)
	TANNIN DEFICENT SEED4		*TDS4*	[+]	(Abrahams et al., 2003)
ANR	*BANYULS*		*BAN*	[+]	(Xie et al., 2003)
UF3GT	*UGT78D2*			[+]	(Tohge et al., 2005)
UF5GT	*UGT75C1*			[+]	(Tohge et al., 2005)
GST	*TRANSPARENT TESTA19*		*TT19*		(Kitamura et al., 2004)
MATE	*TRANSPARENT TESTA12*		*TT12*	[+]	(Debeaujon et al., 2001)
MYB	*TRANSPARENT TESTA2*		*TT2*	[+]	(Nesi et al., 2001)
	PRODUCTION OF ANTHOCYANIN PIGMENT1 PAP1			[+]	(Borevitz et al., 2000)
bHLH	*TRANSPARENT TESTA8*		*TT8*	[+]	(Nesi et al., 2000)
	GLABRA3		*GL3*		(Payne et al., 2000)
	ENHANCER OF GLABRA3		*EGL3*		(Zhang et al., 2003a)
WD40	*TRANSPARENT TESTA GLABRA1*		*TTG1*		(Walker et al., 1999)
Morning Glory					
CHS	*R1*	(N)	*R1*	+	(Iida et al., 2004)
	Anthocyanin	(P)	*A*	+	(Habu et al., 1998)
CHI	*Speckled* or *Cream*	(N)	*Sp, Cr*	+	(Iida et al., 2004)
F3'H	*Magenta*	(N)	*Mg*		(Hoshino et al., 2003)
	Pink	(P)	*P*	+	(Hoshino et al., 2003)
	Fuchsia	(T)	*Fuchsia*		(Hoshino et al., 2003)
DFR	*A3*	(N)	*A3*	+	(Inagaki et al., 1994)
	Pearly	(T)	*Pearly*	+	(Iida et al., 2004)
ANS	*R3*	(N)	*R3*	+	(Y. Morita et al., unpublished)
UF3GT	*Duskish*	(N)	*Dk*	+	(Y. Morita et al., unpublished)
UF3GGT	*Dusky*	(N)	*Dy*		(Morita et al., 2005)
MYB	*C1*	(N)	*C1*		(Y. Morita et al., unpublished)
	W	(P)	*W*		(M. Rauscher, pers. comm.)

Table 6.1 (cont.)

Gene products	Gene name		Gene symbols	TE[a] [T-DNA]	References
bHLH	*Ivory seed*	(P)	*IVS*	+	(K. I. Park et al., unpublished)
	Ivory seed	(T)	*IVS*		(Park et al., 2004)
WD40	*Ca*	(N)	*Ca*		(Y. Morita et al., unpublished)

The symbols + and [+] in the column TE [T-DNA] indicate the presence of an allele generated by insertion of an endogenous transposon among the characterized alleles that may not be cited here and the presence of a T-DNA tagged allele among the first characterized alleles in the literature cited, respectively. The symbol +[a] indicates that the alleles in the original literature cited were generated by insertion of the maize *En-1* transposon. (N), (P), and (T) in the morning glories indicate *I. nil*, *I. purpurea*, and *I. tricolor*, respectively.

3.3. Proanthocyanidins Biosynthesis and Regulation

Proanthocyanins or the condensed tannins in *Arabidopsis* are found only in the seed coat, or testa, and confer brown pigmentation in mature seeds due to their oxidation. Therefore, it is not surprising that most of the genes for anthocyanin biosynthesis were identified on the basis of mutations that abolish or reduce pigmentation in the seed coat, which include *TRANSPARENT TESTA* (*TT*) and *TRANSPARENT TESTA GLABRA* (Winkel-Shirley, 2001a, 2001b). Both structural and regulatory genes were found in the *tt* mutant class, whereas only regulatory genes were identified in the *ttg* class (Table 6.1). Some mutations in the *TT* genes were obtained by screening phenotypes with no pigmentation in stems and leaves (Shikazono et al., 2003). In addition to the regulatory proteins in Table 6.1, transcription factors controlling the proanthocyanidin accumulation at the later stages have been identified: TT1, TT16, and TTG2 proteins belonging to the WIP, MADS, and WRKY family, respectively (Debeaujon et al., 2003). Two alternative pathways lead to proanthocyanidins (Figure 6.1): (1) anthocyanidins are converted into 2,3-*trans*-flavon-3-ols by leucoanthocyanidin reductase (LAR), the cDNA of which was isolated from proanthocyanidins-rich leaves of the legume *Desmodium uncinatum* (Tanner et al., 2003), and (2) anthocyanidins are converted into 2,3-*cis*-flavon-3-ols by anthocyanidin reductase (ANR), which is encoded by the *BAN* gene in *Arabidopsis* (Xie et al., 2003). The seed phenotypes of *tds4* and *ban tds4* double mutants also indicated that anthocyanidin, actually cyanidin, is an intermediate for proanthocyanidin biosynthesis in *Arabidopsis* (Abrahams et al., 2003; Tanner, 2004).

4. FLAVONOID PIGMENTS AS A TOOL TO UNDERSTAND GENETIC AND EPIGENETIC REGULATION

The use of flavonoid pigmentation as a visual and molecular marker has helped in the cloning of genes as well as in unraveling mechanisms of regulation of plant gene expression. The unstable *P1-vv* (variegated pericarp, variegated cob glume) allele was the first mutable allele studied by maize geneticists (Emerson, 1917). Further genetic studies of the *P1-vv* allele led to the identification of a transposable element originally designated as *Mp* (*Modulator of pericarp*) (Brink and Nilan, 1952). Genetic and molecular studies later demonstrated the *Mp* element to be the same as *Ac* (*Activator*). Classical studies of twin sectors on *P1-vv* ears led to the findings that transposition of *Ac* often takes place during DNA replication and the majority of insertions are in the linked sites (Greenblatt and Brink, 1963; Greenblatt, 1984). In recent years, molecular genetic studies have used *P1-vv*-induced variegated phenotypes as a convenient marker to establish transposition mechanisms by easily following the pigmentation of the *Ac* donor site (Chen et al., 1987; Athma et al., 1992; Moreno et al., 1992; Dellaporta and Moreno, 1994; Brutnell, 2002; Kolkman et al., 2005).

Among the plants listed in Table 6.1, maize, snapdragon, petunia, and morning glories have a longer history in breeding and/or classical genetic studies than *Arabidopsis* (Coe and Neuffer, 1977; Iida et al., 1999; Coe, 2001; Schwarz-Sommer et al., 2003; Iida et al., 2004; Gerats and Vandenbussche, 2005), and flower pigmentation is one of the most important traits in breeding of ornamental plants. Because in early times the breeders employed spontaneous mutants as sources of new traits, it is obvious that a significant number of spontaneous mutations were caused by insertion of endogenous transposons. Moreover, mutable alleles conferring variegated phenotypes fascinated many geneticists including Hugo de Vries, who is one of the rediscoverers of the Mendel's laws and systematically studied the inheritance of flower variegations of snapdragon in the 1890s (de Vries, 1903). Indeed, various mutable alleles of maize, snapdragon, petunia, and *Ipomoea* were described and characterized (Nevers et al., 1986), and most of them are believed to be caused by excision of DNA transposons integrated at the mutable loci. Some of these mutants in petunia were isolated by transposon mutagenesis (Koes et al., 1995; van Houwelingen et al., 1998). In addition, several mutations in *Ipomoea* including the *dusky* alleles are likely to be footprints generated by excision of DNA transposable elements (Morita et al., 2005). The situations in these plants clearly contrast with that in *Arabidopsis*, in which many insertion mutations are obtained as T-DNA tagged lines (see Table 6.1).

4.1. Trapping of New Transposons

The cloning and characterization of DNA transposons was first achieved in the mutable *waxy* and *nivea* loci of maize and snapdragon, respectively, and subsequent successful transposon tagging of the maize *Bz1* (*3GT*) and snapdragon *pallida* (*DFR*) loci were performed (Coen et al., 1989; Fedoroff, 1989). In snapdragon, the *nivea* alleles give albino flowers, whereas the *pallida* alleles give ivory flowers with a yellow aurone pigmented area around the lower lip of the corolla, because aurone is derived from chalcone (Figure 6.1). The DNA transposon *Tam3*, residing at the mutable *pallida* locus and which could probably be traced back to de Vries's studies, had been activated and transposed into the already known (or cloned) *nivea* (*CHS*) locus and identified as a newly inserted DNA segment (termed transposon trapping). The mutable *pallida* locus, where the original *Tam3* was integrated, was then identified by using the information of the cloned *Tam3* sequence (termed transposon tagging).

One of the first successful transposon taggings in heterologous plants corresponded to the petunia *Ph6* gene by the maize DNA transposon *Ac*. The resulting insertion displays variegated flower and seed color (Chuck et al., 1993). The mutable *Ph6* allele, deficient in the vacuolar pH control, was later found to be a particular allele of the *AN1* gene encoding an bHLH transcription factor (Spelt et al., 2002). A full description of AN1 and the control of vacuolar pH is found in Chapter 4. Thus, anthocyanin biosynthetic genes played very important roles in establishing trapping of transposons and early demonstration of gene tagging as a powerful molecular genetic tool.

4.2. Identification of Active Transposable Elements in Sorghum and Morning Glories

In sorghum, the high frequency of somatic and germinal reversions of *Y1-cs* to *Y1-rr* led to the hypothesis that the *Y1-cs* phenotype may be the consequence of the presence of a transposable element in the *Y1* gene (Zanta et al., 1994), bearing a marked resemblance to the maize *P1-vv* allele (Chopra et al., 1999). Making use of the similarities in the phlobaphene accumulation patterns, the *Y1-cs* allele allowed the identification of the only known active transposable element *Candystripe1* (*Cs1*) in sorghum. The *Cs1* transposon belongs to the *Enhancer/Suppressor-mutator* (*En/Spm*) or the CACTA super family (Kunze and Weil, 2002) and is 23 kb in size, a property similar to other members of this family. The red-white sectors or the variegated pericarp phenotype result from the somatic excisions of the *Cs1* element from *Y1*, while germinal excisions of the *Cs1* generate heritably functional alleles. Similar to maize *P1-vv*, exploitation of the sorghum variegated pigmentation phenotype as a marker (Figure 6.2) has further allowed the analysis of somatic and germinal excision events of the *Cs1* transposon in sorghum. By applying the traditional approach of monitoring the variegated or full red phenotype of the donor *Cs1* site it has been possible to identify new putative *Cs1*-mutations in sorghum (Carvalho et al., 2005).

About one third of more than 200 spontaneous mutants described in the Japanese morning glory (*I. nil*) exhibit mainly altered flower pigmentation patterns, including several mutable loci conferring flower variegations (Iida et al., 1999). One of them called *flecked* (Figure 6.3) was believed to have appeared in the early 18th century in Japan, and several woodblock prints of its variegated flowers were made in the middle of the 18th century. The mutable *flecked* allele was found to be caused by a 6.4-kb transposon *Tpn1* inserted into the second intron of the *DFR-B* gene (Inagaki et al., 1994). It was later established that the majority of the *I. nil* mutations, including the mutable *speckled* and *purple* alleles (Table 6.1), were caused by insertion of nonautonomous DNA transposons belonging to the *Tpn1* family, in the *En/Spm* superfamily (Hoshino et al., 2001; Iida et al., 2004). Interestingly, all of the *Tpn1* family elements identified so far contain various foreign sequences with genomic exons and introns (Kawasaki and Nitasaka, 2004). Insertion of *Tpn1* into an intron of the *DFR-B* gene resulted in the production of chimeric mRNA molecules consisting of *DFR-B* exons and foreign exon sequences within *Tpn1*, thereby inactivating *DFR-B* (Takahashi et al., 1999). Indeed, several identified mutable alleles conferring flower variegations contain the *Tpn1* family elements inserted into intron sequences of anthocyanin biosynthetic genes (A. Hoshino et al., unpublished). Most of the spontaneous mutations, including mutable alleles giving flower variegations in other morning glories (*I. purpurea* and *I. tricolor*) were shown to be caused by various transposons, most of which belong to the *Ac/Ds* (or hAT) superfamily (Kunze and Weil, 2002) or are *Mu*-related elements (Iida et al., 1999, 2004; Hoshino et al., 2001, 2003; A. Hoshino, unpublished). Interestingly, the mutable allele *ivory seed-variegated* (*ivs-v*) in *I. tricolor*, which exhibits pale-blue flowers with a few fine blue spots and ivory seeds with tiny dark-brown spots (Figure 6.3), is caused by an intragenic tandem duplication within the gene for a bHLH transcriptional regulator, and somatic homologous recombination between the tandem duplication results in the pigmented spots (Park et al., 2004).

4.3. Flavonoid pigments as markers in epigenetic studies

Since Mendel's studies on genetic inheritance, pigmentation by anthocyanins and phlobaphenes has been used as an important trait in elucidating genetic and epigenetic phenomena. Barbara McClintock employed the mutations in the maize pigmentation genes *C1, P1, A1, A2, Pr, and Bz1* (McClintock, 1950, 1951, 1956) to characterize the transposons *Ac, Ds* and *Spm*. She called them "controlling elements," because they were distinct from the chromosomal genes themselves and could alter or modulate gene expression reversibly, independent of their chromosomal locations. The gene regulations she had characterized were later realized to include both genetic and epigenetic gene regulations (Fedoroff, 1989; Feschotte et al., 2002; Lippman and Martienssen, 2004). Epigenetics has been defined as "The study of mitotically and/or meiotically heritable changes in gene function that cannot be explained by changes in DNA sequence" (Russo et al., 1996). Epigenetic modification may be observed as DNA methylation defects and/or chromatin modifications in plants (Lippman and Martienssen, 2004; Matzke and Birchler, 2005). We briefly review the initial observation in alterations of the pigmentation, which have led to the discovery and elucidation of epigenetic gene regulation.

*4.3.1. Transgenes and transposons: Discovery of gene silencing through flavonoid
pigmentation patterns*

Introduction of the maize *DFR* cDNA into a petunia *dfr* mutant resulted in
transgenic plants displaying pelargonidin-based brick-red flowers as well as
variegated flowers with brick-red sectors (Meyer et al., 1987). The flower
variegation was caused by transcriptional gene silencing (TGS) due to DNA
methylation of the promoter of the *DFR* transgene (Meyer, 1995). Subsequently,
posttranscriptional gene silencing (PTGS) of transgenes and endogenous genes were
reported in the petunia *CHS* and *DFR* genes (Napoli et al., 1990; van der Krol et al.,
1990). The generation of these variegated flowers is one of the earliest examples of
sequence-specific RNA-mediated gene silencing, the mechanisms of which have
become one of the central topics in epigenetic gene regulation (Lippman and
Martienssen, 2004; Matzke and Birchler, 2005). Epigenetic modifications have been
shown recently to involve RNA interference and heterochromatin, which in turn are
influenced by invasive DNAs such as transgenes and transposable elements
(Lippman et al., 2004; Lippman and Martienssen, 2004). The flower variegation
patterns caused by the excisions of DNA transposons in *Ipomoea* are often heritable
and they are believed to be regulated epigenetically (Iida et al., 2004). Transposons
and other repetitive sequences are able to regulate adjacent genes epigenetically
(Lippman et al., 2004). A mutant of *I. tricolor* exhibiting variegated flowers was
found to carry a 0.4-kb transposon integrated into the promoter of the *DFR-B* gene,
and the integrated transposon appears to affect DNA methylation of a particular site
within the *DFR-B* promoter (Iida et al., 2004; J.D. Choi and A. Hoshino,
unpublished). Thus, the genetic change caused by the transposon integration results
in a situation in which the epigenetic gene regulation can play important roles in
flower pigmentation (Iida et al., 2004). It is likely that both genetic and epigenetic
regulation associated with transposons can cooperatively act on the generation of
flower pigmentation patterns.

4.3.2. Identification of modifiers that participate in maintenance of gene expression

A classic example of epigenetic regulation is paramutation, which was first
described by R.A. Brink using the expression of the *R* locus in maize kernels (Brink,
1958; Brink et al., 1968). Paramutationlike phenomena now have been described in
several systems including regulatory genes (Chandler and Stam, 2004). In the maize
anthocyanin pathway, recessive *mop1* (*mediator of paramuation1*) and *rmr*
(*required to maintain repression*) mutations are recent demonstrations of the power
of flavonoid pigment-based screens to identify factors participating in the epigenetic
regulation of gene expression (Dorweiler et al., 2000; Hollick et al., 2000). In
addition to anthocyanins, maize phlobaphenes offer a powerful marker to follow
gene expression analysis because of the readily observable pericarp pigmentation in
the kernel or ectopic expression in the plant body (Cocciolone et al., 2000, 2001).
Gene expression studies using *P1-rr* and *P1-wr* alleles have led to the identification
of epigenetic regulatory mechanisms that may play a significant role in generating
tissue specific expression patterns (Chopra et al., 1996, 1998). In 1960, the natural

ectopic expression of phlobaphenes allowed Charles Burnham to identify a spontaneous dominant modifier of maize, which later became known as *Unstable factor for orange1* (*Ufo1*) (Chopra et al., 2003). *Ufo1* modifies the organ-specific expression of the *P1-wr* allele and an enhanced pigmentation state of the tissue is correlated with decrease in DNA methylation of the *P1-wr* sequence (Figure 6.2). The nature of the *Ufo1* mutation is not known but it has now been mapped to maize chromosome 10S (Chopra et al., 2003). Interestingly, *Ufo1* has poor expressivity and may interact differently with single copy versus multicopy hyper- and hypomethylated alleles (R. Sekhon and S. Chopra, unpublished).

5. CONCLUSIONS

It is clear that flower pigmentation attracts not only pollinators but also many plant biologists including geneticists. Because mutations in flavonoid pigmentation genes are nonlethal and confer easily scorable color phenotypes in stems, flowers, and seeds, various flavonoid pigments, including the anthocyanins, proanthocyanidins, and phlobaphenes described here, have served as important traits in elucidating genetic and epigenetic phenomena. Undoubtedly, these secondary metabolites will keep on making important and significant contributions as useful tools for molecular genetic studies in the future.

6. ACKNOWLEDGMENTS

We thank Rajandeep Sekhon and Catherine Svabek for their critical comments and suggestions. The Penn State University research project was supported by grants from USDA-NRI (2002-35318-12676) and NSF (MCB-0416425). The work in Okazaki was supported by grants from the Ministry of Education, Culture, Sports, Science and Technology in Japan. S. I and A. H thanks Yasumasa Morita and Kyeung-Il Park for discussion.

7. REFERENCES

Abrahams, S., Lee, E., Walker, A. R., Tanner, G. J., Larkin, P. J. and Ashton, A. R., 2003, The *Arabidopsis TDS4* gene encodes leucoanthocyanidin dioxygenase (LDOX and is essential for proanthocyanidin synthesis and vacuole development, *Plant J* **35**: 624-636.

Alfenito, M. R., Souer, E., Goodman, C. D., Buell, R., Mol, J., Koes, R. and Walbot, V., 1998, Functional complementation of anthocyanin sequestration in the vacuole by widely divergent glutathione *S*-transferases, *Plant Cell* **10**: 1135-1149.

Athma, P., Grotewold, E. and Peterson, T., 1992, Insertional mutagenesis of the maize *P* gene by intragenic transposition of *Ac*, *Genetics* **131**: 199-209.

Bieza, K. and Lois, R., 2001, An *Arabidopsis* mutant tolerant to lethal ultraviolet-B levels shows constitutively elevated accumulation of flavonoids and other phenolics, *Plant Physiol* **126**: 1105-1115.

Boddu, J., Svabek, C., Sekhon, R., Gevens, A., Nicholson, R., Jones, D., J., P., Gustine, D. and Chopra, S., 2004, Expression of a putative flavonoid 3'-hydroxylase in sorghum mesocotyls synthesizing 3-deoxyanthocyanidin phytoalexins, *Physiol Mol Plant Path* **65**: 101-113.

Boddu, J., Jiang, C., Sangar, V., Olson, T., Peterson, T. and Chopra, S., 2005a, Comparative structural and functional characterization of sorghum and maize duplications containing orthologous Myb transcription regulators of 3-deoxyflavonoid biosynthesis, *Plant Mol Biol*, in press.

Boddu, J., Svabek, C., Ibraheem, F., Jones, A. D. and Chopra, S., 2005b, Characterization of a deletion allele of a sorghum Myb gene *yellow seed1* showing loss of 3-deoxyflavonoids, *Plant Sci*, in press.

Borevitz, J. O., Xia, Y., Blount, J., Dixon, R. A. and Lamb, C., 2000, Activation tagging identifies a conserved MYB regulator of phenylpropanoid biosynthesis, *Plant Cell* **12**: 2383-2394.

Brink, R. A. and Nilan, R. A., 1952, The relation between light variegated and medium variegated pericarp in maize, *Genetics* **37**: 519-544.

Brink, R. A., 1958, Paramutation at the *R* locus in maize, *Cold Spring Harb Symp Quant Biol* **23**: 379-391.

Brink, R. A. and Styles, E. D., 1966, A collection of pericarp factors, *Maize Genet Coop News* **40**: 149-160.

Brink, R. A., Styles, E. D. and Axtell, J. D., 1968, Paramutation: directed genetic change. Paramutation occurs in somatic cells and heritably alters the functional state of a locus, *Science* **159**: 161-170.

Brugliera, F., Holton, T. A., Stevenson, T. W., Farcy, E., Lu, C. Y. and Cornish, E. C., 1994, Isolation and characterization of a cDNA clone corresponding to the *Rt* locus of *Petunia hybrida*, *Plant J* **5**: 81-92.

Brugliera, F., Barri-Rewell, G., Holton, T. A. and Mason, J. G., 1999, Isolation and characterization of a flavonoid 3'-hydroxylase cDNA clone corresponding to the *Ht1* locus of *Petunia hybrida*, *Plant J* **19**: 441-451.

Brutnell, T. P., 2002, Transposon tagging in maize, *Funct Integr Genomics* **2**: 4-12.

Burr, F. A., Burr, B., Scheffler, B. E., Blewitt, M., Wienand, U. and Matz, E. C., 1996, The maize repressor-like gene *intensifier1* shares homology with the *r1/b1* multigene family of transcription factors and exhibits missplicing, *Plant Cell* **8**: 1249-1259.

Byrne, P. F., McMullen, M. D., Snook, M. E., Musket, T. A., Theuri, J. M., Widstrom, N. W., Wiseman, B. R. and Coe, E. H., 1996, Quantitative trait loci and metabolic pathways: genetic control of the concentration of maysin, a corn earworm resistance factor, in maize silks, *Proc Natl Acad Sci USA* **93**: 8820-8825.

Carey, C. C., Strahle, J. T., Selinger, D. A. and Chandler, V. L., 2004, Mutations in the *pale aleurone color1* regulatory gene of the *Zea mays* anthocyanin pathway have distinct phenotypes relative to the functionally similar *TRANSPARENT TESTA GLABRA1* gene in *Arabidopsis thaliana*, *Plant Cell* **16**: 450-464.

Carvalho, C. H. S., Boddu, J., Zehr, U. B., Axtell, J. D., Pedersen, J. F. and Chopra, S., 2005, Genetics and molecular chracterization of *Candystripe1* transposition events in sorghum, *Genetica*, in press.

Chandler, V. L., Radicella, J. P., Robbins, T. P., Chen, J. and Turks, D., 1989, Two regulatory genes of the maize anthocyanin pathway are homologous: isolation of *B* utilizing *R* genomic sequences, *Plant Cell* **1**: 1175-1183.

Chandler, V. L. and Stam, M., 2004, Chromatin conversations: mechanisms and implications of paramutation, *Nat Rev Genet* **5**: 532-544.

Chen, J., Greenblatt, I. M. and Dellaporta, S. L., 1987, Transposition of *Ac* from the *P* locus of maize into unreplicated chromosomal sites, *Genetics* **117**: 109-116.

Chopra, S., Athma, P. and Peterson, T., 1996, Alleles of the maize *P* gene with distinct tissue specificities encode Myb-homologous proteins with C-terminal replacements, *Plant Cell* **8**: 1149-1158.

Chopra, S., Athma, P., Li, X. G. and Peterson, T., 1998, A maize Myb homolog is encoded by a multicopy gene complex, *Mol Gen Genet* **260**: 372-380.

Chopra, S., Brendel, V., Zhang, J., Axtell, J. D. and Peterson, T., 1999, Molecular characterization of a mutable pigmentation phenotype and isolation of the first active transposable element from *Sorghum bicolor*, *Proc Natl Acad Sci USA* **96**: 15330-15335.

Chopra, S., Gevens, A., Svabek, C., Wood, K. V., Peterson, T. and Nicholson, R. L., 2002, Excision of the *Candystripe1* transposon from a hyper-mutable *Y1-cs* allele shows that the sorghum *y1* gene control the biosynthesis of both 3-deoxyanthocyanidin phytoalexins and phlobaphene pigments, *Physiol Mol Plant Path* **60**: 321-330.

Chopra, S., Cocciolone, S. M., Bushman, S., Sangar, V., McMullen, M. D. and Peterson, T., 2003, The maize *Unstable factor for orange1* is a dominant epigenetic modifier of a tissue specifically silent allele of *pericarp color1*, *Genetics* **163**: 1135-1146.

Chuck, G., Robbins, T., Nijjar, C., Ralston, E., Courtney-Gutterson, N. and Dooner, H. K., 1993, Tagging and cloning of a petunia flower color gene with the maize transposable element *Activator*, *Plant Cell* **5**: 371-378.

Cocciolone, S. M., Sidorenko, L. V., Chopra, S., Dixon, P. M. and Peterson, T., 2000, Hierarchical patterns of transgene expression indicate involvement of developmental mechanisms in the regulation of the maize *P1-rr* promoter, *Genetics* **156**: 839-846.

Cocciolone, S. M., Chopra, S., Flint-Garcia, S. A., McMullen, M. D. and Peterson, T., 2001, Tissue-specific patterns of a maize *Myb* transcription factor are epigenetically regulated, *Plant J* **27**: 467-478.

Cocciolone, S. M., Nettleton, D., Snook, M. and Peterson, T., 2005, Transformation of maize with the *p1* transcription factor directs production of silk maysin, a corn earworm resistance factor, in concordance with a hierarchy of floral organ pigmentation, *Plant Biotech* **3**: 225-235.

Coe, E. H. and Neuffer, M. G., 1977, The genetics of corn. In G. F. Sprague (Ed.), *Corn and Corn Improvement* (pp. 111-223). Madison, WI: American Society of Agronomy.

Coe, E. H., Jr., 2001, The origins of maize genetics, *Nat Rev Genet* **2**: 898-905.

Coen, E. S., Robbins, T. P., Almeida, J., Hudson, A. and Carpenter, R., 1989, Consequences and mechanisms of transposition in *Antirrhinum majus*. In D. E. Berg and M. M. Howe, eds, *Mobile DNA* (pp. 413-436). Washington: American Society for Microbiology.

Cone, K. C., Burr, F. A. and Burr, B., 1986, Molecular analysis of the maize anthocyanin regulatory locus *C1*, *Proc Natl Acad Sci. USA* **83**: 9631-9635.

Cone, K. C., Cocciolone, S. M., Burr, F. A. and Burr, B., 1993, Maize anthocyanin regulatory gene *pl* is a duplicate of *c1* that functions in the plant, *Plant Cell* **5**: 1795-1805.

Debeaujon, I., Peeters, A. J. M., Leon-Kloosterziel, K. M. and Koornneef, M., 2001, The *TRANSPARENT TESTA12* gene of *Arabidopsis* encodes a multidrug secondary transporter-like protein required for flavonoid sequestration in vacuoles of the seed coat endothelium, *Plant Cell* **13**: 853-871.

Debeaujon, I., Nesi, N., Perez, P., Devic, M., Grandjean, O., Caboche, M. and Lepiniec, L., 2003, Proanthocyanidin-accumulating cells in *Arabidopsis* testa: regulation of differentiation and role in seed development, *Plant Cell* **15**: 2514-2531.

Dellaporta, S. L., Greenblatt, I., Kermicle, J., Hicks, J. B. and Wessler, S., 1988, Molecular cloning of the maize *R-nj* allele by transposon tagging with *Ac*, *Stadler Genet. Symp* **18**: 263-282.

Dellaporta, S. L. and Moreno, M. A., 1994, Gene tagging with *Ac/Ds* elements in maize. In M. Freeling and V. Walbot, eds, *The Maize Handbook* (pp. 219-233). New York: Springer-Verlag.

de Vetten, N., Quattrocchio, F., Mol, J. and Koes, R., 1997, The *an11* locus controlling flower pigmentation in petunia encodes a novel WD-repeat protein conserved in yeast, plants, and animals, *Genes Dev* **11**: 1422-1434.

de Vetten, N., ter Horst, J., van Schaik, H. P., de Boer, A., Mol, J. and Koes, R., 1999, A cytochrome b_5 is required for full activity of flavonoid 3', 5'-hydroxylase, a cytochrome P450 involved in the formation of blue flower colors, *Proc Natl Acad Sci USA* **96**: 778-783.

de Vlaming, P., Schram, A. W. and Wiering, H., 1983, Genes affecting flower colour and pH of flower limb homogenates in *Petunia hybrida*, *Theor Appl Genet* **66**: 271-278.

de Vries, H., 1903, *Die Mutationstheorie 2*. Leipzig: von Veit u. Co.

Dixon, R. A. and Steele, C. L., 1999, Flavonoids and isoflavonoids—a gold mine for metabolic engineering, *Trends Plant Sci* **4**: 394-400.

Dooner, H. K. and Nelson, O. E., 1977, Genetic control of UDPglucose: flavonol 3-O-glucosyltransferase in the endosperm of maize, *Biochem Genet* **15**: 509-519.

Dooner, H. K., Robbins, T. P. and Jorgensen, R. A., 1991, Genetic and developmental control of anthocyanin biosynthesis, *Annu Rev Genet* **25**: 173-199.

Dorweiler, J. E., Carey, C. C., Kubo, K. M., Hollick, J. B., Kermicle, J. L. and Chandler, V. L., 2000, *Mediator of paramutation1* is required for establishment and maintenance of paramutation at multiple maize loci, *Plant Cell* **12**: 2101-2118.

Emerson, R. A., 1917, Genetical studies of variegated pericarp in maize, *Genetics* **2**: 1-35.

Fedoroff, N. V., Furtek, D. and Nelson, O. E., 1984, Cloning of the *bronze* locus in maize by a simple and generalizable procedure using the transposable controlling element *Activator* (*Ac*), *Proc Natl Acad Sci USA* **81**: 3825-3829.

Fedoroff, N. V., 1989, Maize transposable elements. In D. E. Berg and M. M. Howe, eds, *Mobile DNA* (pp. 375-411). Washington: American Society for Microbiology.

Feschotte, C., Jiang, N. and Wessler, S. R., 2002, Plant transposable elements: where genetics meets genomics, *Nat Rev Genet* 3: 329-341.

Foo, L. Y. and Karchesy, J. J., 1989, Chemical nature of phlopaphene. In R. W. Hemingwa and J. J. Karchesy, eds, *Chemistry and Significance of Condensed Tannins* (pp. 109-118). New York and London: Plenum Press.

Franken, P., Niesbach-Klosgen, U., Weydemann, U., Marechal-Drouard, L., Saedler, H. and Wienand, U., 1991, The duplicated chalcone synthase genes *C2* and *Whp* (*white pollen*) of *Zea mays* are independently regulated; evidence for translational control of *Whp* expression by the anthocyanin intensifying gene *in*, *EMBO J* 10: 2605-2612.

Fukada-Tanaka, S., Inagaki, Y., Yamaguchi, T., Saito, N. and Iida, S., 2000, Colour-enhancing protein in blue petals, *Nature* 407: 581.

Gerats, T. and Vandenbussche, M., 2005, A model system for comparative research: *Petunia, Trends Plant Sci* 10: 251-256.

Goff, S. A., Cone, K. C. and Chandler, V. L., 1992, Functional analysis of the transcriptional activator encoded by the maize *B* gene: evidence for a direct functional interaction between two classes of regulatory proteins, *Genes Dev* 6: 864-875.

Goodman, C. D., Casati, P. and Walbot, V., 2004, A multidrug resistance-associated protein involved in anthocyanin transport in *Zea mays, Plant Cell* 16: 1812-1826.

Goodrich, J., Carpenter, R. and Coen, E. S., 1992, A common gene regulates pigmentation pattern in diverse plant species, *Cell* 68: 955-964.

Goto, T. and Kondo, T., 1991, Structure and molecular stacking of anthocyanins — Flower color variation, *Angrew Chem Int Ed Engl* 30: 17-33.

Greenblatt, I. M. and Brink, R. A., 1963, Transpositions of *Modulator* in maize into divided and undivided chromosome segments, *Nature* 197: 412-413.

Greenblatt, I. M., 1984, A chromosome replication pattrern deduced from pericarp phenotypes resulting from movements of the transposable element *Modulator*, in maize, *Genetics* 108: 471-485.

Grotewold, E., Athma, P. and Peterson, T., 1991, Alternatively spliced products of the maize *P* gene encode proteins with homology to the DNA-binding domain of *myb*-like transcription factors, *Proc Natl Acad Sci USA* 88: 4587-4591.

Grotewold, E., Drummond, B. J., Bowen, B. and Peterson, T., 1994, The *myb*-homologous *P* gene controls phlobaphene pigmentation in maize floral organs by directly activating a flavonoid biosynthetic gene subset, *Cell* 76: 543-553.

Grotewold, E., Chamberlin, M., Snook, M., Siame, B., Butler, L., Swenson, J., Maddock, S., Clair, G. S. and Bowen, B., 1998, Engineering secondary metabolism in maize cells by ectopic expression of transcription factors, *Plant Cell* 10: 721-740.

Habu, Y., Hisatomi, Y. and Iida, S., 1998, Molecular characterization of the mutable *flaked* allele for flower variegation in the common morning glory, *Plant J* 16: 371-376.

Harborne, J. B. and Williams, C. A., 2000, Advances in flavonoid research since 1992, *Phytochem* 55: 481-504.

Hollick, J. B., Patterson, G. I., Asmundsson, I. M. and Chandler, V. L., 2000, Paramutation alters regulatory control of the maize *pl* locus, *Genetics* 154: 1827-1838.

Holton, T. A., Brugliera, F., Lester, D. R., Tanaka, Y., Hyland, C. D., Menting, J. G., Lu, C. Y., Farcy, E., Stevenson, T. W. and Cornish, E. C., 1993, Cloning and expression of cytochrome P450 genes controlling flower colour, *Nature* 366: 276-279.

Holton, T. A., and Cornish, E. C., 1995, Genetics and biochemistry of anthocyanin biosynthesis, *Plant Cell* 7: 1071-1083.

Honda, T. and Saito, N., 2002, Recent progress in the chemistry of polyacylated anthocyanins as flower color pigments, *Heterocycles* 56: 633-692.

Hoshino, A., Abe, Y., Saito, N., Inagaki, Y. and Iida, S., 1997, The gene encoding flavanone 3-hydroxylase is expressed normally in the pale yellow flowers of the Japanese morning glory carrying the *speckled* mutation which produce neither flavonol nor anthocyanin but accumulate chalcone, aurone and flavanone, *Plant Cell Physiol* 38: 970-974.

Hoshino, A., Johzuka-Hisatomi, Y. and Iida, S., 2001, Gene duplication and mobile genetic elements in the morning glories, *Gene* 265: 1-10.

Hoshino, A., Morita, Y., Choi, J. D., Saito, N., Toki, K., Tanaka, Y. and Iida, S., 2003, Spontaneous mutations of the *flavonoid 3'-hydroxylase* gene conferring reddish flowers in the three morning glory species, *Plant Cell Physiol* 44: 990-1001.

Huits, H. S. M., Gerats, A. G. M., Kreike, M. M., Mol, J. N. M. and Koes, R. E., 1994, Genetic control of dihydroflavonol 4-reductase gene expression in *Petunia hybrida*, *Plant J* 6: 295-310.

Iida, S., Hoshino, A., Johzuka-Hisatomi, Y., Habu, Y. and Inagaki, Y., 1999, Floricultural traits and transposable elements in the Japanese and common morning glories, *Annal New York Acad Sci* 870: 265-274.

Iida, S., Morita, Y., Choi, J. D., Park, K. I. and Hoshino, A., 2004, Genetics and epigenetics in flower pigmentation associated with transposable elements in morning glories, *Adv Biophys.* 38: 141-159.

Inagaki, Y., Hisatomi, Y., Suzuki, T., Kasahara, K. and Iida, S., 1994, Isolation of a *Suppressor-mutator/Enhancer*-like transposable element, *Tpn1*, from Japanese morning glory bearing variegated flowers, *Plant Cell* 6: 375-383.

Kambal, A. E. and Bate-Smith, E. C., 1976, A genetic and biochemical study on pericarp pigmentation between two cultivars of grain sorghum, *Sorghum bicolor*, *Heredity* 37: 417-421.

Kawasaki, S. and Nitasaka, E., 2004, Characterization of *Tpn1* family in the Japanese morning glory: *En/Spm*-related transposable elements capturing host genes, *Plant Cell Physiol* 45: 933-944.

Kirby, L. T. and Styles, E. D., 1970, Flavonoids associated with specific gene action in maize aleurone, and the role of light in substituting for the action of a gene. [Corn], *Can J Genet Cytol* 12: 934-940.

Kitamura, S., Shikazono, N. and Tanaka, A., 2004, *TRANSPARENT TESTA 19* is involved in the accumulation of both anthocyanins and proanthocyanidins in *Arabidopsis*, *Plant J* 37: 104-114.

Koes, R. E., Spelt, C. E. and Mol, J. N. M., 1989, The chalcone synthase multigene family of *Petunia hybrida* (V30): differential, light-regulated expression during flower development and UV light induction, *Plant Mol Biol* 12: 213-225.

Koes, R. E., Souer, E., van Houwelingen, A., Mur, L., Spelt, C., Quattrocchio, F., Wing, J., Oppedijk, B., Ahmed, S., Maes, T., Gerats, T., Hoogeneen, P., Meesters, M., Kools, D. and Mol, J. N. M., 1995, Targeted gene inactivation in petunia by PCR-based selection of transposon insertion mutants, *Proc Natl Acad Sci USA.* 92: 8149-8153.

Koes, R. E., Verweij, W. and Quattrocchio, F., 2005, Flavonoids: a colorful model for the regulation and evolution of biochemical pathways, *Trends Plant Sci,* 10: 236-242.

Kolkman, J. M., Conrad, L. J., Farmer, P. R., Hardeman, K., Ahern, K. R., Lewis, P. E., Sawers, R. J. H., Lebejko, S., Chomet, P. and Brutnell, T. P., 2005, Distribution of *Activator* (*Ac*) throughout the maize genome for use in regional mutagenesis, *Genetics* 169: 981-995.

Kreuzaler, F., Ragg, H., Fautz, E., Kuhn, D. N. and Hahlbrock, K., 1983, UV-Induction of chalcone synthase mRNA in cell suspension cultures of *Petroselinum hortense*, *Proc Natl Acad Sci USA* 80: 2591-2593.

Kroon, J., Souer, E., de Graaff, A., Xue, Y., Mol, J. and Koes, R., 1994, Cloning and structural analysis of the anthocyanin pigmentation locus *Rt* of *Petunia hybrida*: characterization of insertion sequences in two mutant alleles, *Plant J* 5: 69-80.

Kunze, R. and Weil, C. F., 2002, The *hAT* and CACTA superfamilies of plant transposons. In N. L. Craig, R. Craigie, M. Gellert and A. M. Lambowitz, eds, *Mobile DNA II* (pp. 565-610). Washington, D. C.: ASM Press.

Larson, R., Bussard, J. B. and Coe, E. H., Jr., 1986, Gene-dependent flavonoid 3'-hydroxylation in maize, *Biochem Genet* 24: 615-624.

Lechelt, C., Peterson, T., Laird, A., Chen, J., Dellaporta, S. L., Dennis, E., Peacock, W. J. and Starlinger, P., 1989, Isolation and molecular analysis of the maize *P* locus, *Mol Gen Genet* 219: 225-234.

Lippman, Z., Gendrel, A. V., Black, M., Vaughn, M. W., Dedhia, N., McCombie, W. R., Lavine, K., Mittal, V., May, B., Kasschau, K. D., Carrington, J. C., Doerge, R. W., Colot, V. and Martienssen, R., 2004, Role of transposable elements in heterochromatin and epigenetic control, *Nature* 430: 471-476.

Lippman, Z. and Martienssen, R., 2004, The role of RNA interference in heterochromatic silencing, *Nature* 431: 364-370.

Lo, S. C. and Nicholson, R. L., 1998, Reduction of light-induced anthocyanin accumulation in inoculated sorghum mesocotyls. Implications for a compensatory role in the defense response, *Plant Physiol* 116: 979-989.

Ludwig, S. R. and Wessler, S. R., 1990, Maize *R* gene family: Tissue-specific helix-loop-helix proteins, *Cell* 62: 849-851.

Martin, C., Carpenter, R., Sommer, H., Saedler, H. and Coen, E. S., 1985, Molecular analysis of instability in flower pigmentation of *Antirrhinum majus*, following isolation of the *pallida* locus by transposon tagging, *EMBO J* 4: 1625-1630.

Martin, C., Prescott, A., Mackay, S., Bartlett, J. and Vrijlandt, E., 1991, Control of anthocyanin biosynthesis in flowers of *Antirrhinum majus*, *Plant J* 1: 37-49.

Martin, C. and Gerats, T., 1993a, Control of pigment biosynthesis genes during petal development, *Plant Cell* 5: 1253-1264.

Martin, C. and Gerats, T., 1993b, The control of flower coloration. In B. R. Jordan, ed, *The Molecular Biology of Flowering* (pp. 219-255). Wallingford, Oxon: C.A.B. International.

Matzke, M. A. and Birchler, J. A., 2005, RNAi-mediated pathways in the nucleus, *Nat. Rev. Genet.* 6: 24-35.

McClintock, B., 1950, The origin and behavior of mutable loci in maize, *Proc Natl Acad Sci USA* 36: 344-355.

McClintock, B., 1951, Chromosome organization and genic expression, *Cold Spring Harb Symp Quant Biol* 16: 13-47.

McClintock, B., 1956, Controlling elements and the gene, *Cold Spring Harbor Symp Quant Biol* 21: 197-216.

McLaughlin, M. and Walbot, V., 1987, Cloning of a mutable *bz2* allele of maize by transposon tagging and differential hybridization, *Genetics* 117: 771-776.

Menssen, A., Hohmann, S., Martin, W., Schnable, P. S., Peterson, P. A., Saedler, H. and Gierl, A., 1990, The *En/Spm* transposable element of *Zea mays* contains splice sites at the termini generating a novel intron from a *dSpm* element in the *A2* gene, *EMBO J* 9: 3051-3057.

Meyer, P., Heidmann, I., Forkmann, G. and Saedler, H., 1987, A new petunia flower colour generated by transformation of a mutant with a maize gene, *Nature* 330: 677-678.

Meyer, P., 1995, DNA methylation and transgene silencing in *Petunia hybrida*, *Curr Top Microbiol Immunol* 197: 15-28.

Mo, Y., Nagel, C. and Taylor, L. P., 1992, Biochemical complementation of chalcone synthase mutants defines a role for flavonols in functional pollen, *Proc Natl Acad Sci USA* 89: 7213-7217.

Mol, J., Grotewold, E. and Koes, R., 1998, How genes paint flowers and seeds, *Trends Plant Sci* 3: 212-217.

Moreno, M. A., Chen, J., Greenblatt, I. and Dellaporta, S. L., 1992, Reconstitutional mutagenesis of the maize *P* gene by short-range *Ac* transpositions, *Genetics* 131: 939-956.

Morita, Y., Hoshino, A., Kikuchi, Y., Okuhara, H., Ono, E., Tanaka, Y., Fukui, Y., Saito, N., Nitasaka, E., Noguchi, H. and Iida, S., 2005, Japanese morning glory *dusky* mutants displaying reddish-brown or purplish-gray flowers are deficient in a novel glycosylation enzyme for anthocyanin biosynthesis, UDP-glucose: anthocyanidin 3-*O*-glucoside-2"-*O*-glucosyltransferase, due to 4-bp insertions in the gene, *Plant J* 42: 353-363.

Napoli, C., Lemieux, C. and Jorgensen, R., 1990, Introduction of a chimeric chalcone synthase gene into petunia results in reversible co-suppression of homologous genes *in trans*, *Plant Cell* 2: 279-289.

Nesi, N., Debeaujon, I., Jond, C., Pelletier, G., Caboche, M. and Lepiniec, L., 2000, The *TT8* gene encodes a basic helix-loop-helix domain protein required for expression of *DFR* and *BAN* genes in *Arabidopsis* siliques, *Plant Cell* 12: 1863-1878.

Nesi, N., Jond, C., Debeaujon, I., Caboche, M. and Lepiniec, L., 2001, The *Arabidopsis TT2* gene encodes an R2R3 MYB domain protein that acts as a key determinant for proanthocyanidin accumulation in developing seed, *Plant Cell* 13: 2099-2114.

Neuffer, M. G., Coe, E. H. and Wessler, S. R., 1997, *Mutants of Maize*. New York: Cold Spring Harbor Laboratory Press.

Nevers, P., Shepherd, N. S. and Saedler, H., 1986, Plant transposable elements, *Adv Bot Res* 12: 103-203.

Nicholson, R. L. and Hammerschmidt, R., 1992, Phenolic compounds and their role in disease resistance, *Ann Rev Phytopathol* 30: 369-389.

O'Reilly, C., Shepherd, N. S., Pereira, A., Schwarz-Sommer, Z., Bertam, I., Robertson, D. S., Peterson, P. A. and Saedler, H., 1985, Molecular cloning of the *a1* locus of *Zea mays* using the transposable elements *En* and *Mu1*, *EMBO J* 4: 877-882.

Ohnishi, M., Fukada-Tanaka, S., Hoshino, A., Takada, J., Inagaki, Y. and Iida, S., 2005, Characterization of a novel Na^+/H^+ antiporter gene *InNHX2* and comparison of *InNHX2* with *InNHX1*, which is responsible for blue flower coloration by increasing the vacuolar pH in the Japanese morning glory, *Plant Cell Physiol* 46: 259-267.

Park, K. I., Choi, J. D., Hoshino, A., Morita, Y. and Iida, S., 2004, An intragenic tandem duplication in a transcriptional regulatory gene for anthocyanin biosynthesis confers pale-colored flowers and seeds with fine spots in *Ipomoea tricolor*, *Plant J* **38**: 840-849.

Payne, C. T., Zhang, F. and Lloyd, A. M., 2000, *GL3* encodes a bHLH protein that regulates trichome development in *Arabidopsis* through interaction with GL1 and TTG1, *Genetics* **156**: 1349-1362.

Paz-Ares, J., Wienand, U., Peterson, P. A. and Saedler, H., 1986, Molecular cloning of the *c* locus of *Zea mays*: a locus regulating the anthocyanin pathway, *EMBO J* **5**: 829-833.

Peterson, T., 1990, Intragenic transposition of *Ac* generates a new allele of the maize *P* gene, *Genetics* **126**: 469-476.

Pooma, W., Gersos, C. and Grotewold, E., 2002, Transposon insertions in the promoter of the *Zea mays a1* gene differentially affect transcription by the Myb factors P and C1, *Genetics* **161**: 793-801.

Quattrocchio, F., Wing, J. F., van der Woude, K., Mol, J. N. M. and Koes, R., 1998, Analysis of bHLH and MYB domain proteins: species-specific regulatory differences are caused by divergent evolution of target anthocyanin genes, *Plant J* **13**: 475-488.

Quattrocchio, F., Wing, J., van der Woude, K., Souer, E., de Vetten, N., Mol, J. and Koes, R., 1999, Molecular analysis of the *anthocyanin2* gene of petunia and its role in the evolution of flower color, *Plant Cell* **11**: 1433-1444.

Radicella, J. P., Turks, D. and Chandler, V. L., 1991, Cloning and nucleotide sequence of a cDNA encoding *B-Peru*, a regulatory protein of the anthocyanin pathway in maize, *Plant Mol Biol* **17**: 127-130.

Ramsay, N. A. and Glover, B. J., 2005, MYB-bHLH-WD40 protein complex and the evolution of cellular diversity, *Trends Plant Sci* **10**: 63-70.

Russo, V. E. A., Martienssen, R. A. and Riggs, A. D., 1996, *Epigenetic Mechanisms of Gene Regulation*. Plainview, N.Y.: Cold Spring Harbor Laboratory Press.

Schijlen, E. G., Ric de Vos, C. H., van Tunen, A. J. and Bovy, A. G., 2004, Modification of flavonoid biosynthesis in crop plants, *Phytochem* **65**: 2631-2648.

Schoenbohm, C., Martens, S., Eder, C., Forkmann, G. and Weisshaar, B., 2000, Identification of the *Arabidopsis thaliana* flavonoid 3'-hydroxylase gene and functional expression of the encoded P450 enzyme, *Biol Chem* **381**: 749-753.

Schwarz-Sommer, Z., Davies, B. and Hudson, A., 2003, An everlasting pioneer: the story of *Antirrhinum* research, *Nat Rev Genet* **4**: 655-664.

Schwinn, K. E. and Davies, K. M., 2004, Flavonoids. In K. Davies, ed, *Plant Pigments and Their Manipulation* (pp. 92-149). Oxford: Blackwell Publishing Ltd.

Shikazono, N., Yokota, Y., Kitamura, S., Suzuki, C., Watanabe, H., Tano, S. and Tanaka, A., 2003, Mutation rate and novel *tt* mutants of *Arabidopsis thaliana* induced by carbon ions, *Genetics* **163**: 1449-1455.

Shirley, B. W., Hanley, S. and Goodman, H. M., 1992, Effects of ionizing radiation on a plant genome: analysis of two *Arabidopsis transparent testa* mutations, *Plant Cell* **4**: 333-347.

Shirley, B. W., Kubasek, W. L., Storz, G., Bruggemann, E., Koornneef, M., Ausubel, F. M. and Goodman, H. M., 1995, Analysis of *Arabidopsis* mutants deficient in flavonoid biosynthesis, *Plant J* **8**: 659-671.

Sidorenko, L. V., Li, X., Cocciolone, S. M., Chopra, S., Tagliani, L., Bowen, B., Daniels, M. and Peterson, T., 2000, Complex structure of a maize *Myb* gene promoter: functional analysis in transgenic plants, *Plant J* **22**: 471-482.

Snyder, B. A. and Nicholson, R. L., 1990, Synthesis of phytoalexins in sorghum as a site-specific response to fungal ingress, *Science* **248**: 1637-1639.

Souer, E., Quattrocchio, F., de Vetten, N., Mol, J. and Koes, R., 1995, A general method to isolate genes tagged by a high copy number transposable element, *Plant J* **7**: 677-685.

Spelt, C., Quattrocchio, F., Mol, J. N. and Koes, R., 2000, *anthocyanin1* of petunia encodes a basic helix-loop-helix protein that directly activates transcription of structural anthocyanin genes, *Plant Cell* **12**: 1619-1631.

Spelt, C., Quattrocchio, F., Mol, J. and Koes, R., 2002, ANTHOCYANIN1 of petunia controls pigment synthesis, vacuolar pH, and seed coat development by genetically distinct mechanisms, *Plant Cell* **14**: 2121-2135.

Springob, K., Nakajima, J., Yamazaki, M. and Saito, K., 2003, Recent advances in the biosynthesis and accumulation of anthocyanins, *Nat Prod Rep* **20**: 288-303.

Styles, E. D. and Ceska, O., 1977, The genetic control of flavonoid synthesis in maize, *Can J Genet Cytol* **19**: 289-302.

Styles, E. D. and Ceska, O., 1981, P and R control of flavonoids in *BRONZE* coleoptiles of maize, *Can J Genet Cytol* **23**: 691-704.

Styles, E. D. and Ceska, O., 1989, Pericarp flavonoids in genetic strains of *Zea mays*, *Maydica* **34**: 227-237.

Takahashi, S., Inagaki, Y., Satoh, H., Hoshino, A. and Iida, S., 1999, Capture of a genomic *HMG* domain sequence by the *En/Spm*-related transposable element *Tpn1* in the Japanese morning glory, *Mol Gen Genet* **261**: 447-451.

Tanaka, Y., Katsumoto, Y., Brugliera, F. and Mason, J., 2005, Genetic engineering in floriculture, *Plant Cell Tiss Org Cult* **80**: 1-24.

Tanner, G. J., Francki, K. T., Abrahams, S., Watson, J. M., Larkin, P. J. and Ashton, A. R., 2003, Proanthocyanidin biosynthesis in plants. Purification of legume leucoanthocyanidin reductase and molecular cloning of its cDNA, *J Biol Chem* **278**: 31647-31656.

Tanner, G. J., 2004, Condensed tannins. In K. Davies, ed, *Plant Pigments and Their Manipulation* (pp. 150-184). Oxford: Blackwell Publishing Ltd.

Tiffin, P., Miller, R. E. and Rausher, M. D., 1998, Control of expression patterns of anthocyanin structural genes by two loci in the common morning glory, *Genes Genet Syst* **73**: 105-110.

Tohge, T., Nishiyama, Y., Hirai, M. Y., Yano, M., Nakajima, J., Awazuhara, M., Inoue, E., Takahashi, H., Goodenowe, D. B., Kitayama, M., Noji, M., Yamazaki, M. and Saito, K., 2005, Functional genomics by integrated analysis of metabolome and transcriptome of *Arabidopsis* plants over-expressing a MYB transcription factor, *Plant J* **42**: 218-235.

van der Krol, A. R., Lenting, P. E., Veenstra, J., van der Meer, I. M., Koes, R. E., Gerats, A. G. M., Mol, J. N. M. and Stuitje, A. R., 1988, An anti-sense chalcone synthase gene in transgenic plants inhibits flower pigmentation, *Nature* **333**: 866-869.

van der Krol, A. R., Mur, L. A., de Lange, P., Mol, J. N. M. and Stuitje, A. R., 1990, Inhibition of flower pigmentation by antisense *CHS* genes: promoter and minimal sequence requirements for the antisense effect, *Plant Mol Biol* **14**: 457-466.

van Houwelingen, A., Souer, E., Spelt, K., Kloos, D., Mol, J. and Koes, R., 1998, Analysis of flower pigmentation mutants generated by random transposon mutagenesis in *Petunia hybrida*, *Plant J* **13**: 39-50.

van Tunen, A. J., Mur, L. A., Recourt, K., Gerats, A. G. M. and Mol, J. N., 1991, Regulation and manipulation of flavonoid gene expression in anthers of petunia: the molecular basis of the *Po* mutation, *Plant Cell* **3**: 39-48.

Walker, A. R., Davison, P. A., Bolognesi-Winfield, A. C., James, C. M., Srinivasan, N., Blundell, T. L., Esch, J. J., Marks, M. D. and Gray, J. C., 1999, The *TRANSPARENT TESTA GLABRA1* locus, which regulates trichome differentiation and anthocyanin biosynthesis in Arabidopsis, encodes a WD40 repeat protein, *Plant Cell* **11**: 1337-1350.

Wienand, U., Sommer, H., Schwarz, Z., Shepherd, N., Saedler, H., Kreuzaler, F., Ragg, H., Fautz, E., Hahlbrock, K., Harrison, B. J. and Peterson, P., 1982, A general method to identify plant structural genes among genomic DNA clones using transposable element induced mutations, *Mol Gen Genet* **187**: 195-201.

Wienand, U., Weydemann, U., Niesbach-Klosgen, U., Peterson, P. A. and Saedler, H., 1986, Molecular cloning of the *c2* locus of *Zea mays*, the gene coding for chalcone synthase, *Mol Gen Genet* **203**: 202-207.

Winkel-Shirley, B., 2001a, Flavonoid biosynthesis. A colorful model for genetics, biochemistry, cell biology, and biotechnology, *Plant Physiol* **126**: 485-493.

Winkel-Shirley, B., 2001b, It takes a garden. How work on diverse plant species has contributed to an understanding of flavonoid metabolism, *Plant Physiol* **127**: 1399-1404.

Wiseman, B. R., Snook, M. and Widstrom, N. W., 1996, Feeding responses of the corn ear worm larvae (Lepidoptera: Noctuidae) on corn silks of varying flavone content, *J Econ Entomol* **89**: 1040-1044.

Wisman, E., Hartmann, U., Sagasser, M., Baumann, E., Palme, K., Hahlbrock, K., Saedler, H. and Weisshaar, B., 1998, Knock-out mutants from an *En-1* mutagenized *Arabidopsis thaliana* population generate phenylpropanoid biosynthesis phenotypes, *Proc Natl Acad Sci USA* **95**: 12432-12437.

Xie, D. Y., Sharma, S. B., Paiva, N. L., Ferreira, D. and Dixon, R. A., 2003, Role of anthocyanidin reductase, encoded by *BANYULS* in plant flavonoid biosynthesis, *Science* **299**: 396-399.

Yamaguchi, T., Fukada-Tanaka, S., Inagaki, Y., Saito, N., Yonekura-Sakakibara, K., Tanaka, Y., Kusumi, T. and Iida, S., 2001, Genes encoding the vacuolar Na$^+$/H$^+$ exchanger and flower coloration, *Plant Cell Physiol* **42**: 451-461.

Yoshida, K., Kondo, T., Okazaki, Y. and Katou, K., 1995, Cause of blue petal colour, *Nature* **373**: 291.

Zanta, C. A., Yang, X., Axtell, J. D. and Bennetzen, J. L., 1994, The candystripe locus, *y-cs*, determines mutable pigmentation of the sorghum leaf, flower, and pericarp, *J Hered* **85**: 23-29.

Zhang, F. and Peterson, T., 2005, Comparisons of maize *pericarp color1* alleles reveal paralogous gene recombination and an organ-specific enhancer region, *Plant Cell* **17**: 903-914.

Zhang, P., Chopra, S. and Peterson, T., 2000, A segmental gene duplication generated differentially expressed *myb*-homologous genes in maize, *Plant Cell* **12**: 2311-2322.

Zhang, F., Gonzalez, A., Zhao, M., Payne, C. T. and Lloyd, A., 2003a, A network of redundant bHLH proteins functions in all TTG1-dependent pathways of *Arabidopsis*, *Development* **130**: 4859-4869.

Zhang, P., Wang, Y., Zhang, J., Maddock, S., Snook, M. and Peterson, T., 2003b, A maize QTL for silk maysin levels contains duplicated *Myb*-homologous genes which jointly regulate flavone biosynthesis, *Plant Mol Biol* **52**: 1-15.

CHAPTER 7

THE EVOLUTION OF FLAVONOIDS AND THEIR GENES

MARK D. RAUSHER

Department of Biology, Box 90338, Duke University, Durham, NC 27708 USA

1. INTRODUCTION

Flavonoids constitute a diverse array of plant secondary compounds that perform a wide variety of physiological and ecological functions (Figure 7.1). The role of anthocyanin pigments as visual signals in angiosperms for attracting pollinators and fruit dispersal agents is well known, but these functions were acquired late in the evolutionary diversification of flavonoids. Less well-known and probably more ancient functions of flavonoids include protection against the detrimental effects of UV radiation; mediation of interactions between pollen and stigma; defense against bacteria, pathogenic fungi, and herbivores; mediation of interactions between plants and mutualistic mycorhizzal fungi; and regulators of hormonal activity. Because recent, thorough reviews of these functions are available elsewhere (Koes et al., 1994; Shirley, 1996), they will not be discussed in detail here. Instead, I focus on the evolutionary processes by which these functions arose and continue to evolve.

As other chapters in this volume make clear, the flavonoid pathway has served as a model system for understanding gene regulation in plants. In a similar way, flavonoid pathway genes, both structural and regulatory, have served as a model system for understanding a variety of evolutionary processes, including the role of gene duplication in facilitating the evolution of novel characters, the causes of evolutionary rate variation among genes, and the relative importance of structural and regulatory genes in evolution of ecologically important characters. In this chapter, I review how examination of patterns of change in flavonoid genes has contributed to our understanding of these and other evolutionary issues, beginning with an examination of our understanding of the historical evolution of the flavonoid biosynthetic pathway.

2.   HISTORICAL EVOLUTION OF THE FLAVONOID PATHWAY

The flavonoid pathway is a classic example of a pathway that has evolved piecemeal, gradually lengthening as new products and new functions were added. Although many details of its evolutionary construction are still hazy, it is possible to reconstruct the main evolutionary events with confidence. The following account is an elaboration of the reconstruction originally suggested by Stafford (1991). As will be seen, a general theme emerges from this reconstruction: gene duplication has provided the raw material for building the pathway. There is a satisfying symmetry to this in that, as we shall see below, once the flavonoid pathway was essentially complete, duplication and divergence of genes from the pathway has led to the emergence of new classes of compounds. Through gene duplication, pathways are not only born themselves, but they spawn additional pathways.

Equally importantly, the evolution of the flavonoid pathway provides an excellent example of how a biochemically complex trait may be built up in stages, with each addition being adaptive. It thus provides an illuminating counterexample to the claim of the "Intelligent Design" movement that complex biochemical adaptations that are irreducibly complex cannot be assembled gradually by natural selection. In this case, the complex biochemical adaptation is flower color produced by anthocyanin pigments. Its complexity lies in the fact that anthocyanin production requires at least six sequential biochemical reactions enabled by six different enzymes. Looking at this character by itself, there is no question that it is irreducibly complex. As numerous knockout mutations in the pathway demonstrate, removal of one enzyme eliminates the production of anthocyanins. It is only because the intermediate products that were invented along the way—the diverse collection of flavonoids produced by land plants—retain their important ecological and physiological functions, and so have not been superseded in these functions by other secondary metabolites, that we are able to recognize that the irreducible complexity of floral pigment production actually evolved gradually.

When considering the evolutionary construction of the flavonoid biochemical pathway, it should be borne in mind that early flavonoid production may have been quite "leaky." Initially, enzymes of primary metabolism already may have produced small quantities of many different simple flavonoids, simply due to lack of complete substrate specificity in those enzymes. In this context, if an environmental change caused the production of one or a subset of these compounds to become advantageous, natural selection would act to enhance its production. At first, the mechanism of enhancement may have been quite crude (e.g., up-regulating a key enzyme of primary metabolism) and probably would have had deleterious pleiotropic effects. If, however, the advantages of increasing the production of flavonoids were large enough, this would evolve despite the pleiotropic effects. Moreover, it would automatically establish selection pressures to lessen the pleiotropic effects. Two particularly effective types of genetic change that could respond to this selection would be specialization of a duplicate gene for carrying out early flavonoid synthesis and the evolution of regulatory control over such a gene.

2.1. Evolution of the flavonoid enzymes

A major clue to the stepwise development of the flavonoid pathway is provided by the distribution of both different types of flavonoids and different flavonoid enzymes among land plant taxa (Figure 7.2). This distribution allows us to recognize a series of successive stages in the evolution of the pathway.

2.1.1. Stage 1: Algae

Land plants are believed to be derived from organisms like green algae (Charales). Although extensive surveys have been conducted (Markham, 1988), flavonoids consistently fail to be found among the algae. Moreover, although several algal genomes have been sequenced, none contain open reading frames that show homology to the coding sequences of any known flavonoid enzymes. The initial evolution of the flavonoid pathway thus probably took place after the colonization of land.

2.1.2. Stage 2: Bryophytes (mosses), liverworts, and hornworts

This paraphyletic group represents the earliest plants to colonize land. It also is the oldest plant group to produce flavonoids. Among the types of flavonoids produced by these plants are chalcones, flavonols, and flavones (Markham, 1988), which are derived from the first three enzymes of the flavonoid pathway (Figure 7.1). In addition, EST (expressed sequence tags) databases report sequences that appear to represent these three enzymes: chalcone synthase, chalcone–flavanone isomerase, and flavanone 3-hydroxylase. All three enzymes appear to have been derived via gene duplication from genes coding for enzymes of primary metabolism. CHS exhibits strong sequence similarity to bacterial genes coding polyketide synthases, particularly those involved in fatty acid synthesis (Verwoert et al., 1992). Moreover, the condensation reactions of these enzymes also are similar, as is the set of substrates used. Based on similarity in sequence and enzymology, F3H belongs to the oxoglutarate-dependent dioxygenase family of enzymes and is presumably derived from a duplication of one of its members (Winkel-Shirley, 2001). The origin of CHI is less clear, since it seems to be unrelated in sequence and tertiary structure to any other plant enzyme (Jez et al., 2000). However, enzymes with similar sequence and secondary structures have been reported from some bacteria and fungi (Gensheimer and Mushegian, 2004), and an enzyme of unknown sequence but exhibiting CHI activity has been isolated from the bacterium *Eubacterium ramulus* (Herles et al., 2004). Interestingly, although this enzyme is capable of converting naringenin chalcone into naringenin, it is believed that its function *in vivo* is the catabolism of naringenin.

Two hypotheses have been put forward regarding the initial advantages associated with producing flavonoids. One suggests that flavonoids evolved as an effective sunscreen protecting against UV radiation as plants began colonizing land

(Markham, 1988; Koes et al. 1994; Shirley 1996). Supporting this suggestion are the observations that even simple flavonoids such as chalcones, aurones, and flavanones absorb UV wavelengths strongly, and that flavonoid knockout mutants often are extremely susceptible to UV damage (Li et al., 1993; Lois and Buchanan, 1994). By contrast, Stafford (1991) has argued that the first function of flavonoids was to regulate or chaperone plant hormones. Stafford argues that this hypothesis is more likely than the UV protection hypothesis because presumably early flavonoid enzymes were not as efficient as current enzymes, and therefore large quantities of flavonoids likely did not accumulate early. Moreover, it has been shown in angiosperms that flavonoids contribute to regulation of auxin transport (Brown et al., 2001; Peer et al., 2004).

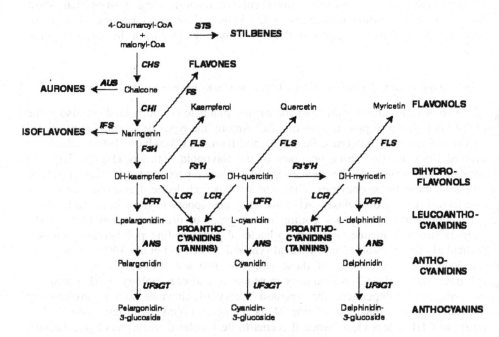

Figure 7.1 Schematic portrayal of the flavonoid pathway. Specific compounds are listed in lower case. Classes of compounds are listed in bold upper case. Enzymes are listed in bold italics. STS: stilbene synthase. CHS: chalcone synthase. CHI: chalcone-flavanone isomerase. F3H: flavanone-3-hydroxylase. DFR: dihydroflavonols 4-reductase. ANS: anthocyanidin synthase. UF3GT: UDP flavonoid glucosyltransferase. F3'H: flavonoid 3' hydroxylase. F3'5'H: flavonoid 3'5' hydroxylase. LCR: leucoanthocyanidins reductase. FS: flavone synthase. FLS: flavonol synthase. IFS: isoflavone synthase. AUS: aureusidin synthase. DH="dihydro." L="leuco."

These three enzymes and the flavonoids produced by them are unlikely to have arisen and evolved their new functions simultaneously. A likely, though currently unproven, sequence of origin is that CHS evolved first, followed by F3H, and then CHI. CHS seems a likely first enzyme because it is the initial enzyme in the pathway. Perhaps equally important, though, is the fact that naringenin chalcone, the product of CHS, spontaneously isomerizes to the flavanone naringenin even in the absence of CHI (Holton and Cornish, 1995). This "leakiness" means that it is likely that early plants with only CHS function could produce a variety of flavonoids that could collectively provide UV protection. The evolutionary incorporation of CHI into the nascent pathway could then have been favored, not because it provided a new function, but because it improved the efficiency of an existing function. The incorporation of F3H into the pathway could have either proceeded or followed CHI, allowing the production of dihydroflavonols and perhaps flavonols. While we can probably never be certain what advantage this conferred, one likely possibility is that these compounds evolved as defenses against pathogenic fungi and bacteria, a function that has been demonstrated in contemporary plants (Kemp and Burden, 1986; Curir et al., 2005).

In addition to the evolution of the CHS-CHI-F3H backbone of the rudimentary flavonoid pathways, this stage also saw the evolutionary development of the three major branches of that pathway (Figure 7.1). These branches correspond to different degrees of hydroxylation of the flavonoid B ring. The first branch produces compounds lacking hydroxylation at either the 3' of 5' position, such as dihydrokaempferol and kaempferol. The second branch produces compounds hydroxylated at the 3' position (dihydroquercetin and quercetin), while compounds produced by the third branch (dihydromyricetin and myricetin) are hydroxylated at both the 3' and 5' positions. Evidence for this claim comes from extensive surveys of flavonoid compounds produced by mosses and liverworts (e.g., Markham, 1988), which reveal the species that produce derivatives of all three branches. Moreover, EST databases from the moss *Physcomitrella patens* contain sequences that are very F3'H like, which in higher plants hydroxylates the 3' position. Sequences similar to F3'5'H, which in higher plants hydroxylates the 5' position, have not yet been reported from mosses or liverworts. Both of these enzymes have apparently been recruited to the pathway through duplication of genes in the cytochrome P450 hydroxylase family (Dixon and Steele, 1999). Although one can only speculate as to the selective advantages associated with addition of branches to the pathway, differential effects of hydroxylated vs. nonhydroxylated flavonoids are known. For example, quercetin seems to be a more effective photoprotectant than kaempferol because it is a more effective antioxidant (Ryan et al., 2001, 2002).

Another enzymatic innovation that appears to have occurred at this stage is the evolution of flavonol synthase. All orders of land plants, including bryophytes, produce flavonols, which are produced from dihydroflavonols by one or more flavonol synthase (FLS) enzymes (Figures 7.1 and 7.2). *FLS* is derived from the 2-oxoglutarate-dependent dioxygenase gene family (Holton et al., 1993). While the presence of flavonols in mosses, liverworts, or ferns indicates that these groups must possess *FLS* genes, they have not yet been detected and characterized, almost certainly because of the little attention that has been paid to the genomes of these groups.

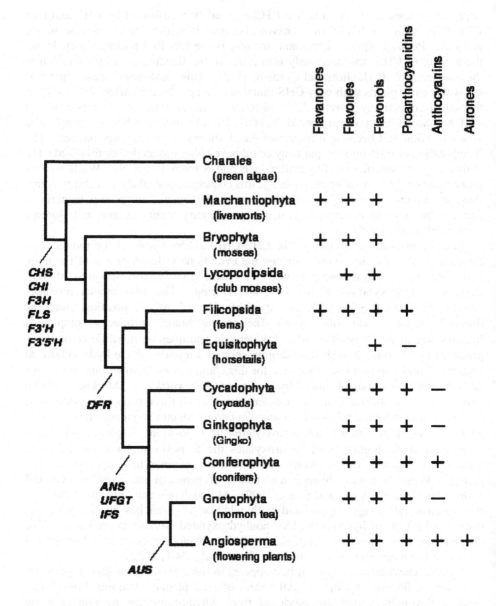

Figure 7.2 *Land plant phylogeny showing the time of origin of flavonoid enzymes.* +, *Documented presence of flavonoid;* —, *possible evolutionary loss of flavonoid. Phylogeny from Savolainen and Chase (2003).*

In contrast to flavonols, isoflavonoids occur only sporadically throughout the land plants (Dewick, 1988). Isoflavonoids have been reported in the moss *Bryum capillare* (Anhut et al., 1984), suggesting that the key enzyme for producing these compounds, isoflavone synthase (IFS) may have originated at this stage of plant evolution. However, the isoflavonoids have not been reported from any other plant outside of the gymnosperms and angiosperms (Dewick, 1988). This pattern suggests that the production of isoflavonoids may have evolved independently in *Byrum* and in the seed plants. Moreover, until isoflavonoids are detected in other bryophytes, the acquisition of the ability of *Bryum* to synthesize these compounds may represent a relatively recent evolutionary event. Since *IFS* has not yet been cloned from *Bryum*, it is not currently possible to evaluate this hypothesis.

2.1.3. Stage 3: Filicophyta (ferns)

The ferns and allies (lycopsids and equisitopsids) are believed to be derived from bryophytelike ancestors. They are the oldest group of plants known to produce proanthocyanidins (also called leucoanthocyanidins). Proanthocyanidins are commonly found in polymerized form, molecules known as tannins. In higher plants, proanthocyanidins are produced by the enzyme dihydroflavonol-4-reductase (DFR), which uses dihydroflavonols as substrates. Although this enzyme has not been reported in ferns, the genomic characterization of ferns is in its infancy. It would be very surprising if DFR were not eventually found in these plants.

One of the primary functions of tannins in plants is believed to be defense against bacterial and fungal pathogens, as well as herbivores (Feeny, 1970). This is likely to be one of the primary advantages that drove the evolution of the capability to produce proanthocyanidins in the lineage leading to ferns. As with other enzymes, DFR seems to have been derived from a duplicated gene associated with primary metabolism, in this case NADPH-dependent reductases associated with steroid metabolism (Baker and Blasco, 1992). Additionally, the three branches of the flavonoid pathway established in the bryophytes are maintained in the ferns, with procyanidin and prodelphinidin, as well as the flavonols kaempferol, quercetin, and myricetin being produced by species in many different fern families (Markham, 1988).

2.1.4. Stage 4: Gymnosperms and angiosperms

It is within these two groups of land plants that anthocyanins finally made their appearance (Timberlake and Bridle, 1980; Niemann, 1988). The key addition to the flavonoid pathway involved in this step was the recruitment of the enzyme anthocyanidin synthase (ANS), presumably as a result of gene duplication from the family of 2-oxo-glutarate-dependent oxygenases. This enzyme catalyzes the production of colored anthocyanins from colorless leucoanthocyanidins. It is not clear, however, whether it was the production of color per se that constituted the original function of anthocyanins. Although color signaling, especially to pollinators and fruit dispersal agents, is clearly a primary function of anthocyanins in angiosperms, such signaling is rare if not absent in gymnosperms from which the angiosperms arose.

In gymnosperms and angiosperms, anthocyanins are typically sequestered in vacuoles (see Chapter 5 for a detailed description of the vacuolar transport of flavonoids). This requires two additional enzymes, one for the addition of sugar moieties to the anthocyanidin skeleton to make the compound more soluble and one to transfer the resulting anthocyanin across the vacuolar membrane. In angiosperms, these two functions are accomplished, respectively, by enzymes such as UF3GT, which glycosylates anthocyanidins, and glutathione-*S*-transferase (GST). The former are derived from the large family of sugar transferases (e.g., UDP glucosyltransferases) (Mackenzie et al., 1997). The latter has apparently been derived independently at least twice in angiosperms from the large glutathione transferase family (Alfenito et al., 1998).

As in the case of the first three flavonoid enzymes, it is not likely that ANS, UFGT, and GST were recruited simultaneously. Once anthocyanidins began to be produced, it is likely that nonspecific glycosyltrasferases provided sufficient activity to convert them to anthocyanins. Similarly, nonspecific GSTs may have initially allowed some degree of vacuolar sequestration. Indeed, experimental transformation of anthocyanin-GST deficient maize kernels with GSTs not associated with anthocyanin conjugation produced vacuolar accumulation of anthocyanins (Alfenito et al., 1998). The efficiency of these functions then could have been improved through gene duplication and specialization on anthocyanin substrates.

Two other classes of flavonoids may have made their appearance in the seed plants. Except for the moss *Bryum*, isoflavonoids appear to be restricted to gymnosperms and angiosperms (Dewick, 1988), suggesting two separate origins of these compounds and the enzymes that produce them (see above). While isoflavonoids are widely distributed throughout the seed plants, their occurrence is sporadic (Figure 7.3). This taxonomic distribution may represent either multiple origins of isoflavone synthase and related enzymes or a single origin with multiple secondary losses of the ability to produce these compounds. These hypotheses can be distinguished in principle by constructing a phylogeny of *IFS* and related cytochrome P450 hydroxylases, as has been done for the origin of stilbene synthases from chalcone synthases (see below). To date, however, only one *IFS* gene from a plant outside the Fabaceae (legumes) has been characterized definitively, preventing at present a phylogenetic reconstruction.

Aurones are the second class of compounds that appear to have originated at this stage. They have never been found in ferns, fern allies (horsetails, lycopods, whisk ferns), or in gymnosperms (Bohm, 1975; Markham, 1988). By contrast, they have been found in at least seven orders of eudicots, where they typically occur as yellow pigments in flowers and in sedges (Bohm, 1975). They also have been reported from one species of bryophyte and one genus of liverworts (Markham, 1988), though it is unclear whether these represent mistaken chemical identification or independent origins. The gene coding the enzyme aureusidin synthase (AUS), which catalyzes the conversion of naringenin chalcone into the aurone aureusidin, recently has been cloned from *Antirrhinum* (Nakayama et al., 2000, 2001) and is derived by duplication from a plant polyphenol oxidase gene. As with *IFS*, the hypothesis of independent origins in angiosperms and bryophytes cannot be tested at this point because *AUS* has not been cloned from any other species.

2.2. Historic evolution of flavonoid gene regulation

Evolution of regulatory control of the flavonoid structural genes is poorly understood, but there are some intriguing patterns. In most dicots that have been examined, the core structural genes are coordinately regulated in two sets: a group of "early" genes and a group of "late" genes. In *Arabidopsis*, for example, light activation of flavonoid production first results in the production of transcripts of *CHS*, *CHI*, *F3H*, and *FLS*. Somewhat later, *DFR* and *ANS* are activated (Kubasek et al., 1992; Pelletier and Shirley, 1996; Pelletier et al., 1997). Moreover, activation of the late genes requires the product of the *TTG1* gene, whereas activation of the early genes does not, indicating that these two sets of genes regulated in different fashion. The patterns in *Antirrhinum* and *Petunia* are similar, although in *Antirrhinum*, *F3H* belongs to the set of "late" genes (Martin et al., 1991)

It is intriguing that the "early" genes in *Arabidopsis* correspond to those that evolved in early land plants (bryophytes and liverworts), while the "late" genes evolved in ferns and seed plants. This correspondence may indicate that a system of regulatory control over flavonoid production was established early in the assembly of the flavonoid pathway and has remained intact for more than 100 million years. Activation of this control system is induced by UV radiation (van Tunen et al., 1988; Kubasek et al., 1992; Pelletier et al., 1997), exactly what would be expected if flavonoids evolved originally to provide photoprotection as plants moved onto land. Under this scenario, a second regulatory control system would have been added as additional enzymes were incorporated into the flavonoid pathway, giving rise to the coordinated control over the "late" genes. Since the products of these "late" genes provide functions other than photoprotection, and since these products were needed in other tissues, it is not surprising that a separate regulatory system evolved to control them.

Although the regulatory genes that activate the "early" genes have not yet been identified, circumstantial evidence points to the involvement of genes from the MYB family of transcription factors. In particular, Moyano et al. (1996) demonstrated that two MYB proteins from *Antirrhinum* flowers activate *CHS* and *CHI* when expressed in yeast. Homologous MYB proteins also control expression of "late" genes in angiosperms, although these require the coparticipation of bHLH transcription factors (Mol et al., 1998) (see Chapter 4 for a description of the factors involved in the transcriptional regulation of flavonoid genes).

The observation that biosynthetic genes that presumably evolved at different times are regulated coordinately raises the question of how coordinate regulation arose. Although we currently do not have the information needed to answer this question, it is useful to speculate on how coordinate regulation *may* have evolved, if for no other reason than to generate concrete hypotheses that can be tested empirically.

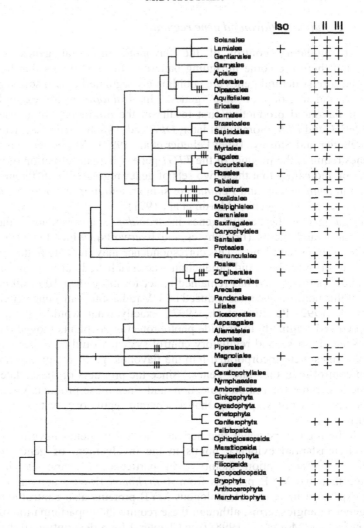

Figure 7.3 *Distributions of isoflavonoids and flavonoids from the three main branches of the flavonoid pathway in land plants. Pluses in the "Iso" column indicate orders for which isoflavonoids have been reported. Pluses in the "I, II, and III" columns indicate presence of flavonoids in the pelargonidin, cyanidin, and delplhinidin branches, respectively, of the pathway. Minuses indicate putative absence. Hash marks on phylogeny indicate most parsimonious locations of putative losses of flavonoids from the three branches. | , ||, and |||, respectively, represent loss of pelargonidin, cyanidin, and delphinidin branches. Phylogeny from Savolainen and Chase (2003). Data for taxonomic distribution of flavonoids are from Markham (1988), Giannasi (1988), Niemann (1988), and Williams and Harborne (1988). A more detailed analysis of distributions in plant families would reveal many more cases of probable loss of function in one or more of the pathway branches.*

In this spirit, I offer the following model for the evolution of coordinate regulation (Figure 7.4). Imagine a time very early in the evolution of land plants in which chalcone synthase was the only flavonoid enzyme that had evolved. Presumably it was regulated by a MYB transcription factor (*MYB1* in Figure 7.4). When it became advantageous to incorporate another step into the pathway (e.g., flavonone-3-hydroxylase; this may have preceded the incorporation of CHI since the reaction catalyzed by CHI can proceed without the enzyme, albeit more slowly), this enzyme (F3H in Figure 7.4B) was derived via duplication from an oxoglutarate-dependent oxygenase (OGDO), which presumably was already regulated by its own transcription factor(s) (*TF2* in Figure 7.4A). Because timing and spatial pattern of expression controlled by *MYB1* and *TF2* probably were dissimilar, natural selection would probably have favored a more coordinated expression of *CHS* and *F3H* to increase the efficiency at which dihydroflavonols were produced. This could be accomplished easily by adding MYB1-protein binding sites to the promoter region of *F3H* (Figure 7.4C). At this point, *F3H* would be activated by either *MYB1* or *TF2*. Subsequently, selection also might favor the elimination of *tf2*-protein binding sites, if activation of *F3H* by *TF2* were deleterious (Figure 7.4D). This sequence of events thus would establish coordinated regulation of *CHS* and *F3H* by *MYB1*.

In a similar manner, additional enzymes could be brought under the control of *MYB1* as they evolved to produce novel flavonoids. In addition, the independent coordinated regulation of "late" biosynthetic genes could be established in similar fashion. In this case, the first enzyme, DFR, was derived by duplication from an NADH-dependent reductase, which presumably was already regulated by its own transcription factor(s), possibly a MYB protein (*MYB2* in Figure 7.4E). Duplication of the reductase, followed by divergence, created the DFR enzyme, which presumably still was regulated by *MYB2* (Figure 7.4F).

Efficient production of leucoanthocyanidins requires the coordinated expression of *DFR* with the "early" genes. Because at this stage *DFR* is regulated independently of the "early" genes, coordination is likely to be poor. Consequently, selection would probably favor the evolution of a mechanism that would produce increased coordination. One way of accomplishing this would involve duplication of *MYB2* to produce *MYB2a*, which could then evolve to activate *DFR* in a way that was better coordinated with the "early" genes (Figure 7.4G). This would establish a regulatory control system of *DFR* expression, to which then could be added subsequent "late" enzymes (*ANS, UF3GT*) as they evolved, by the same mechanism that established the coordinated regulation of the early enzymes.

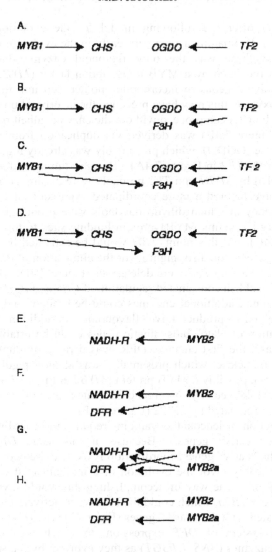

Figure 7.4 *A model for the evolution of coordinated regulation. Arrows indicate regulatory control of transcription factor (lower case) over a biosynthetic gene (upper case). (A–D) The capture of regulatory control. (A) CHS and oxoglutarate-dependent oxygenase (OGDO) controlled by separate transcription factors. (B) Duplication of OGDO and subsequent divergence gives rise to F3H. (C) Change in cis-regulatory region of F3H brings it under joint control of MYB1 and TF2. (D) Selection eliminates control of F3H by TF2. (E–H) Establishment of novel control. (E) NADH-dependent reductase (NADH-R) controlled by transcription factor MYB2. (F) Duplication of NADH-R and subsequent divergence gives rise to DFR. (G) Duplication of MYB2 and divergence gives rise to MYB2a, which causes DFR to be expressed with "early" genes in a more coordinated manner. (H) Selective elimination of the control of MYB2 over DFR and of MYBa over NADH-R.*

While this scenario is hypothetical, it does make two very strong predictions: (1) Coordination of expression within a set of biosynthetic genes should have been brought about by evolutionary change in the promoter modules of the biosynthetic genes themselves, rather than by change in transcription factors to "capture" control over additional enzymes; and (2) the transcription factor(s) controlling "late" genes should be more closely related to those controlling the expression of the NADH-dependent reductase gene that gave rise to *DFR* than to the transcription factor controlling the expression of the "early" genes (i.e., *MYB2a* should be more closely related to *MYB2* than to *MYB1*) (Figure 7.4).

Although evaluating the second prediction must await further characterization of the actual transcription factors associated with the flavonoid pathway, some evidence already supports prediction (1). For example, during the production of anthocyanins in maize, both the "early" and "late" biosynthetic genes are coordinately activated by a single MYB transcription factor, acting with a single bHLH factor. Transient expression experiments indicate that the "late" genes were brought under the control of the "early" transcription factor by changes in the *cis*-regulatory regions of the "late" genes. In these experiments, transcription factors from *Petunia* and maize were reciprocally bombarded into tissues of each of these two species. The *Petunia* transcription factors used were those that activate the "late" genes in *Petunia*. Both sets of factors activated the "early" genes in maize, while neither set of factors activated the "early" genes in *Petunia*, demonstrating that the difference between maize and *Petunia* in control over "early" genes is due to changes in the promoter regions of those genes (Quattrocchio et al., 1993, 1998).

The hypothetical scenario described above assumes that control over "early" genes by the same transcription factor that controls "late" genes is a derived character in maize. This assumption is also testable. Currently, we have information about control of "early" and "late" genes during anthocyanin synthesis in five taxa. In three of the taxa (*Arabidopsis*, *Antirrhinum*, and *Petunia*), the two sets of genes are under separate control, while in maize and *Ipomoea* there is unified control of both sets. The phylogenetic relationships among these taxa are shown in Figure 7.5, along with the present character states. With this information, inferring that the ancestral state was either separate or unified, control is equally parsimonious, since in either case only two transitions are required to produce the observed character states. It should be possible, however, to distinguish between these possibilities when regulatory control is deciphered for additional taxa.

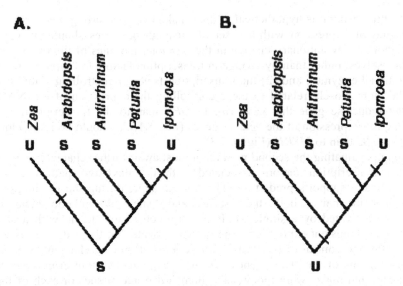

Figure 7.5 *Evolution of coordinate regulation of flavonoid biosynthetic genes. Character states indicated by U or S. U: Anthocyanin regulator has unified control over "early" and "late" genes. S: Control of "early" genes separate from control of "late" genes. A and B show that a minimum of two character state changes (hash marks) whether the ancestral state was S or U. Phylogeny from Savolainen and Chase (2003).*

3. IS PHENOTYPIC EVOLUTION MEDIATED BY STRUCTURAL OR REGULATORY GENES?

A central issue in evolutionary biology regards the nature of genes involved in adaptive evolutionary change. Substantial evidence is accruing to indicate that regulatory sequences, including both *cis*-regulatory regions as well as transcription factors, play a major role in morphological evolution in plants and animals (King and Wilson, 1975; Britten and Davidson, 1969, 1971; Dickinson, 1991; Doebley, 1993; Doebley and Lukens, 1998; Wray et al., 2003). By contrast, much physiological adaptation clearly involves changes in structural genes (Watt, 1977, 1983; Hochachka and Somero, 1984; Crawford and Powers, 1989; Gillespie, 1991; Yokoyama et al., 1993; Newcomb et al., 1997; Purrington and Bergelson, 1997). It still is not clear, however, whether this apparent contrast represents a fundamental difference in the genetic underpinnings of morphological and physiological evolution or instead perhaps reflects the recent simultaneous burgeoning of interest in gene regulation and in evolution of development. More generally, evolutionary biology currently lacks a coherent framework for predicting whether any particular type of trait is likely to evolve primarily by changes in structural genes or changes in regulatory genes.

Investigations of the evolution of flower color are beginning to provide some insight into these evolutionary issues. Flower color is a particularly appropriate trait for addressing these issues for several reasons: (1) the biochemical pathway that produces anthocyanins, probably the most important floral pigments, is relatively simple, and thus the connection between genotype and phenotype is straightforward and easily interpreted; (2) both structural genes and regulatory genes of the pathway have been extensively characterized in many different plants; (3) there is a tremendous amount of naturally occurring variation in flower color, both within as well as among species, that can be used to dissect the genetic factors responsible for evolutionary change; and (4) much is understood about the ecological significance of differences in flower color, which provides a context for interpreting evolutionary change in this character.

Evolutionary transitions in flower color frequently accompany or are accompanied by, changes in floral morphology that are believed to enhance the efficiency of interactions with new pollinators. Indeed, this is such a widespread phenomenon that "pollinator syndromes" have been recognized by plant evolutionary biologists for decades (Faegri and van der Pijl, 1966). For example, bee-pollinated flowers typically are blue-purple, have relatively short, broad tubes, broad limbs that serve as landing platforms, small amounts of concentrated nectar, and inserted anthers and stigmas. By contrast, hummingbird-pollinated flowers typically have reddish flowers, long narrow tubes, small limbs, copious dilute nectar, and exserted anthers and stigmas. Moth- and bat-pollinated flowers tend to be white, fragrant, and open at night. Many evolutionary changes in flower color thus seem to be adaptations associated with pollinator attraction.

Interestingly, many floral color transitions appear to be unidirectional. In the genus *Ipomoea*, for example, ancestral flowers were blue-violet and adapted to bee pollination. From this state, there have been numerous independent transitions to red, white and yellow flowers, but no documented cases of a transition from these colors back to blue-violet (Figure 7.6). Similarly, in the genus *Penstemon* there have been many independent transitions from blue, bee-pollinated flowers to red, hummingbird-pollinated flowers, without any evidence of the reverse transition (Wilson et al., 2004).

A simple hypothesis for this pattern is that these evolutionary transitions result from loss-of-function (LOF) mutations, for which reversions are unlikely. Several cases of naturally occurring LOF mutations of this type are known. For example, in the normally blue-flowered common morning glory, *Ipomoea purpurea*, red-flowered variants result from inactivation of the enzyme F3'H caused by a transpositional insertion into the *F3'H* gene that produces a premature stop codon (Zufall and Rausher, 2003; Hoshino et al., 2003). In the same species, natural white-flowered variants result from a similar transpositional inactivation of the enzyme CHS (Habu et al., 1998; Coberly, 2003) and from a deletion and frameshift in the coding region of a MYB transcription factor *IpMYB1* (Chang et al., 2005). In *Petunia axillaris*, white flowers have evolved from purple ancestors. The primary cause seems to have been a transposon-mediated deletion/frameshift in *AN2*, encoding a MYB transcription factor homologous to *IpMYB1*, although changes in several anthocyanin structural genes also contribute to lack of pigmentation (Quattrocchio et al. 1999). In the red-flowered, hummingbird-pollinated morning glory *Ipomoea quamoclit*, which has evolved from blue-flowered, bee-pollinated ancestors, the transition from blue to red pigmentation is caused by production of pelargonidin-based rather than cyanidin-based anthocyanins. This shift has resulted from some combination of three different LOF mutations that block the cyanidin branch of the pathway: an almost complete down-regulation of F3'H, a knockout of F3'H function *in planta*, and a loss in the enzyme DFR of the ability to metabolize dihydroquercetin (Zufall and Rausher, 2004). Finally, in the white-flowered *I. alba* and *I. aquatica*, expression of the structural genes *CHS* and *DFR* is markedly reduced, though it is not known whether changes in other anthocyanin genes also may contribute to lack of floral pigmentation in these species (Durbin et al., 2003).

The recognition that LOF mutations often may contribute to flower color evolution suggests that there may be a relatively straightforward conceptual framework for predicting and explaining whether flower color evolution is more likely to involve structural or regulatory genes. Two extreme possibilities can be envisioned, depending on the magnitude of deleterious pleiotropy associated with LOF mutations. One possibility is that mutations in regulatory sequences incur substantially lower levels of deleterious pleiotropy than mutations in structural genes. Under this hypothesis, whenever selection favors a novel flower color, regulatory mutations will have a higher net selection coefficient than structural gene mutations, because they experience lower deleterious pleiotropy. If the magnitude of pleiotropy is large enough, it is possible that only the regulatory mutations will have a positive net selection coefficient, and thus contribute to flower color evolution. Even if the magnitude of deleterious pleiotropy is not sufficient to completely offset the flower color advantage in structural gene mutations, the net selection coefficient will be smaller for those mutations than for mutations in regulatory sequences. Because the probability that an advantageous mutation actually will become fixed is roughly proportional to its selection coefficient (Hartl and Clark, 1989), this means that it will still be more likely for flower color evolution to be cause by changes in regulatory sequences.

The second extreme possibility is that when a new flower color becomes advantageous, the associated fitness benefit dwarfs the fitness cost of pleiotropy in any anthocyanin gene, structural or regulatory. Under this circumstance, selective coefficients of LOF mutations will not differ substantially among genes and all will have a roughly equal probability of fixation. Chance will predominate and determine whether regulatory or structural gene mutations ultimately give rise to a novel flower color.

3.1. How common is differential pleiotropy?

Substantial differences in the magnitude of deleterious pleiotropy between structural and regulatory LOF mutations may be expected for several reasons. First, anthocyanin structural genes often have broad expression domains (e.g., Durbin et al., 2003). For example, all structural genes in *Arabidopsis thaliana*, except flavonol synthase, are single-copy genes, and thus are expressed in all tissues that accumulate anthocyanins (Winkel-Shirley, 2001). Moreover, the upstream genes are expressed in all tissues that accumulate any type of flavonoid. In other species, some structural genes exist as small gene families (Koes et al., 1989; Durbin et al., 1995; Inagaki et al., 1999). Nevertheless, in many cases, one member of the gene family appears to be the primary copy expressed in most tissues (Koes et al., 1989; Inagaki et al., 1999; Durbin et al., 2003). Because of this broad expression domain, a knockout of a structural gene typically would be expected not only to produce altered flower color but also reduce or eliminate production of some anthocyanins and other flavonoids in many other tissues (Quattrocchio et al., 1993; Huits et al., 1994; van Houwelingen et al., 1998). Because these compounds serve a variety of often-crucial physiological and ecological functions in plants (Koes et al., 1994; Shirley, 1996), such a reduction is likely to substantially counteract any fitness advantages associated with the production of altered flower color.

By contrast, LOF in anthocyanin regulatory sequences are expected *a priori* to have fewer deleterious consequences for at least two reasons. First, anthocyanin transcription factors often exist as multigene families, with each member of the family having a restricted domain of expression (Ludwig and Wessler, 1990; Cone et al., 1993; Quattrocchio et al., 1993; Huits et al., 1994). As a consequence, knockouts of these genes can eliminate pigment production in flowers (or even in parts of flowers), without affecting anthocyanin or general flavonoid production in other tissues. Second, as is expected because of the general modularity of *cis*-regulatory sequences, expression of anthocyanin genes is commonly regulated independently in different tissues (Ludwig et al., 1990; Radicella et al., 1992; Patterson et al., 1995). In particular, mutations in *cis*-regulatory regions can differentially alter anthocyanin production in different tissues (e.g., Hansen et al., 1996). These observations suggest that mutations in *cis*-regulatory regions can greatly down-regulate anthocyanin production in flowers without necessarily reducing flavonoid production in other tissues. Such tissue-specific structural gene down-regulation is correlated with absence of floral pigmentation in several *Ipomoea* species (Durbin et al. 2003). Although in this case it has not been directly demonstrated that the changes in expression level are due to changes in *cis*-regulation rather than alteration of a transcription factor, the former seems likely because coordinate down-regulation of blocks of structural genes is not observed. In the white-flowered *I. alba*, for instance, *CHS* is down-regulated but *CHI* is not. Our own unpublished data indicate that only *CHS* and *DFR* among the seven core structural genes are down-regulated in *I. alba* flowers. Similarly, in the red-flowered *I. quamoclit*, only *F3'H* is down-regulated; absence of F3'H results in the absence of blue cyanidin-based pigments (Zufall and Rausher, 2003, 2004).

Experimental examination of the fitness consequences of anthocyanin structural and regulatory gene LOF in *Ipomoea purpurea* supports the hypothesis that structural gene mutants that produce white flowers experience substantially greater negative pleiotropy than regulatory mutants. In natural populations of this species, mutations at two different loci (*A* and *W*) produce very similar white-flower phenotypes (Ennos and Clegg, 1983; Epperson and Clegg, 1987).

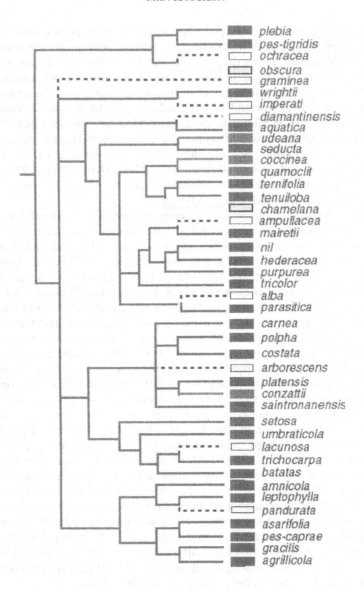

Figure 7.6 *Unidirectional evolution of flower color in Ipomoea. Phylogeny and ancestral state reconstruction based on Miller et al. (1999, 2004) and Zufall (2003). Boxes indicate flower color. All species are in the genus Ipomoea. See Color Section for figure in colors.*

At the *A* locus, white flowers result from a mutation in the structural gene *CHS-D* (Habu et al., 1998; Coberly, 2003), the primary copy of chalcone synthase that is expressed in most plant tissues (Fukada-Tanaka et al., 1997; Durbin et al., 2003). This mutation is caused by a transposon insertion into the sole intron that results in a truncated transcript. By contrast, white flowers at the *W* locus are a consequence of an out-of-frame deletion that introduces a premature stop codon into *IpMYB1* (Chang et al., 2005), which coordinately activates many of the anthocyanin structural genes in *I. purpurea* (Tiffin et al., 1998).

The direct effect of the white flowers produced by these two mutations on the mating system, and thus the direct fitness effects of the white phenotype, are very similar (Brown and Clegg, 1984; Rausher et al., 1983; Fehr and Rausher, 2004). Pleiotropic effects differ substantially, however. Compared to "wild-type" plants with blue flowers (*AAWW* genotype), white-flowered *aa* plants exhibit reduced germination and early survival, increased susceptibility to insect herbivores and fungal pathogens, and reduced seed set and pollination success at high (though natural) temperatures (Coberly, 2003; Coberley and Rausher, 2003). By contrast, white-flowered *ww* plants exhibit similar germination rates, viability, and seed set compared to blue-flowered plants (Rausher and Fry, 1993; Mojonnier and Rausher, 1997) and show no differences in susceptibility to herbivores or pathogens (Fineblum and Rausher, 1997). This difference in the magnitude of the deleterious pleiotropy is not only consistent with the expectation of greater negative effects of LOF mutations in structural genes, but may explain why the *w* allele is much more common in natural populations than the *a* allele (Coberly and Rausher, 2003; Fehr and Rausher, 2004).

Whether this pattern of differential pleiotropy holds in general remains to be explored. Numerous examples of pleiotropic effects of both anthocyanin structural and regulatory gene mutations have been documented (e.g., Coe et al., 1981; Mo et al., 1992; Van der Meer et al. 1992; Li et al., 1993), but the relative magnitude of pleiotropy is seldom assessed in the same species in fitness units under natural conditions, so it is unclear whether the fitness effects of negative pleiotropy tend to be higher in structural genes. Nor is it clear whether floral color adaptation is accomplished more frequently through modification of regulatory genes than of structural genes. Only a handful of attempts have been made to discern the genetic changes responsible for evolutionary change in flower color (Quattrocchio et al., 1999; Zufall and Rausher, 2004; Durbin et al., 2003), and even in these cases it is unclear whether redundant LOF changes in genes not examined might contribute substantially to the observed change in floral hue.

4. EVOLUTIONARY RATE VARIATION AMONG ANTHOCYANIN GENES

Variation in evolutionary rates among proteins is ubiquitous (Li, 1997). Rates of amino acid substitution vary over several orders of magnitude. Some proteins, such as histones, have undergone essentially no amino acid substitutions over hundreds of millions of years. By contrast, the *Drosophila* gene *OdsH* has undergone 10 amino acid replacements in approximately 1 million years (Ting et al., 1998). Two explanations have commonly been proposed to account for this variation: variation in selective constraint and variation in the frequency of advantageous substitutions.

At the extremes, these explanations are probably correct. Histones interact directly with DNA and other histones in the formation of the nucleosome. Because most amino acids in histone molecules participate in these interactions, it is likely that the optimal amino acid configuration was achieved early in the history of life and that there have been few subsequent replacements because virtually any amino acid substitution would disrupt function. In other words, selective constraint is almost complete in these proteins. In the case of *OdsH*, repeated episodes of advantageous substitutions are indicated by the very high Ka/Ks ratio.

For the large majority of proteins that do not exhibit such extremely high or low substitution rates, however, the causes of rate variation are less clear. Although population genetic tools that would permit a determination of the relative importance of variation in constraint and variation in positive selection have been available for nearly a decade, they seldom have been applied to addressing the causes of rate variation. Recently, however, application of these techniques to substitution patterns in anthocyanin pathway genes has begun to illuminate the causes of rate variation among these genes. In addition, they have begun to suggest that the rate at which particular genes evolve may in part be determined by the position and role that their protein products in the anthocyanin biochemical pathway.

4.1. Rate variation among structural genes

Rausher et al. (1999) quantified rates of amino acid substitution in six core anthocyanin structural genes (*CHS, CHI, F3H, DFR, ANS,* and *UF3GT*) by comparing maize sequences to *Antirrhinum* and *Ipomoea* sequences. They found more than a fivefold difference in the nonsynonymous substitution rate between the most rapidly evolving gene (*UF3GT*) and the most slowly evolving gene (*CHS*). Because of the broad taxonomic comparison in this study, substitutions at synonymous sites were saturated, and it thus was not possible to distinguish between alternative explanations for rate variation.

In a subsequent taxonomically more restricted study, Lu and Rausher (2003) quantified both synonymous and nonsynonymous substitution rates in *CHS, ANS,* and *UF3GT* by examining six species within the genus *Ipomoea*. As in the previous study, *CHS* exhibited the lowest amino acid substitution rate, *UF3GT* the highest, and *ANS* an intermediate rate. Differences in the underlying mutation rates were ruled out as a cause of this variation because the gene with the lowest non-synonymous substitution rate had the highest synonymous substitution rate. In addition, codon-based analyses of Ka/Ks ratios (Yang et al., 2000) revealed very little evidence that positive selection had caused extensive amino acid substitution in any of the three genes: no positively selected sites were detected for either *CHS* or *UF3GT*, and only two positively selected sites (out of 349) were detected for *ANS*. Together, these results suggest that variation in evolutionary rates among anthocyanin structural genes is most likely due to a variation in selective constraint. This inference must be viewed as tentative, however, because the analyses used to detect positive selection can be very conservative and can fail to detect selection if advantageous substitutions are scattered throughout the gene rather than being concentrated in certain sites.

A striking pattern that emerged from the Rausher et al. (1999) and Lu and Rausher (2003) analyses of rate variation is that downstream enzymes evolve faster than upstream enzymes. This pattern implies that genes corresponding to the downstream enzymes are under lower selective constraint than upstream genes. Although this pattern could be due simply to chance, there is another intriguing explanation: the position of an enzyme in the pathway determines in part the degree of selective constraint it experiences.

Upstream anthocyanin pathway enzymes also are required for the production of other flavonoid compounds, whereas downstream enzymes are not (Figure 7.1). Slightly deleterious mutations in upstream genes that reduce catalytic efficiency are thus likely to reduce flavonoid production in addition to anthocyanin production. By contrast, a similar mutation in a downstream gene potentially will affect only anthocyanin production. In other words, equivalent mutations are likely to have greater deleterious pleiotropy in upstream enzymes. This greater pleiotropy is in turn likely to increase the magnitude of the selection coefficient associated with the mutation, making it less likely that the mutation will be fixed by genetic drift. This scenario thus envisions that at evolutionary equilibrium most nonsynonymous substitutions represent chance fixation of slightly deleterious mutations and that the rate of such fixation is higher in downstream enzymes. While this hypothesis is consistent with observed patterns of rate variation, there is currently very little direct evidence supporting it.

4.2. Rapid evolution of regulatory genes

Because of their role in controlling developmental processes, it has been hypothesized that plant transcription factors may play a central role in the evolution of plant morphology (Doebley, 1993; Doebley and Lukens, 1998; Purugganan, 1998). An intriguing observation that is in accordance with this hypothesis is that plant transcription factors often evolve at elevated rates compared to the structural genes they regulate (Purugganan et al., 1995; Purugganan, 1998; Barrier et al., 2001; Remington and Purugganan, 2002; Dias et al., 2003), a pattern that also appears to hold for anthocyanin gene transcription factors (Purugganan and Wessler, 1994; Rausher et al. 1999). Usually, this elevated rate of nonsynonymous substitutions is concentrated in certain protein domains, while other domains (e.g., DNA-binding domains) are highly constrained. In the rapidly evolving domains, Ka/Ks ratios often approach 1. Moreover, high rates of insertion/deletion often are found in these rapidly evolving domains (Dias et al., 2003; Chang et al., 2005; Shavorskaya and Lagercrantz, unpublished manuscript). As is the case for the rapid evolution of structural genes, this rapid rate of evolution can be explained in two ways: by greatly relaxed selective constraint or by substantially enhanced, repeated positive selection. Confirmation of the latter explanation would constitute support for the importance of transcription factors in morphological evolution.

Analysis of substitution patterns in the *IpMYB1* anthocyanin transcription factor in *Ipomoea* does not provide such support (Chang et al., 2005). This gene, which is a member of the R2R3-MYB family of transcription factors, can be divided into two domains: an approximately 350 bp encoding the DNA-binding domain and an approximately 550 bp encoding the non-DNA-binding domain region. While the binding domain is believed to function in both DNA sequence recognition and binding to partner bHLH transcription factors (Goff et al., 1992; Williams and Grotewold, 1997; Sainz et al., 1997; Grotewold et al., 2000), little is known about the function of the variable region other than that it contains a transcription activation domain (Goff et al., 1991).

As expected, the DNA binding domain of this gene is highly conserved, exhibiting a relatively low Ka/Ks ratio of 0.235. The nonbinding domain, on the other hand, exhibits a Ka/Ks ratio of approximately 0.75. A series of independent analyses indicate that approximately 25% of the amino acid sites in the nonbinding domain are very highly conserved, while the remaining sites are evolving effectively neutrally. At these sites, Ka/Ks = 1, the proportions of radical and conservative amino acid substitutions do not differ from neutral expectations, and neither inter-specific nor intraspecific patterns of variation reveal any evidence that positive selection has contributed to the elevated replacement rate. It thus appears that the rapid evolution of this transcription factor is caused by greatly reduced selective constraints. Interestingly, the highly conserved sites are scattered randomly through the non-DNA-binding domain region, suggesting that the domain as a whole serves one or more crucial functions. Moreover, these conserved sites contain a disproportionate number of negatively charged, acidic amino acids. These observations suggest that the known function of transcriptional activation may be the primary function of nonbinding domain. The important functional feature of transcription activation domains of eukaryotic transcription factors appear to consist of negatively charged amino acids aligned along the same edge or face of alpha helices or other secondary structures, while the composition and three-dimensional structure of the remainder of the domain appears to be largely irrelevant (Ptashne, 1988). One thus would expect to see strong conservation at sites where acidic amino acids are present, as well as at sites that maintain the proper orientation of those sites but little constraint on other sites, a pattern that is consistent with what is observed in *IpMYB1*.

The absence of detectable positive selection in *IpMYB1* is consistent with results of other investigations of selection on rapidly evolving plant transcription factors. Remmington and Purugganan (2002) failed to detect positive selection on growth-regulating transcription factors in Hawaiian Silverswords, and Shavorskaya and Lagercrantz (unpublished results) similarly detected no positive selection on *CoI1* in the *Brassicaceae*. Although it is dangerous to generalize from a small number of investigations, current evidence suggests that rapidly evolving domains of plant transcription factors do not contribute to adaptive phenotypic evolution.

The results of two recent investigations of DNA-binding domain evolutionary diversification in members of the R2R3-MYB transcription factor family (Jia et al., 2003, 2004), which includes several anthocyanin regulatory genes, may at first seem difficult to reconcile with this conclusion (but see Dias et al., 2003). In both cases, positive selection was demonstrated to be responsible for DNA-binding domain substitutions. However, these investigations examined the divergence of paralogs within the same species. This divergence presumably occurred early in the evolutionary history of the gene family, when different paralogs were acquiring different regulatory functions. By contrast, the three studies referred to in the previous paragraph all examined evolutionary divergence of orthologous copies. During the divergence of these orthologs, a common function for the binding domains (e.g., activation of anthocyanin structural genes, in the case of *IpMYB1*) has been maintained. One thus would not expect to see major functional divergence and thus substantial positive selection in the binding domains of these genes, and one does not.

One of the most remarkable aspects of rapidly evolving anthocyanin gene transcription factors is that function is highly conserved. Ectopic expression of *C1* (a *MYB* gene) and *R* (a *bHLH* gene) from maize elicits anthocyanin production in *Petunia*, *Arabidopsis*, and *Nicotiana* (all dicots) (Lloyd et al., 1992; Quattrocchio et al., 1993), despite the common ancestors of maize and these species having lived approximately 100–125 million years ago (Savolainen and Chase, 2003). This common function has been maintained despite, in some cases, the absence of recognizable homology in the rapidly evolving non-DNA-binding domains.

Explaining how an essentially neutrally evolving region at the same time can maintain its function over long periods of evolutionary time is a challenging evolutionary mystery. One possibility, alluded above, is that for some regions the only functional requirement is the presence of certain types of amino acid (e.g., acidic). Another possibility is that only the three-dimensional configuration, rather than the specific amino acid composition, of the region is of primary importance to function (e.g., in mediating contact between transcription factors). If this hypothesis is correct, then function may be quite tolerant to virtually any individual amino acid substitution at most sites, although at any one time there may be a minority of sites that are crucial for maintaining configuration, and thus at least over a relatively short evolutionary period, are under severe constraint, as seen in *IpMYB1*. Over longer time periods, however, turnover may be possible at constrained sites also if, for example, substitutions at other previously unconstrained sites relieve constraint by introducing new foci for stabilization of the three-dimensional configuration.

Although, as suggested above, much flower color evolution appears to involve loss-of-function mutations in the anthocyanin pathway, it also is clear that this pathway has contributed frequently to the evolution of novel characters. Arguably, the most important process yielding new function is gene duplication followed by the evolution of novel function in one of the duplicated copies (neofunctionalization). Evolutionary analyses of flavonoid gene families suggests that this process has repeatedly given rise to novel classes of secondary compounds in plants.

5. GENE DUPLICATION AND THE EVOLUTION OF NOVEL FUNCTION

5.1. Duplication and divergence in chalcone synthase

A striking example of the evolution of new function is provided by stilbene synthases (Tropf et al., 1994). Stilbene phytoalexins are produced sporadically throughout the plant kingdom as defenses against fungal pathogens (Figure 7.7). The backbone of these compounds is synthesized by the enzyme stilbene synthase.

Stilbene synthases exhibit moderate to high amino acid sequence similarity to chalcone synthases and catalyze a similar condensation reaction. Using a phylogenetic analysis of chalcone and stilbene synthases, Tropf et al. (1994) showed that stilbene synthases have been derived independently several times from chalcone synthase, presumably following *CHS* duplication. They also demonstrated, using site-directed mutagenesis and *in vitro* enzyme assays, that modification of only three or four amino acids in chalcone synthase is needed to produce stilbene synthase activity.

Duplication and divergence of chalcone synthases apparently has produced a variety of other novel enzymes. For example, the common morning glory, *Ipomoea purpurea*, contains a six-member gene family exhibiting sequence similarity to chalcone synthases from other plants. Five of these copies appear to produce functional enzymes, while the sixth copy is apparently a pseudogene (Clegg and Durbin, 2003). Of the five functional copies, *CHS-D* is the primary transcript in most tissues, including flowers (Fukada-Tanaka et al., 1997; Durbin et al., 2003). This copy, along with *CHS-E*, has been shown to be capable of catalyzing the canonical chalcone synthase reaction: condensation of 4-coumaroyl-CoA and malonyl-CoA to form naringenin chalcone, the first step of the flavonoid pathway (Clegg and Durbin, 2003). By contrast, three of the copies, CHS-A, -B, and -C, are not able to catalyze this reaction, suggesting that they have evolved alternate as yet unknown functions. This suggestion is supported by the observation of an accelerated nonsynonymous substitution rate in the lineage leading to these copies (Durbin et al., 1995; Rausher et al., 1999), as well as the demonstration that at least some of the substitutions on this branch were positively selected (Yang et al., 2004).

Helariutta et al. (1996) described a similar situation in the Asterids. They isolated a gene with deduced amino acid similarity to asterid chalcone synthases and stilbene synthases. But the expression domain of this gene differs from the typical expression domains of chalcone and stilbene synthases, as does its substrate specificity *in vitro*, indicating that this gene has evolved some novel unknown function. A survey of several Asterid species revealed that this gene itself had duplicated and formed a small gene family, the members of which have diverged in both expression domain and substrate specificity. Nevertheless, the gene tree they obtained indicated that this novel family was derived from chalcone synthase. In addition to these novel enzymes, duplicates of chalcone synthase have evolved into resveratrol synthases and bibenzyl synthases (Figure 7.7) (Clegg and Durbin, 2003).

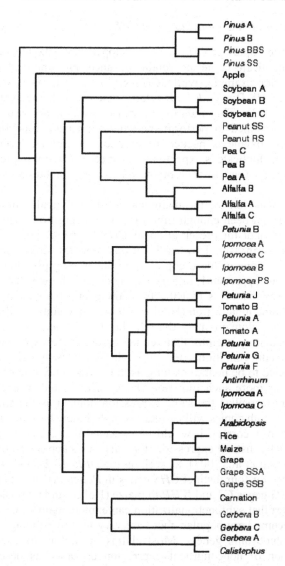

Figure 7.7 *Evolution of stilbene synthases (SS), bibenzyl synthases (BBS), and resveratrol synthases (RS) from chalcone synthases. Black: chalcone synthases. Red: modified chalcone synthases. In cases in which the modified chalcone synthase is not listed as another enzyme (Ipomoea A, B, C, Gerbera B), these enzymes have been shown to preferentially utilize substrates other than 4-coumaroyl-CoA and/or malonyl-CoA, the normal chalcone synthase substrates. Modified from Tropf et al. (1994) and Clegg and Durbin (2003). See Color Section for figure in colors.*

5.1. Duplication and subfunctionalization in CHS

While some duplicate copies of *CHS* have developed new enzyme functions, it is also common for duplicate copies to retain the ancestral canonical function. In *Ipomoea purpurea*, for example, while *CHS-A*, *-B*, and *-C* have acquired novel catalytic properties, *CHS-D* and *-E* both retain standard chalcone synthase activity (Clegg and Durbin, 2003). This situation raises the evolutionary issue of why both of these copies are retained to perform the same enzymatic function. Unless a double dose of this enzyme is needed to produce an optimal level of flux, an unlikely scenario, one duplicate is expected in most cases (except in presumably rare cases of neofunctionalization) to accumulate inactivating mutations and become a pseudogene.

One possible explanation for the maintenance of catalytically similar copies is simply that there has not been enough time for pseudogenization. This explanation seems unlikely to account for preservation of *CHS-D* and *-E* in *Ipomoea* because selective constraint, as measured by Ka/Ks ratios, remains strong on both copies (Yang et al., 2004). If one of these copies were undergoing pseudogenization, the Ka/Ks ratio would tend approach 1, rather than the observed low value of 0.055.

An alternative explanation is that the two copies are maintained because they have undergone subfunctionalization (Hughes, 1994; Force et al., 1999). Under this process, the expression domains may diverge through the accumulation of neutral mutations in *cis*-regulatory regions that reduce or eliminate expression in certain tissues or under certain conditions (e.g., drought stress, pathogen attack). Such mutations are expected to be neutral as long as they affect only one of the duplicate copies. Once one or more such mutations have accumulated in each copy, both copies are necessary to realize the same expression domain as the single ancestral copy. Consequently, both copies will be retained indefinitely by purifying selection.

If this explanation is correct, it would be expected that duplicate copies would exhibit at least partially distinct expression domains. This appears to be the case for *CHS-D* and *-E* (Durbin et al., 2003), as well as for *DFR-A*, *-B*, and *-C* in *Ipomoea* (Inagaki et al., 1999), for four distinct *CHS* genes in *Petunia* (Koes et al., 1989), and for the duplicate *CHS* genes *C2* and *WHP* in maize (Holton and Cornish, 1995).

Although the idea that subfunctionalization can explain maintenance of duplicate gene copies has become very popular, there are very few convincingly documented cases, including anthocyanin genes. Most claims of subfunctionalization rest on simple demonstration of noncongruent expression domains, as described above. However, noncongruent domains also can result from acquiring novel expression domains in different copies (e.g., Hansen et al., 1996). The only way to distinguish between these two causes of noncongruence is to carefully reconstruct phylogenetically the ancestral, single-copy expression domain and determine if the domains of the current duplicate copies are subsets of that ancestral domain. Flavonoid structural genes are an excellent candidate for this approach.

5.2. Duplication and divergence in flavonoid regulatory genes

Gene duplication and divergence also has played a significant role in the evolution of flavonoid regulation. Four classes of transcription factors have been implicated across angiosperms as contributing to regulation (Mol et al., 1998). Two of these, belonging to the *MYB* and *bHLH* gene families, themselves are found in multiple copies in plants. In maize, for example, MYB transcription factors include C1 and PL1. The two exhibit greater than 90% amino acid identity in their amino- and carboxyl-terminal domains, which participate in regulatory function. These two duplicates have diverged greatly, however, in their expression domains: *Pl1* activates anthocyanin production in vegetative tissues and flowers, while *C1* activates anthocyanin biosynthesis in the aleurone layer of kernels (Cone et al., 1993).

bHLH-Type flavonoid transcription factors are represented in maize by a half-dozen paralogs (Ludwig and Wessler, 1990) grouped in two sets. One set of apparent tandem duplicates is found on chromosome 10, while a second set is found on chromosome 2. These copies have largely nonoverlapping expression domains, and thus activate anthocyanin production in different tissues (Ludwig and Wessler, 1990). Nevertheless, particle bombardment experiments demonstrate that any of these copies, when driven by a constitutive promoter, will induce pigment formation in virtually any tissue to which it is introduced (Ludwig et al., 1990; Goff et al., 1990). Multiple copies of these transcription factors also are found in dicots. For example, in *Petunia*, at least three distinct MYB factors control pigmentation in different parts of the flower (Kroon, 2004), while *Arabidopsis* also has at least three such copies (Nesi et al., 2001; Borevitz et al., 2000).

Transient and stable transformation of anthocyanin regulatory genes from one species into corresponding LOF mutants in distantly related species repeatedly has demonstrated that copies from different species are functionally exchangeable (Quattrocchio et al., 1993, 1998; Lloyd et al., 1992). This observation, coupled with differentiation of expression domains among duplicated copies suggests that maintenance of multiple copies of these transcription factors may be the result of subfunctionalization, though the caveats described above apply here also. Regardless of the evolutionary processes that maintain the duplicate copies, however, their noncongruent expression domains presumably allow pigmentation patterns to evolve somewhat independently in different tissues, including different floral parts. Divergence among species in complex floral color patterning thus presumably has been rendered possible by prior duplication of anthocyanin regulatory genes.

6. PATHWAY DEGENERATION AND EVOLUTIONARY POTENTIAL

Anthocyanins are a highly diverse group of flavonoids. Several hundred different anthocyanins have been described from the gymnosperms and angiosperms. Nevertheless, this great diversity is built upon structural elaboration of a small number of anthocyanidins, which form the backbone of every anthocyanin molecule. There are six commonly occurring anthocyanidins, plus another ten or so that occur sporadically in these plant groups (Timberlake and Bridle, 1980). The diversity of anthocyanins results from the myriad of ways these few anthocyanidins can be decorated by the addition of sugars.

The six common anthocyanidins are the product of three different branches of the anthocyanin pathway (Figure 7.1). One branch gives rise to a single unmethylated anthocyanidin, pelargonidin, which tends to produce red or orange anthocyanins. A second branch gives rise to two anthocyanidins, the unmethylated cyanidin and the singly methylated peonidin. Anthocyanins derived from these compounds tend to be blue or magenta. The branching enzyme leading to this branch is F3′H, which adds a hydroxyl group to the 3′ carbon of the anthocyanidin skeleton. Finally, a third branch gives rise to three anthocyanidins, the unmethylated delphinidin, the singly methylated petunidin, and the doubly methylated malvidin, which tend to produce blue-purple pigments. The branching enzyme associated with this part of the pathway is F3′5′H, which adds hydroxyl groups to the 3′ and 5′ carbons of the skeleton.

Flower color is in large part determined by which of these branches is most active in a particular species. In *Ipomoea*, for example, blue- and purple-flowered species tend to produce almost exclusively cyanidin-based anthocyanins. Pathway flux in these species is almost entirely down the second branch. However, mutations that knock out the enzyme F3′H redirect flux down the pelargonidin branch and result in red flowers (Zufall and Rausher, 2003; Hoshino et al., 2003). Red-flowered *Ipomoea* species also almost always produce pelargonidin-based rather than cyanidin-based, anthocyanins (Zufall and Rausher, 2004; Zufall, 2003). A similar pattern is seen in *Penstemon* (Scrophulariaceae), in which blue/purple, bee-pollinated flowers tend to produce delphinidin-derived anthocyanins, while red, hummingbird-pollinated flowers tend to produce pelargonidins (Scogin and Freeman, 1987). Adjusting the relative amounts of flux down the different pathway branches thus seems to be a common way of altering flower color.

Several lines of evidence indicate that these three branches evolved very early in the diversification of plants. First, the ability to produce all three classes of anthocyanidins is exhibited by most angiosperm orders as well as many gymnosperms (Figure 7.3). Second, the hydroxylating enzymes F3′H and F3′5′H, exhibit high-sequence homology throughout these two groups of plants, indicating that they were each recruited into flavonoid metabolism once very early in the diversification of the land plants. Flavonols derived from all three of the branches have been reported in ferns (Wollenweber and Schneider, 2000) and from the first two branches in bryophytes (Webby et al., 1996).

Despite the near taxonomic ubiquity at the level of plant orders of these three branches of the flavonoid pathway there are numerous known instances in which one or more of the branches has been inactivated (Table 7.1). At least two types of genetic changes have been associated with this inactivation. In some genera (*Arabidopsis*, *Petunia*, *Cymbidium*), enzymes below the branch point, such as DFR, have evolved to be substrate specialists. For example, in *Petunia* no pelargonidin-derived anthocyanins are produced because the *Petunia* DFR does not metabolize dihydrokaempferol, a precursor of pelargonidin. In other genera (*Ipomoea, Chrysanthemum, Rosa, Dianthus*), the branching enzymes themselves have been inactivated. Thus, morning glories (*Ipomoea*) do not produce derivatives of delphinidin because they lack the enzyme F3'5'H. Extensive surveys of land plants for the types of flavonoids present (Markham, 1988; Giannasi, 1988; Niemann, 1988; Williams and Harborne, 1988) also suggest that knockouts of one or more of these three branches may be common, since there are numerous plant orders in which compounds associated with particular branches are apparently absent (Figure 7.3).

Table 7.1 *Taxa for which a branch of the anthocyanidins pathway has been inactivated*

Taxon	Family	Branch Inactivated	Implicated Enzyme
Arabidopsis	Brassicaceae	pelargonidin	DFR substrate specialization
Chrysanthemum	Asteridae	delphinidin	F3'5'H activity absent
Cymbidium	Orchidaceae	pelargonidin	DFR substrate specialization
Dianthus	Caryophyllaceae	delphinidin	F3'5'H activity absent
Ipomoea	Convolvulaceae	delphinidin	F3'5'H activity absent
Petunia	Solanaceae	pelargonidin	DFR substrate specialization
Rosa	Rosaceae	delphinidin	F3'5'H activity absent

On theoretical grounds, the evolution of redundant knockouts in a branch of the pathway may be expected, because once a branch has been inactivated by a knockout in one enzyme it is unlikely that purifying selection will act on other enzyme functions associated with that pathway. Thus, in the absence of a DFR that metabolizes dihydrokaempferol, selection on ANS and UF3GT to maintain the ability to utilize leucopelargonidin and pelargonidin will be removed. The genes coding for these enzymes therefore will tend to accumulate mutations that inhibit these activities.

Although it is clear that in the species listed in Table 7.1, at least one enzyme has been inactivated, for most of these species it is not known whether other enzymes associated with the same pathway branch also may have been redundantly inactivated. In *Petunia*, however, transformation of a strain lacking DFR, F3'H, and F3'5'H with a *Gerbera* DFR, which efficiently metabolizes both DHK and DHQ, produces copious pelargonidin-based anthocyanins (Johnson et al., 1999), indicating that enzymes downstream of DFR have not degenerated in *Petunia*. By contrast, transformation of maize *DFR* into a DFR- and F3'H-deficient *Arabidopsis* strain produces only minimal amounts of pelargonidin-derived anthocyanins (Dong et al., 2001), suggesting that one or more downstream genes in *Arabidopsis* may be at least partially degenerate with respect to metabolizing pelargonidin precursors or derivatives.

This type of pathway "degeneration" is likely to have important implications for future evolutionary potential. When a branch of the pathway has been inactivated because of a change in only one enzyme, it is possible for back-mutation to restore that branch if it becomes advantageous to do so (e.g., if a different flower color becomes advantageous). However, once two or more enzymes acquire inactivating changes, the chances of restoring the lost branch become much more remote for the simple reason that the chances of two or more back mutations occurring simultaneously is exceedingly small. Branch inactivation thus is likely to be permanent, thus restricting the evolutionary possibilities for ecologically important characters such as flower color. This phenomenon is likely to reinforce the apparent tendency for unidirectionality of flower color change (Figure 7.6) (see above).

A recent investigation of changes in anthocyanin pathway enzymes in the morning glory *Ipomoea quamoclit* provides an apparent example of this type of pathway degeneration (Zufall and Rausher, 2004). *I. quamoclit* is a member of a small clade of species, all of which have red or orange flowers pollinated by hummingbirds. In addition to color, flowers of these species exhibit several morphological features characteristic of the bird pollination syndrome (Faegeri and van der Pijl, 1996). Their closest known relatives, *I. ternifolia* and *I. tenuiloba*, exhibit the ancestral bee pollination syndrome, which includes blue/purple flowers. It thus appears that the evolutionary transition from blue to red flowers in the ancestral lineage leading to the *I. quamoclit* group constituted an adaptation to exploit a novel pollinator.

As described above, three different genetic changes in the anthocyanin pathway have been identified in *I. quamoclit*. In flowers, the branching enzyme F3'H has been significantly down-regulated. When transformed using a constitutive promoter into an *Arabidopsis* strain lacking F3'H function, the *F3'H* gene from *I. quamoclit* fails to restore anthocyanin production, whereas the same gene from the blue-flowered *I. purpurea* successfully complements the *Arabidopsis* knockout. This result suggests that the functioning of *I. quamoclit* F3'H *in planta* also is impaired. Finally, as demonstrated both by enzyme assays and by complementation tests in *Arabidopsis*, DFR from *I. quamoclit* has lost the ability to metabolize dihydroquercetin, the precursor of cyanidin.

Although it has not yet been definitively proven, it is likely that any one of these changes is sufficient to inactivate the cyanidin branch of the anthocyanin pathway in *I. quamoclit*, thus forcing flux down the pelargonidin branch to produce red pigments. This conclusion, in turn, suggests that the evolutionary transition from blue to red flowers was caused by one of these genetic changes, while the other two represent subsequent degeneration of that branch of the pathway. Moreover, it seems unlikely that species in the *I. quamoclit* group would be able to reinstate a functional cyanidin branch because it would require multiple simultaneous mutations. If in the future ecological conditions make it advantageous for one of these species to utilize bees rather than hummingbirds as pollinators, prior adaptation and accompanying pathway degeneration may make this nearly impossible. It will be of interest to determine whether the other species listed in Table 7.1 have similarly undergone pathway degeneration, and thus have a reduced evolutionary potential.

7. CONCLUSIONS

Several important themes regarding the evolution of flavonoids and their genes emerge from the above considerations. First, the flavonoid metabolic network has been and continues to be evolutionarily very fluid, both in terms of its structure and of its regulation. Gene duplications have been co-opted repeatedly for the production of novel enzymes that produce novel flavonoids. In many cases (e.g., stilbene synthases, glutathione S-transferases, and probably isoflavone synthase), enzymes with the same function have been recruited independently two or more times. While this creative assembly of the network has been occurring constantly since plants colonized land, loss of components of the network apparently also has occurred frequently (e.g., parallel inactivation of pelargonidin, cyanidin, and/or delphinidin branches in different taxa, perhaps loss of ability to produce isoflavonoids). These two processes, working in tandem, are in large part responsible for the enormous variation in types of flavonoids produced by different plants.

A second important theme is that pleiotropy is likely to play a major role in determining which flavonoid genes participate in adaptive evolutionary change and how rapidly those genes evolve. While evolutionary biologists have long recognized that pleiotropic costs theoretically can constrain adaptive evolutionary change, it has been difficult to study this phenomenon empirically. The flavonoid pathway holds great promise as a model system for examining the evolutionary consequences of pleiotropy because in several plant groups there have been repeated parallel instances of shifts in flower color. Nature thus has provided us with a natural experiment from which we should be able to determine whether evolution has repeatedly used the same genes to create similar phenotypes, and if so whether it has done so because changes in those genes incur relatively small pleiotropic fitness decrements.

A final theme emerging from this review is the importance of loss-of-function mutations in adaptive evolutionary change. Often biologists tend to think of adaptation as a constructive process that adds novel characters to a preexisting phenotype. As should be clear, the assembly of the flavonoid pathway reflects just this type of process and demonstrates the importance of gene duplication and neofunctionalization in creating new characters. However, the numerous examples described above of inactivation of branches of the flavonoid pathway suggest that evolutionary change often involve destructive processes. In a few cases, such as the evolution of red flowers in hummingbird-pollinated *Ipomoea*, it is clear that pathway inactivation was the mechanism of natural selection used to produce an adaptive phenotype. In most other cases, however, it remains to be determined whether pathway inactivation arises from fixation of neutral inactivating mutations by genetic drift when certain flavonoids are no longer needed or whether selection actually turns off the pathway. In either case, however, loss of function is likely to constrain the possible directions in which future evolutionary change can occur.

8. REFERENCES

Alfenito, M. R., Souer, E., Goodman, C. D., Buell, R., Mol, J., Koes, R., and Walbot, V., 1998, Functional complementation of anthocyanin sequestration in the vacuole by widely divergent glutathione S-transferases, *Plant Cell* **10**: 1135-1149.

Anhut, S., Zinsmeister, H. D., Mues, R., Barz, W., Mackenbrock, K., Koster, J., and Markham, K. R., 1984, The first identification of isoflavones from a bryophyte, *Phytochem* **23**: 1073-1075.

Baker, M. E., and Blasco, R., 1992, Expansion of the mammalian 3β-hydroxysteroid dehydrogenase/plant dihydroflavonols reductase superfamily to include bacterial cholesterol dehydrogenase, a bacterial UDP-galactose-4-epimerase and open reading frames in vaccinia virus and fish lymphocystis disease virus, *FEBS Lett* **301**: 89-93.

Barrier, M., Robichaux, R. H., and Purugganan, M. D., 2001, Accelerated regulatory gene evolution in an adaptive radiation, *Proc Natl Acad Sci USA* **98**: 10208-10213.

Bohm, B. A., 1975, Chalcones, aurones and dihydrochalcones, in *The Flavonoids*, Harborne, J. B., Mabry, T. J., and Mabry, H., eds. Academic Press, New York, pp. 442-504.

Britten, R., and Davidson, E., 1969, Gene regulation for higher cells: A theory, *Science* **165**: 349-357.

Britten, R., and Davidson, E., 1971, Repetitive and non-repetitive DNA sequences and a speculation on the origins of evolutionary novelty, *Quart Rev Biol* **46**: 111-133.

Brown, B. A., and Clegg, M. T., 1984, Influence of flower color polymorphism on genetic transmission in a natural population of the common morning glory, *Ipomoea purpurea*, *Evolution* **38**: 796-803.

Brown, D. E., Rashotte, A. M., Murphy, A. S., Normanly, J., Tague, B. W., Peer, W. A., Taiz, L., and Muday, G. K., 2001, Flavonoids act as negative regulatorys of auxin transport *in vivo* in *Arabidopsis*, *Plant Physiol* **126**: 524-535.

Borevitz, J. O., Xia, Y., Blount, J., Dixon, R. A., and Lamb, C., 2000, Activation tagging identifies a conserved MYB regulator of phenylpropanoid biosynthesis, *Plant Cell* **12**: 2383-2394.

Chang, S-M., Lu, Y., and Rausher, M. D., 2005, Neutral evolution of the non-binding region of the anthocyanin regulatory gene *Ipmyb1* in *Ipomoea*, *Genetics* **170**: 1967–1978.

Clegg, M. T., and Durbin, M. L., 2003, Tracing floral adaptations from ecology to molecules, *Nat Rev Gen* **4**: 206-215.

Coberly, L. S., 2003, The cost of white flosers: pleiotropy and the evolution of flower color, Ph.D. dissertation, Duke University, Durham, NC.

Coberly, L. C., and Rausher, M. D., 2003, Analysis of a chalcone synthase mutant in *Ipomoea purpurea* reveals a novel function for flavonoids: amelioration of heat stress, *Mol Ecol* **12**: 1113-1124.

Coe, E. H., McCormick, S., and Modena, S. A., 1981, White pollen in maize, *J Heredity* **72**: 318-320.

Cone, K. C., Cocciolone, S. M., Burr, F. A., and Burr, B., 1993, Maize anthocyanin regulatory gene *pl* is a duplicate of *c1* that functions in the plant, *Plant Cell* **5**: 1795-1805.

Crawford, D. L., and Powers, D. A., 1989, Molecular basis of evolutionary adaptation at the lactate dehydrogenase-B locus in the fish *Fundulus heteroclitus*, *Proc Natl Acad Sci USA* **86**: 9365-9369

Curir, P., Dolci, M., and Galeotti, F., 2005, A phytoalexin-like flavonol involved in the carnation *(Dianthus caryopohyllus)-Fusarium oxysporum* f. Sp *dianthi* pathosystem, *J Phytopath* **153**: 65-67.

Dewick, P. M., 1988, Isoflavonoids, in *The Flavonoids. Advances in Research Since 1980*, Harborne, J. B., ed., Chapman and Hall, London, pp. 125-209.

Dias, A. P., Braun, E. L., McMullen, M. D., and Grotewold, E., 2003, Recently duplicated maize *R2R3 Myb* genes provide evidence for distinct mechanisms of evolutionary divergence after duplication, *Plant Physiol* **131**: 610-620.

Dickinson, W., 1991, The evolution of regulatory genes and patterns in *Drosophila*, *Evol Biol* **25**: 127-174.

Dixon, R. A., and Steele, C. L., 1999, Flavonoids and isoflavonoids—a gold mine for metabolic engineering, *Trends Plant Sci* **10**: 394-400.

Doebley, J., 1993, Genetics, development and plant evolution, *Curr Opin Genet Dev* **3**: 865-872.

Doebley, J., and Lukens, L., 1998, Transcriptional regulators and the evolution of plant form, *Plant Cell* **10**: 1075-1082.

Dong, X., Braun, E.L., and Grotewold, E. (2001). Functional conservation of plant secondary metabolic enzymes revealed by complementation of Arabidopsis flavonoid mutants with maize genes. *Plant Physiol.* **127**, 46-57.

Durbin, M. L., Learn, G. H., Huttley, G. A., and Clegg, M. T., 1995, Evolution of the chalcone synthase gene family in the genus *Ipomoea*, *Proc Natl Acad Sci USA* **92**: 3338-3342.

Durbin, M. L., Lundy, K. E., Morrell, P. L., Torres-Martinez, C. L., and Clegg, M. T., 2003, Genes that determine flower color: the role of regulatory changes in the evolution of phenotypic adaptations, *Mol Phyl Evol* **29**: 507-518.

Ennos, R. A., and Clegg, M. T., 1983, Flower color variation in the morning glory, *Ipomoea purpurea*, *J. Heredity* **74**: 247-250.

Epperson, B. K., and Clegg, M. T., 1987, Instability at a flower color locus in the morning glory, *J Heredity* **78**: 346-352.

Faegri, D.T., and van der Pijl, L., 1966, *The Principles of Pollination Ecology*, Pergamon, Oxford.

Feeny, P., 1970, Seasonal changes in oak leaf tannins and nutrients as a cause of spring feeding by winter moth caterpillars, *Ecology* **51**: 565-581.

Fehr, C., and Rausher, M. D., 2004, Effects of variation at the flower-color *A* locus on mating system parameters in *Ipomoea purpure*, *Mol Ecol* **13**: 1839-1847.

Fineblum, W. L., and Rausher, M. D., 1997, Do genes influencing floral pigmentation also influence resistance to herbivores and pathogens? The *W* locus in *Ipomoea purpurea*, *Ecology* **78**: 1646-1654.

Force, A., Lynch, M., Pickett, F. B., Amores, A., Yan, Y., and Postlethwait, J., 1999, Preservation of duplicate genes by complementary, degenerative mutations, *Genetics* **151**: 1531-1545.

Fukada-Tanaka, S., Hoshino, A., Hisatomi, Y., Habu, Y., Hasebe, M., and Iida, S., 1997, Identification of new chalcone synthase genes for flower pigmentation in the Japanese and common morning glories, *Plant Cell Physiol* **38**: 88-96.

Gensheimer, M., and Mushegian, A., 2004, Chalcone isomerase family and fold: no longer unique to plants, *Protein Sci* **13**: 540-544

Giannasi, D. E., 1988, Flavonoids and evolution in the dicotyledons, in *The Flavonoids. Advances in Research Since 1980*, Harborne, J. B., ed., Chapman and Hall, London, pp. 479-504.

Gillespie, J. H., 1991, *The Causes of Molecular Evolution*, Oxford Univ. Press, Oxford, UK.

Goff, S. A., Cone, K. C., and Chandler, V. L., 1992, Functional analysis of the transcription activator encoded by the maize B-gene: evidence for a direct functional interaction between two classes of regulatory proteins, *Genes Dev* **6**: 864-875.

Goff, S. A., Cone, K. C., and. Fromm, M. E, 1991, Identification of functional domains in the maize transcriptional activator C1: comparison of wild-type and dominant inhibitor proteins, *Genes Dev* **5**: 298-309.

Goff, S. A., Klein, T. M., Roth, B. A., Fromm, M. E., Cone, K. C., Radicella, J. P., and Chandler, V. L., 1990, Transactivation of anthocyanin biosynthetic genes following transfer of *B*-regulatory genes into maize tissues, *EMBO J* **9**: 2517-2522.

Grotewold, E., Sainz, M.B., Tagliani, L., Hernandez, J.M., Bowen, B., and Chandler, V.L., 2000, Identification of the residues in the Myb domain of maize C1 that specify the interaction with the bHLH cofactor *R. Proc Natl Acad Sci* USA **97**, 13579-13584.

Habu, Y., Hisatomi, Y., and Iida, S., 1998, Molecular characterization of the mutable *flaked* allele for flower variegation in the common morning glory, *Plant J* **16**: 371-376.

Hansen, M. A., Gaut, B. S., Stec, A. O., Fuerstenberg, S. I., Goodman, M. M., Coe, E. H., and Doebley, J. F., 1996, Evolution of anthocyanin biosynthesis in maize kernels: The role of regulatory and enzymatic loci, *Genetics* **143**: 1395-1407.

Hartl, D. L., and Clark, A. G., 1989, *Principles of Population Genetics, 2nd Edition*, Sinauer, Sunderland, MA.

Helariutta, Y., Kotilainen, M., Elomaa P., Kalkkinen, N., Bremer, K., Teeri, T. H. and Albert, V. A., 1996, Duplication and functional divergence in the chalcone synthase gene family of Asteraceae: Evolution with substrate change and catalytic simplification, *Proc Natl Acad Sci USA* **93**: 9033-9038.

Herles, C., Braune, A., and Blaut, M., 2004, First bacterial chalcone isomerase isolated from Eubacterium ramulus, *Arch Microbiol* **181**: 428-434

Hochachka, P. W., and Somero, G. N., 1984, *Biochemical Adaptation*, Princeton Univ. Press, Princeton.

Holton, T. A., and Cornish, E. C., 1995, Genetics and biochemistry of anthocyanin biosynthesis, *Plant Cell* **7**: 1071-1083.

Holton, T. A., Brugliera, F., and Tanaka, Y., 1993, Cloning and expression of flavonol synthase from *Petunia hybrid, Plant J* **4**: 1003-1010.

Hoshino, A., Moriga, Y., Choi, J.-D., Saito, N., Toki, K., Tanaka, Y., and Iida, S., 2003, Spontaneous mutations of the flavonoid 3'-hydroxylase gene conferring reddish flowers in the three morning glory species, *Plant Cell Physiol* **44**: 990-1001.

Hughes, A. L. 1994, The evolution of functionally novel proteins after gene duplication, *Proc Roy Soc Lond B.* **256**: 119-124.

Huits, H.S.M., Gerats, A.G.M., Kreike, M.M., Mol, J.N.M., and Koes, R.E., 1994, Genetic control of dihydroflavonol 4-reductase gene expression in *Petunia hybrida*, *Plant J* **6**: 295-310.

Inagaki, Y., Johzuka-Hisatomi, Y., Mori, T., Takahashi, S., Hayakawa, Y., Peyachoknagul, S., Ozeki, Y., and Iida, S., 1999, Genomic organization of the genes encoding dihydroflavonols 4-reductase for flower pigmentation in the Japanese and common morning glories, *Gene* **226**: 181-188.

Jez, J. M., Bowman, M. E., Dixon, R. A., and Noel, J. P., 2000, Structure and mechanism of chalcone isomerase: an evolutionarily unique enzyme in plants, *Nat Struct Biol* **7**: 786-791.

Jia, L, Clegg, M. T., and Jiang, T., 2003, Excess nonsynonymous substitutions suggest that positive selection episodes operated in the DNA-binding domain evolution of *Arabidopsis* R2R3-MYB genes, *Plant Mol Biol* **52**: 627-642.

Jia, L, Clegg, M. T., and Jiang, T., 2004, Evolutionary dynamics of the DNA-binding domains in putative *R2R3-myb* genes identified from rice subspecies *indica* and *japonica* genomes, *Plant Physiol* **134**: 575-585.

Johnson, E. T., Hankuil, Y., Shin, B., Oh, B-J., Cheong, H., and Choi, G., 1999, *Cymbidium hybrida* dihydroflavonols 4-reductase does not efficiently reduce dihydrokaempferol to produce orange pelargonidin-type anthocyanins, *Plant J* **19**: 81-85.

Kemp, M. S., and Burden, R. S., 1986, Phytoalexins and stress metabolites in the sapwood of trees, *Phytochemistry* **25**: 1261-1269.

King, M.-C., and Wilson, A. C., 1975, Evolution at two levels in humans and chimpanzees, *Science* **188**: 107-116.

Koes, R.E., Spelt, C.E., and Mol, J.N., 1989, The chalcone synthase multigene family of *Petunia hybrida* (V30): Differential, light-regulated expression during flower development and UV light induction, *Plant Mol Biol* **12**: 213-225.

Koes, R. E., Quattrocchio, R., and Mol, J. N. M., 1994, The flavonoid biosynthetic pathway in plants: function and evolution, *BioEssays* **16**: 123-132.

Kroon, A. R., 2004, *Transcription Regulation of the Anthocyanin Pathway in Petunia hybrida,* Ph.D. dissertation, Vrije Universiteit, Amsterdam, The Netherlands.

Kubasek, W. L., Shirley, B. W., McKillop, A., Goodman, H. M., Briggs, W., and Ausubel, F. M., 1992, Regulation of flavonoid biosynthetic genes in germinating *Arabidopsis* seedlings, *Plant Cell* **4**: 1229-1236.

Li, W-H., 1997, *Molecular Evolution*, Sinauer, Sunderland, MA.

Li, J., Ou-Lee, T. M., Raba, R., Amundson, R. G., and Last, R. L., 1993, *Arabidopsis* flavonoid mutants are hypersensitive to UV-B radiation, *Plant Cell* **5**: 171-179.

Lloyd, A. M., Walbot, V., and Davis, R. W., 1992, *Arabidopsis* and *Nicotiana* anthocyanin production activated by maize regulators *R* and *C1*, *Science* **258**: 1773-1775.

Lois, R., and Buchanan, B. B., 1994, Severe sensitivity to ultraviolet radiation in an *Arabidopsis* mutant deficient in flavonoid accumulation, *Planta* **194**: 504-509.

Lu, Y., and Rausher, M. D., 2003, Evolutionary rate variation in anthocyanin pathway genes, *Mol Biol Evol* **20**: 1844-1853.

Ludwig, S., Bowen, B., Beach, L. and Wessler, S., 1990, A regulatory gene as a novel visible marker for maize transformation, *Science* **247**: 449-450.

Ludwig, S.R., and Wessler, S.R., 1990, Maize *R* gene family: Tissue specific helix-loop-helix proteins, *Cell* **62**: 849-851.

Mackenzie, P.I., Owens, I. S., Burchell, B., Bock, K. W., Bairoch, A., Belanger, A., FournelGigleux, S., Green, M., Hum, D. W., Iyanagi, T., Lancet, D., Louisot, P., Magdalou, J., Chowdhury, J. R., Ritter, J. K., Schachter, H., Tephly, T. R., Tipton, K. F., and Nebert, D. W., 1997, The UDP glycosyltransferase gene superfamily: Recommended nomenclature update based on evolutionary divergence, *Pharmacogenetics* **7**: 255-269.

Markham, K. R., 1988, Distribution of flavonoids in the lower plants and its evolutionary significance, in *The Flavonoids. Advances in Research Since 1980*, Harborne, J. B., ed., Chapman and Hall, London, pp. 427-468.

Martin, C., Prescott, A., Mackay, S., Bartlett, J., and Vrijlandt, E., 1991, Control of anthocyanin biosynthesis in flowers of *Antirrhinum majus, Plant J* 1: 37-49

Miller, R. E., McDonald, J. A., and Manos, P. S., 2004, Systematics of *Ipomoea* subgenus *Quamoclit* (Convolvulaceae) based on ITS sequence data and a Bayesian phylogenetic analysis, *Am J Bot* **91**: 1208-1218.

Miller, R. E., Rausher, M. D., and Manos, P. S., 1999, Phylogenetic systematics of *Ipomoea* (Convolvulaceae) based on ITS and *waxy* sequences, *Syst Bot* **24**:209-227.

Mo, Y., Nagel, C., and Taylor, L. P., 1992, Biochemical complementation of chalcone synthase mutants defines a role for flavonols in functional pollen, *Proc Nat Acad Sci USA* **89**:7213-7217.

Mojonnier, L., and Rausher, M. D., 1997, Selection on a floral color polymorphism in the common morning glory (*Ipomoea purpurea*): The effects of overdominance in seed size, *Evolution* **51**: 608-614.

Mol, J., Grotewold, E., and Koes, R., 1998, How genes paint flowers and seeds. *Trends Plant Sci.* **3**, 212-217.

Moyano, E., Mart'nez-Garcia, J.F. and Martin, C., 1996, Apparent redundancy in *myb* gene function provides gearing for the control of flavonoid biosynthesis in *Antirrhinum* flowers, *Plant Cell* **8**: 1519-1532.

Nakayama, T., Yonekura-Sakakibara, K., Sato, T., Kikuchi, S., Fukui, Y., Fukuchi-Mizutani, M., Ueda, T., Nakao, M., Tanaka, Y., Kusumi, T., and Nishino, T., 2000, Aureusidin synthase: a polyphenol oxidase homolog responsible for flower coloration, *Science* **290**: 1163-1166.

Nakayama, T., Sato, T., Fukui, Y., Yonekura-Sakakibara, K., Hayashi, H., Tanaka, Y., Kusumi, T., and Nishino, T., 2001, Specificity analysis and mechanism of aurone synthesis catalyzed by aureusidin synthase, a polyphenol oxidase homolog responsible for flower coloration, *FEBS Lett.* **499**: 107-111.

Nesi, N., Jond., C., Debeaujon, I., Caboche, M., and Lepiniec, L., 2000, The *TT8* gene encodes a basic helix-loop-helix domain protein required for expression of *DFR* and *BAN* genes in *Arabidopsis* siliques, *Plant Cell* **13**: 2099-2114.

Nesi, N., Jond, C., Debeaujon, I., Caboche, M., and Lepiniec, L., 2001, The Arabidopsis TT2 gene encodes an R2R3 MYB domain protein that acts as a key determinant for proanthocyanidin accumulation in developing seed. *Plant Cell* **13**, 2099-2114.

Newcomb, R. D., Campbell, P. M., Ollis, D. L., Cheah, E., Russell, R. J., and Oakeshott, J. G., 1997, A single amino acid substitution converts a carboxylesterase to an organophosphorus hydrolase and confers insecticide resistance on a blowfly, *Proc Nat Acad Sci USA* **94**: 7464-7468.

Niemann, G. J., 1988, Distribution and evolution of the flavonoids in gymnosperms, in *The Flavonoids. Advances in Research Since 1980*, Harborne, J. B., ed., Chapman and Hall, London, pp. 469-478.

Patterson, G. I., Kubo, K. M., Shrover, T. and Chandler, V. L., 1995, Sequences required for paramutation of the maize *b* gene map to a region containing the promoter and upstream sequences, *Genetics* **140**: 1389-1406.

Pelletier, M. K., Murrell, J. R., and Shirley, B. W., 1997, Characterization of flavonol synthase and leucoanthocyanidins dioxygenase genes in Arabidopsis, *Plant Physiol* **113**: 1437-1445.

Pelletier, M.K., and Shirley, B.W., 1996, Analysis of flavanone 3-hydroxylase in *Arabidopsis* seedlings. Coordinate regulation with chalcone synthase and chalcone isomerase, *Plant Physiol* **111**: 339-345.

Peer, W. A., Bandyopadhyay, A., Blakeslee, J. JU., Makam, S. N., Chen, R. J., Masson, P. H., and Murphy, A. S., 2004, Variation in expression and protein localization of the PIN family of auxin efflux facilitator proteins in flavonoid mutants with altered auxin transport in *Arabidopsis thaliana, Plant Cell* **16**: 1898-1911.

Ptashne, M., 1988 How eukaryotic transcriptional activators work, *Nature* **335**: 683-689.

Purrington, C. B., and Bergelson, J., 1997, Fitness consequences of genetically engineered herbicide and antibiotic resistance in *Arabidopsis thaliana, Genetics* **145**: 807-814.

Purugganan, M. D., 1998, The molecular evolution of development, *BioEssays* **20**: 700-711.

Purugganan, M. D., Rounsley, S. D., Schmidt, R. J., and Yanofsky, M. F., 1995, Molecular evolution of flower development: diversification of the plant MADS-box regulatory gene family, *Genetics* **140**: 345-356.

Purugganan, M. D., and Wessler, S. R., 1994, Molecular evolution of the plant *R* regulatory gene family, *Genetics* **138**: 849-854.

Quattrocchio, F., Wing, J.F., Leppen, H.T.C., Mol, J.N.M., and Koes, R.E., 1993, Regulatory genes controlling anthocyanin pigmentation are functionally conserved among plant species and have distinct sets of target genes, *Plant Cell* **5**: 1497-1512.

Quattrocchio, F., Wing, J.F., van der Woude, K., Mol, J.N.M., and Koes, R., 1998, Analysis of bHLH and MYB-domain proteins: Species-specific regulatory differences are caused by divergent evolution of target anthocyanin genes, *Plant J* **13**: 475–488.

Quattrocchio, F., Wing, J., van der Woude, K., Souer, E., de Vetten, N., Mol, J., and Koes, R., 1999, Molecular analysis of the *anthocyanin2* gene of Petunia and its role in the evolution of flower color, *Plant Cell* **11**: 1433-1444.

Radicella, J. P., Brown D., Tolar, L. A. and Chandler, V. L., 1992, Allelic diversity of the maize *B* regulatory gene: different leader and promoter sequences of two *B* alleles determine distinct tissue specificities of anthocyanin production, *Genes Dev* **6**: 2152-2164.

Rausher, M.D., Augustine, D., and Vanderkoi, A., 1983, Absence of pollen discounting in a genotype of *Ipomoea purpurea* exhibiting increased selfing, *Evolution* **47**: 1688–1695.

Rausher, M. D., and Fry, J. D., 1993, Effects of a locus affecting floral pigmentation in *Ipomoea purpurea* on female fitness components, *Genetics* **134**: 1237-1247.

Rausher, M. D., Miller, R. E., and Tiffin, P., 1999, Patterns of evolutionary rate variation among genes of the anthocyanin biosynthetic pathway, *Mol Biol Evol* **16**: 266-274.

Remington, D. L., and Purugganan, M. D., 2002, *GAI* homologues in the Hawaiian Silversword alliance (Asteraceae-Madiinae): molecular evolution of growth regulators in a rapidly diversifying plant lineage, *Mol Biol Evol* **19**: 1563-1574.

Ryan, K. G., Swinny, E. E., Winefield, C., and Markham, K. R., 2001, Flavonoids and UV photoprotection in *Arabidopsis* mutants, *Zeitschrift fur Naturforschung (Section C)* **56**: 745-754.

Ryan, K. G., Swinny, E. E., Markham, K. R. , and Winefield, C., 2002, Flavonoid gene expression and UV photoprotection in transgenic and mutant *Petunia* leaves, *Phytochem* **2002**: 59:23-32.

Sainz, M. B., Goff, S. A., and Chandler, V. L., 1997, Extensive mutagenesis of a transcriptional activation domain identifies single hydrophobic and acidic amino acids important for activation *in vivo*, *Mol Cell Biol* **17**: 115-122.

Savolainen, V., and Chase, M. W., 2003, A decade of progress in plant molecular phylogenetics, *Trends Genet* **19**: 717- 724.

Scogin, R., and Freeman, C. E., 1987, Floral anthocyanins of the genus *Penstemon*: Correlations with taxonomy and pollination, *Bioch Syst Ecol* **15**: 355-360.

Shirley, B. W., 1996, Flavonoid biosynthesis: "new" functions for an "old" pathway, *Trends Plant Sci* **1**: 377-382.

Stafford, H. A., 1991, Flavonoid evolution: an enzymatic approach, *Plant Physiol* **96**: 680-685.

Tiffin, P., Miller, R. E., and Rausher, M. D., 1998, Control of expression patterns of anthocyanin structural genes by two loci in the common morning glory, *Genes Gen Syst* **73**: 105-110.

Timberlake, C. F., and Bridle, P., 1980, The anthocyanins, in *The Flavonoids*, Harborne, J. B., Mabry, T. J., and Mabry, H., eds., Academic Press, New York, pp. 214-266.

Ting, C. T., Tsaur, S. C., Wu, M. L., and Wu, C.-I., 1998, A rapidly evolving homeobox at the site of a hybrid sterility gene, *Science* **282**: 1501-1504.

Tropf, S., Lanz, T., Rensing, S. A., Schroeder, J., and Schroeder, G., 1994, Evidence that stilbene synthases have developed from chalcone synthases several times in the course of evolution, *J Mol Evol* **38**: 610-618.

Van der Meer, I. M., Stam, M., van Tunen, A. J., Mol, J. N. M., and Stuitje, A. R., 1992, Antisense inhibition of flavonoid biosynthesis in Petunia anthers results in male sterility, *Plant Cell* **4**: 253-262.

van Houwelingen, A., Souer, E., Spelt, C., Kloos, D., Mol, J., and Koes, R., 1998, Analysis of flower pigmentation mutants generated by random transposon mutagenesis in *Petunia hybrida*, *Plant J* **13**: 39–50.

Van Tunen, A. J., Koes, R. E., Spelt, C. E., van der Krol, A. R., Stuitje, A. R., and Mol, J. N. M., 1988, Cloning of the two chalcone flavanone isomerase genes from *Petunia hybrida*: coordinate, light-regulated and differential expression of flavonoid genes, *EMBO J* **7**: 1257-1263.

Verwoert, I. I. G. S., Verbree, E. C., Van der Linden, K. H., Nijkamp, H. J. J., and Stuitje, A. R., 1992, Cloning, nucleotide sequence and expression of the *Escherichi coli fabD* gene, encoding malonyl coenzyme A-acyl carrier protein transacylase, *J Bact* **174**: 2851-2857.

Watt, W. B., 1977, Adaptation at specific loci 1. Natural selection on phosphoglucose isomerase of *Colias* butterflies—biochemical and population aspects, *Genetics* **87**: 177–194.

Watt, W.B., 1983, Adaptation at specific loci. 2. Demographic and biochemical-elements in the maintenance of the *Colias* PGI polymorphism, *Genetics* **103**: 691–724.

Webby, R. F., Markham, K. R., and Smith, R. I. L., 1996, Chemotypes of the Antarctic moss *Bryum algens* delineated by their flavonoid constituents, *Bioch Syst Ecol* **24**: 469-475.

Williams, C. E., and Grotewold, E., 1997, Differences between plant and animal Myb domains are fundamental for DNA binding activity and chimeric Myb domains have novel DNA-binding specificities, *J Biol Chem* **272**: 563-571.

Williams, C. A., and Harborne, J. B., 1988, Distribution and evolution of flavonoids in the monocotyledons, in *The Flavonoids. Advances in Research Since 1980*, Harborne, J. B., ed., Chapman and Hall, London, pp. 505-524.

Wilson, P., Castellanos, M. C., Hogue, J. N., Thomson, J. D. and Armbruster, W. S., 2004, A multivariate search for pollination syndromes among penstemons, *Oikos* **104**: 345-361.

Winkel-Shirley, B., 2001, Flavonoid biosynthesis. A colorful model for genetics, biochemistry, cell biology and biotechnology, *Plant Physiol* **126**: 485-492.

Wollenweber, E., and Schneider, H., 2000, Lipophilic exudates of Pteridaceae—chemistry and chemotaxonomy, *Bioch Syst Ecol* **28**: 751-777.

Wray, G.A., Hahn, M. W., Abouheif, E., Balhoff, J. P., Pizer, M., Rockman, M. V., and Romano, L. A., 2003, The evolution of transcriptional regulation in eukaryotes, *Mol Biol Evol* **20**: 1377- 1419.

Yang, J., Gu, H., and Yang, Z., 2004, Likelihood analysis of the chalcone synthase genes suggests the role of positive selection in morning glories (*Ipomoea*), *J Mol Evol* **58**: 54-63.

Yang, Z., Nielsen, R., Goldman, N., and Pedersen, A. K., 2000, Codon-substitution models for heterogeneous selection pressure at amino acid sites, *Genetics* **155**: 431–449.

Yokoyama, S., Starmer, W. T., and Yokoyama, R., 1993, Paralogous origin of the red- and green-sensitive visual pigment genes in vertebrates, *Mol Biol Evol* **10**: 527-538.

Zufall, R. A., 2003, *Evolution of Red Flowers in Ipomoea*, Ph.D. dissertation, Duke University, Durham, NC.

Zufall, R. A., and Rausher, M. D., 2003, The genetic basis of a flower-color polymorphism in the common morning glory, *Ipomoea purpurea*, *J Heredity* **94**: 442-448.

Zufall, R. A., and Rausher, M. D., 2004, Genetic changes associated with floral adaptation restrict future evolutionary potential, *Nature* **428**: 847-850.

CHAPTER 8

FLAVONOIDS AS NUTRACEUTICALS

JEN-KUN LIN[1] AND MENG-SHIH WENG

Institute of Biochemistry and Molecular Biology, College of Medicine, National Taiwan University, No. 1, Section 1, Jen-ai Road, Taipei, Taiwan
[1]*Corresponding author; E-mail: jklin@ha.mc.ntu.edu.tw*

1. INTRODUCTION

"Nutraceutical" is a term coined in 1979 by Stephen DeFelice (DeFelice, 1992). It is defined "as a food or parts of food that provide medical or health benefits, including the prevention and treatment of disease." Subsequently, several other terms (medical food, functional food, and nutritional supplements) were used. A nutraceutical is any nontoxic food extract supplement that has scientifically proven health benefits for both the treatment and prevention of disease (Dillard and German, 2000). Nutraceuticals may range from isolated nutrients, dietary supplements, and diets to genetically engineered "designer" food, herbal products, and processed products, such as cereals, soups, and beverages. The increasing interest in nutraceuticals reflects the fact that consumers hear about epidemiological studies indicating that a specific diet or component of the diet is associated with a lower risk for a certain disease.

The major active nutraceutical ingredients in plants are flavonoids. The flavonoids are a group of organic molecules ubiquitously distributed in vascular plants. Approximately 2000 individual members of the flavonoids group of compounds have been described. As is typical for phenolic compounds, they can act as potent antioxidants and metal chelators. They also appear to be effective at influencing the risk of cancer. Overall, several of these flavonoids appear to be effective anticancer promoters and cancer chemopreventive agents. The presentation in this chapter is designed to provide the reader the tools to understand the biological and molecular role of plant flavonoids, including their antioxidant and antiproliferative activities and their role in intracellular signaling cascades.

The flavonoids, ubiquitous in plants, are the largest class of polyphenols, with a common structure of diphenylpropanes (C6-C3-C6), consisting of two aromatic rings linked through three carbons. The six major subclasses of flavonoids include the flavones (e.g., apigenin, luteolin), flavonols (e.g., quercetin, myricetin), flavanones (e.g., naringenin and hesperidin) (Figure 8.1), flavanols (or catechins) (e.g., epicatechin and gallocatechin) (Figure 8.2), anthocyanidins (e.g., cyanidin and pelargonidin) (Figure 8.3), and isoflavones (e.g., genistein and daidezin) (Figure 8.1) (Ross and Kasum, 2002).

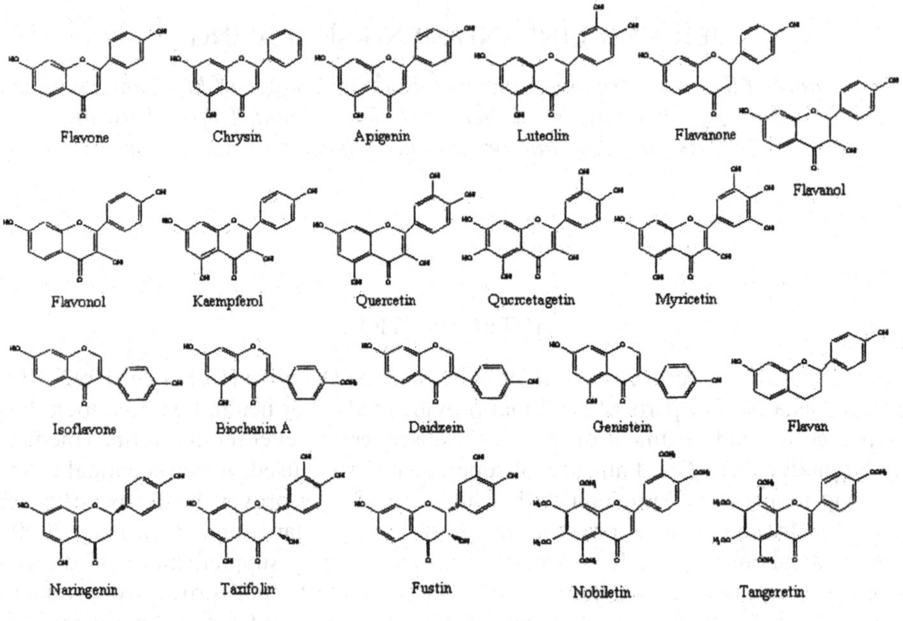

Figure 8.1 *Structure of select flavones, flavonols, flavanones, flavanols, isoflavones, and flavans.*

They are widely distributed in foods and beverages of plant origin, such as fruits, vegetables, tea, cocoa, and wine. Numerous publications report their content in various foods (Kelm et al., 2005; Schreier, 2005). Within the subgroups of the flavonols and the flavones, the flavonol quercetin is the most frequently occurring compound in foods. Also common are kaempferol, myricetin, and the flavones apigenin and luteolin. Tea and onions are the main dietary sources of flavonols and flavones. In this review, we describe the recent developments on the biological activities of flavonoids that have provided the important basis for their nutraceutical functions.

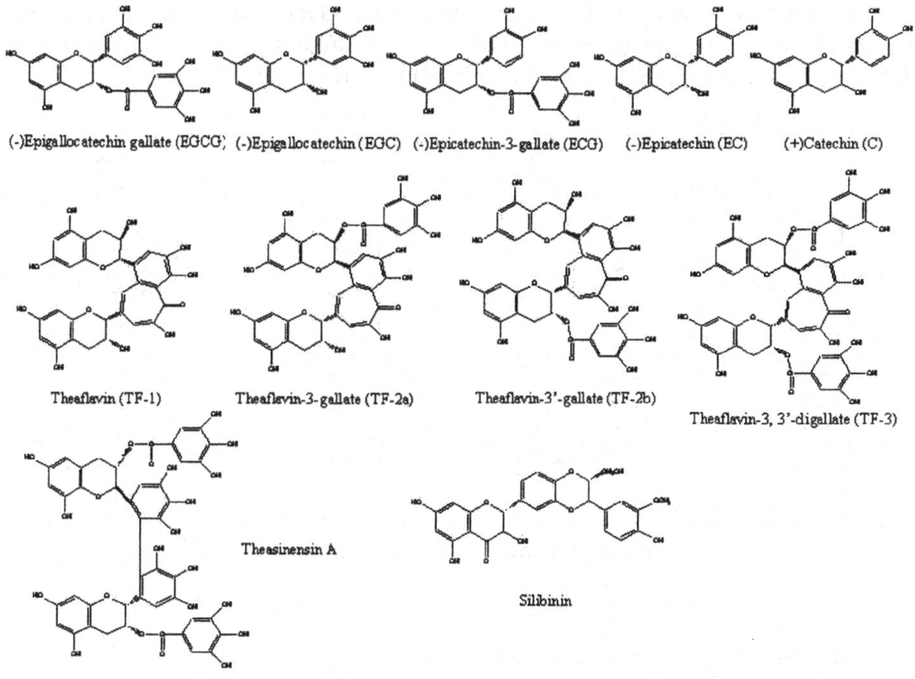

(-)Epigallocatechin gallate (EGCG)   (-)Epigallocatechin (EGC)   (-)Epicatechin-3-gallate (ECG)   (-)Epicatechin (EC)   (+)Catechin (C)

Theaflavin (TF-1)   Theaflavin-3-gallate (TF-2a)   Theaflavin-3'-gallate (TF-2b)

Theaflavin-3, 3'-digallate (TF-3)

Theasinensin A

Silibinin

Figure 8.2 *Structure of tea polyphenols.*

2. ANTIOXIDANT ACTIVITY

2.1. Reactive oxygen species and antioxidant systems

Diets high in flavonoids, fruits, and vegetables are protective against a variety of diseases, particularly cardiovascular disease and some types of cancer (Ness and Powles, 1997). Antioxidants and dietary fiber are believed to be the principal nutrients responsible for these protective effects. Reactive oxygen species (ROS) are formed *in vivo* during normal aerobic metabolism and can cause damage to DNA, proteins, and lipids, despite the natural antioxidant defense system of all organisms (Bors and Saran, 1987). ROS contribute to cellular aging (Sastre et al., 2000), mutagenesis (Takabe et al., 2001), carcinogenesis (Kawanishi et al., 2001), and coronary heart disease (Khan and Baseer, 2000) possibly through the destabilization of membranes (Takabe et al., 2001), DNA damage, and oxidation of low-density lipoprotein (LDL). Many *in vitro* studies have demonstrated the potent peroxyl radical scavenging abilities of flavonoids, which contribute to inhibiting lipid peroxidation and oxidation of LDL (Castelluccio et al., 1995; Salah et al., 1995). Since oxidation of LDL is implicated in the pathogenesis of coronary heart diseases (Witztum and Steinberg, 1991) through its ability to decrease the susceptibility of

LDL to oxidation, a number of researches have undertaken investigations examining the activity of dietary agents rich in flavonoids in inhibiting LDL oxidation *ex vivo* (Ishikawa et al., 1997; Nigdikar et al., 1998; van het Hof et al., 1997).

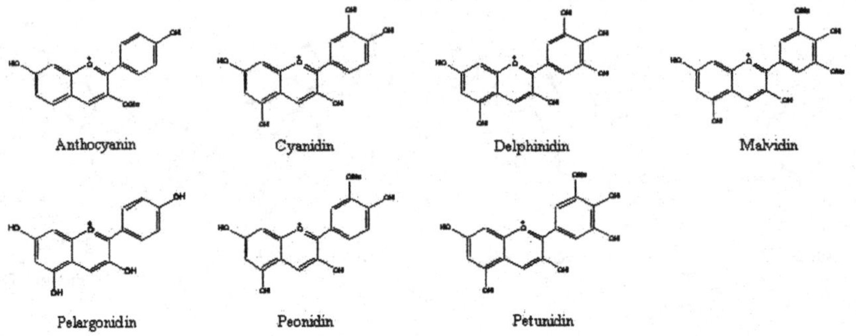

Figure 8.3 Structure of anthocyanidins.

2.2. Reactive nitrogen species and inducible nitric oxide synthase

Reactive nitrogen species (RNS) also appear to contribute to the pathology of cardiovascular diseases. NO is one RNS produced by the action of nitric oxide synthase in endothelial cells, neurons, and other cell types. At the sites of inflammation, inducible nitric oxide synthase (iNOS) is also augmented, and NO synthesis is further activated. Peroxynitrite, a potent oxidant generated by the reaction of nitric oxide (NO) with superoxide in the vascular endothelium, induces LDL oxidation (Leeuwenburgh et al., 1997; Moore, et al., 1995) and pro-inflammatory cytokine-mediated myocardial dysfunction (Ferdinandy et al., 2000; W. Wang et al., 2002). Another potential source of RNS derives from dietary nitrite, which reacts with the acidic gastric juice to produce nitrous acid, which decomposes to oxides of nitrogen. Nitrous acid and its products are able to nitrosate amines, deaminate DNA bases, and nitrate aromatic compounds including tyrosine. Several flavonoids and phenolic compounds, including the epicatechin/gallate family of flavanols, are powerful inhibitors of nitrous acid-dependent nitration and DNA deamination *in vitro* (Oldreive et al., 1998).

2.3. Protective effects of flavonoids

The protective effects of flavonoids in biological systems are ascribed to their capacity to transfer free radical electrons, chelate metal catalysts (Ferrali et al., 1997), activate antioxidant enzymes (Elliott et al., 1992), reduce alpha-tocopherol radicals (Hirano et al., 2001), and inhibit oxidases (Cos et al., 1998).

Green tea is a rich source of flavonoids, primarily catechins and flavonols. In black tea, as a consequence of the fermentation process, catechins are converted to complex condensation products, the theaflavins (Figure 8.2). Tea polyphenols show strong antioxidative effects and provide powerful scavengers against superoxide, hydrogen peroxide, hydroxyl radicals, nitric oxide, and peroxynitrite produced by various chemicals and biological systems. With regard to *in vitro* LDL oxidation, gallate esters were found to be less efficient than the respective free forms in inhibiting the oxidation catalyzed by Cu(II). Their activity follows the order epigallocatechin gallate (EGCG) > epicatechin gallate (ECG) > catechin (C) > epicatechin (C) (Miura et al., 1995). Anderson and collaborators reported that green tea polyphenols partially protect DNA from •OH radical-induced strand breaks and base damage (Anderson et al., 2001). Pulse radiolysis results support the mechanism of electron transfer (or H-transfer) from catechins to radical sites on DNA (Anderson et al., 2001). In black tea, all the theaflavins showed the same capacity to inhibit the production of superoxide. Green tea, black tea, and EGCG were shown to block the production of oxygen free radicals derived from the cooked meat mutagen 2-amino-3-methylimidazo[4,5-f]quinoline (IQ) in the presence of a NADPH-cytochrome P450 reductase (Hasaniya et al., 1997). These results support an antioxidant role of catechins in their direct interaction with DNA radicals. Catechin polyphenols could also decrease the peroxynitrite-induced nitration of tyrosine and protect the apolipoprotein B-100 of LDL from peroxynitrite-induced modification of critical amino acids, which contribute to its surface charge (Pannala et al., 1997). Recently, our laboratory found that oral feeding of green tea leaves to rats resulted in enhanced SOD activity in serum and catalase activity in liver and an increased concentration of glutathione in the liver (Lin et al., 1998). We also established that theaflavins and EGCG inhibit xanthine oxidase (XO). They inhibit XO to produce uric acid and also act as scavengers of superoxides. Theaflavin 3,3'-digallate (TF-3) inhibited the superoxide production in HL-60 cells. Therefore, the antioxidative activity of tea polyphenols may be due not only to their ability to scavenge superoxides, but also because of their ability to block XO and relative oxidative signal transducers (Lin et al., 2000). Other flavonoids such as quercetin, kaempferol, myristin, apigenin, and leuteolin also have antioxidative activity in many *in vitro* studies (Dwyer 1995).

The Caerphilly Study began in 1979 with the overall objective of examining the determinants and predictive ability of new and classic risk factors for incident CHD. During the initial recruitment phase (1979–83) 2512 men aged 45–59 years were examined, representing 90% of the population of men in this age group from the town of Caerphilly, South Wales, UK, and its surrounding villages (total population 40,000). Since then they have been examined at 5-year intervals. At the first reexamination between 1984 and 1988, when the men were aged 49–64 years, men of the same age who had moved into the defined geographical area also were deemed to be eligible. A total of 2398 men were recruited into the reconstructed cohort and they form the baseline population for the current study. This Caerphilly Study, which investigated flavonols and ischemic heart disease in Welsh men, found that quercetin can inhibit LDL oxidation and therefore inhibit platelet aggregation *in vitro* (Hertog et al., 1997). The positive effect of red wine phenolics on the modulation of human LDL resistance against oxidative modification has been demonstrated *in vitro* (Frankel et al., 1993). These authors showed that catechin oligomers, anthocyanidin dimers and trimers, as well as myricetin were main antioxidant components in red wine (Frankel et al., 1993). Ghiselli and collaborators observed that anthocyanins were the most effective, both in scavenging ROS and in inhibiting lipoprotein oxidation (Ghiselli et al., 1998). In soybean, the isoflavone genistin inhibited the oxidation of LDL by metal anions, superoxide/nitric oxide, and endothelial cells (Kapiotis et al., 1997). The feeding of a soy protein high-fat diet (21g isoflavone in 100 g protein) to C57BL/6 mice result in the reduction of plasma cholesterol levels, increased the resistance of LDL against oxidation, and decreased the atherosclerotic lesion area (Kirk et al., 1998). Humans who eat three soy bars (containing 12 mg genistein and 7 mg daidzein each) daily for 2 weeks could increase the plasma isoflavone level and the resistance of LDL to Cu-catalyzed oxidation (Tikkanen et al., 1998). Other studies in humans found that chronic red wine consumption (400 ml/day) reduced the susceptibility of LDL to lipid peroxidation catalyzed by Cu (Fuhrman et al., 1995). Similarly, a small but significant increase in the lag time of LDL oxidation was observed after 4 weeks of black tea consumption (600 ml/day) (Ishikawa et al., 1997).

3. ANTICARCINOGENESIS

Studies on cancer prevention have assessed the impact of a wide variety of flavonoids and a selected few isoflavones for their efficacy in inhibiting cancer in a number of animal models. These studies demonstrated that flavonoids inhibit carcinogenesis *in vitro* and substantial evidence indicates that they also do so *in vivo* (Caltagirone et al., 2000; Miyagi et al., 2000). Flavonoids may inhibit carcinogenesis by affecting the molecular events in the initiation, promotion, and progression stages. Animal studies and investigations using different cellular models suggested that certain flavonoids could inhibit tumor initiation as well as tumor progression (Deschner et al., 1991; Makita et al., 1996; Tanaka et al., 1997, 1999).

3.1. Quercetin

Dietary quercetin inhibited DMBA-induced carcinogenesis in hamster buccal pouch (Balasubramanian and Govindasamy, 1996) and in rat mammary gland (Verma et al., 1988). When given during the initiation stage, quercetin also inhibited DEN-induced lung tumorgenesis in mice (Khanduja et al., 1999). In a medium-term multi-organ carcinogenesis model in rats, quercetin (1% in the diet) inhibited tumor promotion in the small intestine (Akagi et al., 1995). Feeding rats with quercetin, during either the initiation or promotion stage, inhibited 4-NQO-induced carcinoma formation in the tongue (Makita et al., 1996). Siess and coworkers investigated the effects of feeding rats with flavone, flavanone, tangeretin, and quercetin on two steps of aflatoxin B1 (AFB1)-induced hepatocarcinogenesis (initiation and promotion) and found that flavones, flavanones, and tangeretin administered through the initiation period decreased the number of gamma-glutamyl transpeptidase-preneoplastic foci (Siess et al., 2000). Quercetin decreased oxidative stress-induced neuronal cell membrane damage more than vitamin C. These results suggest that quercetin, in addition to many other biological benefits, contributes significantly to the protective effects of neuronal cells from oxidative stress-induced neurotoxicity, such as Alzheimer's disease (Heo and Lee, 2004). On the other hand, the suppressive effects of flavones, such as chrysin and apigenin, on the expression of the high affinity IgE receptor FcεRI, which plays a central role in the IgE-mediated allergic response (Yano et al., 2005) has been demonstrate.

3.2. Genistein and daidzein

Genistein and daidzein (isoflavones derived from soybeans) have been shown to inhibit the development of both hormone- and non-hormone-related cancers, including mouse models of breast, prostate, and skin cancer. Treatment of TRAMP mice with 100–500 mg genistein/kg diet reduced the incidence of advanced-stage prostate tumors, in a dose-dependent manner (Mentor-Marcel et al., 2001). A high-isoflavone diet also was shown to inhibit methylnitrosourea-induced prostate tumor in Lobund-Wistar rats (J. Wang et al., 2002). Topically applied genistein reduces the incidence and multiplicity of skin tumors in the DMBA-initiated and TPA-promoted multiplicity of skin mouse model by 20% and 50%, respectively (J. Wang et al., 2002). In the UVB light-induced complete carcinogenesis model, topical pretreatment of SKH-1 mice with 10 μM genistein significantly reduced the formation of H_2O_2 and 8-bydroxy-2'-deoxyguanosine, but not of pyrimidine dimers in the epidermis (J. Wang et al., 2002).

3.3. Anthocyanins

Only a few studies have been performed to elucidate a potential anticarcinogenic activity of anthocyanins, despite their presence and importance in the human diet. Hagiwara et al. performed an animal experiment in 1, 2-dimethylhydrazine (DMH)-initiated f344/DuCrj rats. Anthocyanins from purple sweet potato and red cabbage were given at a dietary level of 5.0% in combination with 0.02% 2-amino-1-methyl-6-phenylimidazo [4,5-b] pyridine (PhIP), a promoter in the diet until 36 weeks. Lesion development induced by DMH and PhIP was suppressed by the anthocyanins. The marked inhibitory effects on colon carcinogenesis were apparent for the anthocyanins comprising cyanidin, but not peonidin, as the main constituent, suggesting that the inhibition of anthocyanins on carcinogenesis may be related to the number of hydroxyl groups on the B-ring (Hagiwara et al., 2002). Hou and co-workers used JB6 mouse epidermal cells to study the molecular mechanisms of anticarcinogenesis by anthocyanins (Hou et al., 2004). Their data indicated that both TPA-induced cell transformation and AP-1 transactivation were significantly inhibited by delphinidin, petunidin, and cyanidin, but not by pelargonidin, peonidin, or malvidin. These results suggest that the orthodihydroxyphenyl structure on the B-ring of anthocyanins may be essential for the inhibitory action, because pelargonidin, peonidin, and malvidin, having no such orthodihydroxyphenyl structure, failed to show the inhibitory effects. The molecular mechanism of delphinidin inhibition of TPA-induced AP-1 transactivation was due to a suppression of TPA-induced phosphorylation of ERK and JNK/SAPK.

Because strawberries are shown to contain higher concentrations of phytochemicals, including anthocyanins, and have a antioxidant capacity when compared with other common fruits, their neuroprotective activity was tested *in vitro* on PC12 cells treated with H_2O_2. Their protective effect and antioxidant capacity also were compared with those of banana and orange, which are the fresh fruits consumed at highest levels in the United States. The overall relative neuronal cell protective activity of these three fruits followed the decreasing order strawberry> banana > orange. The protective effects appeared to be due to the higher phenolic contents including anthocyanins, which are the major contributors in strawberries (Heo and Lee, 2005b).

3.4. Tea and tea polyphenols

The anticarcinogenesis effects of EGCG, green tea, and black tea extracts on various organs and animal model have been reported. Studies by Khan and collaborators showed that green tea polyphenols have a potent inhibitory effect on skin tumorigenicity in Sencar mice (Khan et al., 1988). In recent years, many studies demonstrated that topical application or oral feeding of a polyphenolic fraction from tea extract or of individual catechin derivatives had anticarcinogenesis effects in animal skins and other organs (Yang and Wang, 1993). Tea extracts were found to be effective in inhibiting 4-(methylnitrosamino)-1-(3-pyridyl)-1-butanone (NNK)-induced lung tumorigenesis in A/J mice with any of three dosing schedules (Yang et al., 2002). In the NNK-induced lung tumorigenesis model, administration of black tea extracts to adenoma-breeding mice significantly inhibited tumor cell proliferation and the progression of adenoma to carcinoma (Yang et al., 1997a and b). Inhibition of tumor invasion and metastasis in transplanted and spontaneous metastasis models by orally administered green tea infusion of EGCG also was reported (Liu et al., 2001; Sazuka et al., 1995). Most of the studies were conducted with chemical- or ultraviolet light-induced tumorigenesis models. For example, administration of 1% or 2% freshly brewed green or black tea significantly inhibited the spontaneous development of lung adenoma and rhabdomyosarcoma in A/J mice (Landau et al., 1998). The bioavailability of tea constituents is apparently a key factor determining the effectiveness of tea in inhibiting tumor formation. In this respect, the oral cavity and digestive tract, which have direct contact with orally administered tea, may represent good targets for chemoprevention. In the 7,12-dimethylbenz[a]anthracene (DMBA)-induced oral carcinogenesis hamster model, treatment with 0.6% green tea as the sole source of drinking fluid reduced the number of visible tumors by 35% and reduced tumor volume by 57%. In addition, immunohistochemical analyses showed that tea increased the apoptotic index of the tumors while decreasing the proliferation index and microvessel density (Li et al., 2002). Purified tea constituents, EGCG and theaflavins, also have been reported to inhibit tumorigenesis. For example, EGCG inhibited lung tumorigenesis in A/J mice induced by NNK and cisplatin (Xu et al., 1992). Theaflavins (a mixture of theaflavin, theaflavin-3-gallate, theaflavin-3'-gallate, and theaflavin-3,3'-digallate) reduced NNK-induced lung tumor multiplicity and volume in A/J mice (Yang et al., 1997a and b). These findings are interesting, given the extremely poor bioavailability of theaflavins, and may suggest that the theaflavins are metabolized to a more-bioavailable active metabolite.

The membrane protective effects of the phenolics determined by LDH release and trypan blue exclusion assays demonstrated that epicatechin, catechin, and their mixture protect cellular membrane from β-amyloid-induced cytotoxicity. It has been demonstrated that the major flavonoids of cocoa, epicatechin and catechin, protect PC12 cells from Aβ-induced neurotoxicity, and suggest that cocoa may have an anti-neurodegenerative effect in addition to other known chemopreventive effects (Heo and Lee, 2005a).

An ester of catechin, 3-O-octanoyl-(+)-catechin (OC), was synthesized from (+)-catechin by the incorporation of an octanoyl chain into catechin in the light of (-)-epicatechin gallate (ECG) and (-)-epigallocatechin-3-gallate (EGCG). OC was found to inhibit the response of ionotropic GABA receptors and Na+/glucose co-transporters expressed in *Xenopus* oocytes in a noncompetitive manner, more efficiently than catechin. OC also induced a nonspecific membrane current and decreased the membrane potential of the oocyte. This newly synthesized catechin derivative OC possibly binds to the lipid membrane more strongly than does catechin, ECG, or EGCG and as a result perturbs the membrane structure (Aoshima et al., 2005).

4. SUPPRESSION OF CANCER GROWTH

4.1. Antiproliferative effects

Deregulated proliferation appears to be a hallmark of increased susceptibility to neoplasia. Cancer prevention generally is associated with inhibition, reversion, or delay of cellular hyperproliferation. Most flavonoids have been demonstrated to inhibit proliferation in many types of cultured human cancer cell lines, whereas they have little or no toxicity to normal human cells. For example, Kandaswami and co-workers reported antiproliferative effects of four citrus flavonoids (quercetin, taxifolin, nobiletin, and tangeretin, at 2-8 µg/ml for 3–7 days) on squamous cell carcinoma HTB43 (Kandaswami et al., 1991). Kuo showed antiproliferative potency of five flavonoids and two isoflavonoids (0–100 µM) on colon carcinoma HY29 and Caco-2 cell lines, with the induction of apoptosis (Kuo, 1996). Le Bail and coworkers suggested antiproliferative activity of certain flavonoids and genistein at high concentrations (50 µM) on breast cancer MCF-7 through a mechanism independent of the estrogen receptor (Le Bail et al., 1998). Twenty-seven citrus flavonoids were investigated for antiproliferative activities, at 40 µM, on several tumor cell lines, including lung carcinoma A549 and gastric

TGBC11TKB cancer cells and were found to inhibit proliferation of cancer cell lines. However, they did not significantly affect proliferation of normal human cell lines (Kawaii et al., 1999). Studies on the inhibition of cell proliferation and angiogenesis by flavonoids in six different cancer cell lines had been reported and noted that the IC50 of active flavonoids were in the low micromolar range, physiologically available concentrations (Fotsis et al., 1997). Genistein and synthetic isoflavone analogues, at 0.1–25 µg/ml, inhibited intestinal epithelia cell proliferation and induced apoptosis *in vitro* (Booth et al., 1999). An inhibition of human breast cancer cell proliferation and delay of mammary tumorigenesis by flavonoids also was found (So et al., 1996). Over 30 flavonoids had been screened for their effects on cell proliferation and potential cytotoxicity in two human colon cancer cell lines. All compounds tested, including specific flavones, flavonols, flavanones, and isoflavonones, demonstrated antiproliferative activity in the absence of cell cytotoxicity (Kuntz et al, 1999). No notable structure–activity relationships were found on the basis of flavonoid subclass. Other studies evaluated how the core structure of the flavones, 2-phenyl-4H-1-benzopyran-4-one, affects proliferation, differentiation, and apoptosis in a human colon cancer cell line (Wenzel et al., 2000). In particular, this study evaluated the effect of the flavone on the expression of cell cycle and apoptosis-related genes in the cell line, reporting dramatic changes in mRNA levels of specific genes including cyclooxygenase-2, NF-κ, and BCL-X. Further, there was a high selectively for apoptosis of the transformed cells. The authors concluded that flavones could furnish a new chemopreventive agent.

4.2. *Inhibition of cell cycle progression*

Perturbations in the progression of the cell cycle may account for the anti-carcinogenetic effects of many flavonoids. Mitogenic signals commit cells to entry into a series of regulated steps allowing the progression of the cell cycle. Synthesis of DNA (S phase) and separation of two daughter cells (M phase) are main features of cell cycle progression. The time between the S and M phases is known as the G2 phase. This phase is important to allow cells to repair errors that occur during DNA duplication, preventing the propagation of these errors to daughter cells. In contrast, the G1 phase represents the period of commitment to cell cycle progression that separates M and S phases as cells prepare for DNA duplication upon mitogenic signals.

CDKs and CDKIs have been recognized as key regulators of cell cycle progression. Alternation and deregulation of CDK activity are pathogenic hallmarks of neoplasia. A number of cancers are associated with hyperactivation of CDKs as a result of mutations of the CDK or CDKI genes. Therefore, compounds that function as inhibitors or modulators of these enzymes are of interest as novel potential therapeutic agents in cancer (Senderowicz, 2002, 2003). Flavonoids and tea polyphenols have been found to perturb the cell cycle in certain cancer cell lines. Genistein produces cell cycle arrest at both G1/S and G2/M phase in the human myelogenous leukemia HL60 cells and in the lymphocytic leukemia MOLT-4 cells (Traganos et al., 1992). Furthermore, it has been shown that genistein induced G2/M arrest through p21Waf1/Cip1 up-regulation and apoptosis induction in a non-small-cell lung cancer cell lines (Lian et al., 1998). Isoflavonones (genistein, genistin, daidzein, and biochanin A) also inhibit growth of murine and human bladder cancer cell lines by inducing cell cycle arrest, apoptosis, and angiogenesis (Zhou et al., 1998). Quercetin blocks the cell cycle at G1/S phase in human colonic COLO320 DM cells (Hosokawa et al., 1990) and leukemic T cells (Yoshida, et al., 1992). Another widely distributed flavonoid, apigenin, significantly induced a reversible G2/M arrest in keratinocyte, fibroblasts, and colonic carcinoma cell lines (Lepley et al., 1996; Lepley and Pelling, 1997; Wang et al., 2000). Studies on silibinin found that perturbations in cell cycle progression may account for the anticarcinogenic effects on human prostate carcinoma cell lines and colon carcinoma cell line (Agarwal et al., 2003; Tyagi et al., 2002a, 2002b). Anthocyanins block the cell cycle progression through CDK inhibition in many cancer cell lines (Favot et al., 2003; Lazze et al., 2004; Martin et al., 2003). Tea polyphenols inhibit cell proliferation and suppress tumor growth activity. Our laboratory has investigated the effects of EGCG and other catechins on the cell cycle progression (Liang et al., 1999). The results suggest that EGCG either exerts its growth-inhibitory effects through modulation of the activities of several key G1 regulatory proteins such as CDK2 and CDK4 or mediates the induction of the CDK inhibitors p21Waf1/Cip1 and p27Kip1.

5. MOLECULAR MECHANISMS OF CANCER CHEMOPREVENTION BY FLAVONOIDS

5.1. Epidermal growth factor receptor family

The epidermal growth factor receptor (EGFR) family consists of four transmembrane receptor tyrosine kinases, EGFR (HER1), HER2 (ErbB2, neu), HER3 (ErbB3), and HER4 (ErbB4), whose function is to transmit extracellular cues to intercellular signal transduction pathways that regulate proliferation, survival, and differentiation responses. At least two members of the family, EGFR and HER2, are frequently deregulated in human epithelial tumors, as a consequence of autocrine stimulation, overexpression, or mutation. Deregulation is often associated with an adverse prognosis in an array of tumor types, including the central nervous system, head and neck, GI, and breast cancer (Arteaga, 2001). EGFR heterodimerizes with other family members, which are more potent in terms of receptor stability as well as both mitogenic and cell survival signaling, compared to its own homodimer. EGFR-mediated signaling cascades activate Shc-Grb2-Ras-Raf, and ultimately mitogen-activated protein kinases (MAPK) (Seger and Krebs, 1995).

EGCG strongly inhibits the protein tyrosine kinase activities of EGFR, PDGFR, and FGFR, exhibiting an IC$_{50}$ value of 0.5–1 μg/ml. In an *in vivo* assay, EGCG reduced the autophosphorylation levels of EGFR by EGF and phosphoamino acid analyses of EGFR revealed that EGCG inhibited the EGF-stimulated increase in phosphotyrosine level in A431 cells (Liang et al., 1997). The theaflavins have an antiproliferative activity on tumor cells and the molecular mechanisms of antiproliferation may involve a block of the growth factor binding to its receptor, and thus suppress mitogenic signal transduction (Liang et al., 1999a and c). Tea polyphenols (EGCG and theaflavins) also inhibit fatty acid synthase through EGFR/PI-3K/Akt/SP-1 pathway and further inhibited cell proliferation (Yeh et al, 2003). Flavonoids, including quercetin, luteolin, and apigenin, are potent inhibitors of EGFR family of tyrosine kinases (Lee et al., 2004). Apigenin-induced apoptosis of breast cancer cells is through the proteasome-mediated degradation of Her-2/neu, which results in the inhibition of the PI3K/Akt survival pathway (Way et al., 2004a). These studies also examined the structure–activity relationship of flavonoids on Her-2/neu, indicating that (1) the position of B ring; and (2) the existence of the 3',4'-hydroxyl group on the 2-phenyl group was important for the degradation of HER2/neu protein by flavonoids (Way et al., 2005). The anthocyanins cyanidin and delphinidin also have potent inhibitory effects on the tyrosine kinase activity of EGFR (Meiers et al., 2001). Structure–activity studies showed that the presence of vicinal hydroxy substituents at the phenyl ring at the 2-position (B-ring) is crucial for target interaction. The presence of a single hydroxy group or the introduction of methoxy substituents in the B-ring results in a substantial loss of inhibitory activity (Marko et al., 2004).

5.2. Mitogen-activated protein kinases (MAPKs) pathways

Three classes of MAPKs are known and include the c-Jun N-terminal kinases/stress activated protein kinase (JNKs/ SAPKs), the p38 kinase and the extracellular signal-regulated protein kinase (ERKs) (Boulton et al., 1991; Davis, 1994; Kallunki et al., 1994; Kyriakis et al., 1994). JNKs/SAPKs and p38 kinases are activated by various forms of stress, including ultraviolet (UV) irradiation (Kallunki et al., 1994). In contrast, ERKs are strongly activated and play a critical role in transmitting signals initiated by tumor promoters such as TPA and growth factors, including EGF and platelet-derived growth factor (PDGF) (Cowley et al., 1994; Minden et al., 1994). However, the activation of these pathways is not mutually exclusive. For example, heat shock and UV irradiation partially activate the ERKs cascade and EGF partially activates the JNKs pathway (Davis, 1994; Minden et al., 1994). Evidence strongly indicates that the activation of MAPKs by tumor-promoting agents plays a functional role in tumor promoter-induced malignant transformation (Dong et al., 1997a, 1997b; Huang et al., 1997, 1998). MAPK are activated by translocation (Bonni et al., 1999; Kharbanda et al., 2000) to the nucleus, where they influence target transcription factors. ERK1/2 are usually associated with prosurvival signaling (Kharbanda et al., 2000) through mechanisms that may involve activation of the cyclic AMP regulatory binding protein (CREB) (Bonni et al., 1999; Kaplan and Miller, 2000), the up-regulation of the antiapoptotic protein Bcl-2, and nontranscriptional inhibition of BAD. On the other hand, JNK has been strongly linked to a transcription-dependent apoptotic signaling, possibly through the activation of c-Jun (Behrens et al., 1999) and other AP-1 proteins including JunB, JunD, and ATF-2 (Davis, 2000).

Many of the molecular alterations that are associated with carcinogenesis occur in cell-signaling pathways responsible for regulating cellular proliferation or apoptosis. Tea polyphenols and flavonoids have been shown to interact with MAP kinase pathways and mediate signaling by influencing activation and phosphorylation of these molecules (Anter et al., 2004; Arakaki et al., 2004; Chen, et al., 1999; Ko et al., 2005; Mallikarjuna et al., 2004; Miyata et al., 2004; Zhang et al., 2005). Evidence indicates that tea inhibits tumor promoter- or growth factor-induced cell transformation and AP-1 activation. AP-1 is a well-characterized transcription factor composed of homodimers and/or heterodimers of the Jun and Fos gene families (Cohen and Curran, 1988; Halazonetis et al., 1988; Hirai et al., 1989) and it regulates the transcription of various genes associated with cellular inflammation, proliferation, and apoptosis (Angel and Karin, 1991). In cell culture and animal models, AP-1 was shown to be involved in tumor progression and metastasis (Crawford and Matrisian, 1996) and to play a key role in the preneoplastic-to-neoplastic transformation (Barthelman et al., 1998; Bernstein & Colburn, 1989). Significantly, when tumor promoter-induced AP-1 activity was blocked, neoplastic transformation was inhibited (Dong et al., 1994). AP-1 thus appears to be a key target for chemopreventive agents such as flavonoids and tea polyphenols (McCarty, 1998). EGCG has been shown to effectively inhibit cell transformation in A172 and transfected NIH 3T3 cells (Ahn et al., 1999). In addition, both EGCG and theaflavins inhibit EGF- or TPA-induced cell transformation in a dose-dependent manner (Dong et al., 1997b). At a dose range similar to that which inhibited cell transformation, EGCG and theaflavins repress AP-1-dependent transcriptional activity and AP-1 DNA-binding activity induced by TPA. Furthermore, these tea compounds inhibited TPA- or EGF-induced c-Jun phosphorylation and JNKs activation, but not ERKs phosphorylation. Based on these results and what is known about tumor-promoted activation of AP-1, EGCG and theaflavins appear to exert their chemopreventive effects primarily through the inhibition of AP-1 transactivation and subsequent AP-1 DNA-binding activity (Dong et al., 1997).

5.3. Phosphatidylinositol-3-kinase (PI3K) pathways

Members of the three classes of the phosphoinositide 3-kinase (PI3K) family of enzymes have a central role in many cellular functions, including proliferation, differentiation, cell migration, survival, glucose homeostasis, and control of cell growth (Katso et al., 2001). PI3Ks are lipid kinases whose biological responses are evoked through their ability to phosphorylate phosphoinositides (Cantley, 2002; Katso et al., 2001). PI3Ks are heterodimers, consisting of a variable catalytic subunit (p110) and a regulatory subunit (p85, p55, or p50) that transduce signals from activated receptor tyrosine kinases (RTKs) (Katso et al., 2001). After stimulation by various different growth factors and cytokines, PI3K is recruited either directly or indirectly to the membrane; this association activates the enzyme and brings it into close proximity with its lipid substrate, phosphatidylinositol (4,5)-bisphosphate [PtdIns(4,5)P2] (White, 2003), thereby generating PtdIns(3,4,5)P3. The tumor suppressor protein PTEN reverses the action of PI3K by dephosphorylating PtdIns(3,4,5)P3 at the D-3 position, and thus providing an essential suppressor of PI3K signaling whose function is lost in various advanced stage cancers (Katso et al., 2001). PtdIns(3,4,5)P3 generated in the membrane by PI3K recruits members of the intracellular signaling pathways containing pleckstrin homology (PH) domains to the plasma membrane, thereby coupling PI3K signals to downstream effector molecules. Activation of one particular effector, protein kinase B (PKB; also known as AKT), seems to be essential not only in mediating the effects of insulin on glucose homeostasis but also in regulating the profound effects of insulin and IGFs on mTOR signaling and cell growth. PKB phosphorylates and inactivates various substrates involved in diverse processes including cell survival (the proapoptotic protein BAD), glycogen synthesis (glycogen synthase kinase-3), and gene transcription (FOXO transcription factors) (Cantley, 2002; Fresno Vara et al., 2004; White, 2003). However, PKB is also known to promote cell and organismal growth downstream of PI3K.

Recently, significant additional signaling molecules have been implicated in the mechanism(s) responsible for the antitumor effects of the flavonoids and tea polyphenols. PI3K is an important factor in carcinogenesis and an inhibitory effect of flavonoids, EGCG, and theaflavins on the activation of PI3K and its downstream effectors may further explain the antitumor promotion action of these flavonoids (Bagli et al., 2004; Harmon and Patel, 2004; Koh et al., 2004; Miyata et al., 2004; Way et al., 2004a and b, 2005).

5.4. Suppression of NF-κB activation

The transcription factor NF-κB-induced signaling is well known for its important roles in the control of processes including cell growth, apoptosis, inflammation, and stress response (Karin et al., 2002; Lin and Karin, 2003; Storz and Toker, 2003). It has been known that NF-κB is inactive when bound to I-κB in the cytosol and could be activated by phosphorylation and degradation of I-κB (Silverman and Maniatis, 2001). I-κB is phosphorylated by activated I-κB kinase (IKK), and IKK is phosphorylated and activated by mitogen activated kinase kinase 1 (MEKK1), one of the kinases in the MAPK pathway (Gupta et al., 2004; Kucharczak et al., 2003). The phosphorylation of I-κB by IKKs leads to proteasome-dependent degradation of I-κB, setting NF-κB free; NF-κB then can translocate into the nucleus to activate the expression of a set of NF-κB-responsive genes. However, NF-κB is the key protein in the pathway and has been described as a major culprit and a therapeutic target in cancer (Bharti and Agarwal, 2002; Biswas et al., 2001; Orlowski and Baldwin, 2002).

Studies have demonstrated that flavonoids inhibit activation of NF-κB in different cancer cell lines, suggesting a possible explanation for the inhibitory effects of these agents on cancer cells (Davis et al., 1999; Gupta et al., 2002; Liang et al., 1999a-c; Yang et al., 2001). Recently, our laboratory has investigated the inhibition of IKK activity in LPS-activated murine macrophages by various polyphenols including EGCG and theaflavins (Pan et al., 2000). TF-3 inhibited IKK activity more strongly than did the other polyphenols. TF-3 strongly inhibited both IKK1 and IKK2 activity and prevented the degradation of I-κB and I-κB in activated macrophage cells. These results suggest that TF-3 and other polyphenols may exert their antiinflammatory and cancer chemoprevention actions by suppressing the activation of NF-κB through inhibition of IKK activity. EGCG has been shown to inhibit the constitutive activation of NF-κB in H891 head and neck carcinoma cells and MDA-MB-231 breast carcinoma cells (Masuda et al., 2002). In murine RAW 264.7 macrophage cells, EGCG could inhibit LPS-induced iNOS activation through inhibited I-κB degradation and NF-κB nuclear translocation and activity (Lin and Lin, 1997). In A431 epidermoid carcinoma cells, treatment of EGCG dose- and time-dependently increased I-κB level and inhibited NF-κB nuclear translocation (Gupta et al., 2004). The UVB-induced NF-κB activation of normal human epidermal keratinocytes was associated with increased I-κB phosphorylation and degradation and EGCG was shown to block NF-κB activation and nuclear translocation (Afaq et al., 2003). Although ROS have been suggested to be involved in the activation of the NF-κB signaling system and that its inhibition by EGCG is due to the antioxidant activity, direct evidence for this mechanism is lacking. It has been proposed that the inhibition of NF-κB signaling by EGCG can be interpreted simply by the inhibition of IKK-catalyzed phosphorylation of I-κB. In addition, EGCG had a concurrent effect on two important transcription factors p53 (stabilization of p53) and NF-κB (negative regulation of NF-κB activity), causing a change in the ratio of Bax/Bcl-2 in a manner that favors apoptosis (Hastak et al., 2003).

5.4. Androgen and estrogen receptor

It has been found that the androgen receptor (AR) signaling pathway plays important roles in carcinogenesis and cancer progression through regulation of transcription of androgen-responsive genes (Luke and Coffey, 1994). Genistein and theaflavins regulate the molecules involved in the AR signaling pathway when they were used to inhibit growth of cancer cells (Davis et al., 2000; Lee et al., 2004). Our experiments showed that TF3 inhibits rat liver microsomal 5-α-reductase activity and significantly reduced androgen-responsive LNCaP prostate cancer cell growth, suppressed expression of the AR, and lowered androgen-induced prostate-specific antigen secretion and fatty acid synthase protein level. Our result suggests that TF3 might be useful as a chemoprevention agent for prostate cancer through suppressing the function of androgen and its receptor (Lee et al., 2004).

Many environmental chemicals have been found to be estrogenic and stimulate the growth of ER-positive human breast cancer cells (Welshons et al., 2003). It is important to develop dietary strategies to prevent the stimulated growth of breast tumors by environmental estrogens because it is difficult to avoid human exposure to them. Isoflavonoids, including genistein, daidzein, and formononetin, have been shown to be estrogenic agonists in various animal models (Farnsworth et al., 1975). By using an estrogen receptor (ER)-dependent transcriptional response assay, Miksicek reported that commonly occurring flavonoids also had estrogenic activity (Miksicek, 1993). Of the flavonoids and isoflavonoids tested in this sensitive assay, the order of estrogenicity was genistein > kaempherol > naringenin > apigenin > daidzein > biochanin A > formononetin > luteolin > fisetin > catechin/taxifolin > hesperetin (Miksicek, 1995). Some properties of soy isoflavones, such as preventing osteoporosis (Anderson, 1999) and lowering cholesterol levels (Lichtenstein, 1998), may suggest an estrogen-related mechanism.

Isoflavones are believed to exert their effects through ER signaling pathway because of the structural similarity to estrogen. However, experimental studies have found that isoflavones at different concentrations may exhibit different effects (Martin et al., 1978). Genistein at concentrations lower than 1 µM may induce breast cancer cell proliferation by estrogenic agonistic properties, while genistein at concentrations higer than 5 µM may prevent hormone-dependent growth of breast cancer cells by potential estrogen-antagonistic activity. Recent studies indicate that genistein at 50 and 100 µM significantly down-regulate mRNA expression of ER and arrest the growth of MCF-7 cells at G2/M phase, suggesting that the inhibitory action of genistein on human breast cancer cells is partially mediated by the alteration of ER-dependent pathways (Chen et al., 2003). However, experimental studies also showed that isoflavones exert their inhibitory effects on ER-negative MDA-MB-231 breast cancer cells (Li et al., 1999) and hormone-independent cancer cells (Alhasan et al., 2001; Lian et al., 1998, 1999).

The inhibitory effects of black tea polyphenols on aromatase activities has been investigated in our laboratory. We found that black tea polyphenols, TF-1, TF-2, and TF-3, significantly inhibited rat ovarian and human placental aromatase activities. In *in vivo* models, these black tea polyphenols also inhibited the proliferation induced by 100 nM dehydroepiandrosterone (DHEA) in MCF-7 cells. Interestingly, black tea polyphenols had antiproliferative effects in breast cancer cells with hormonal resistance. The inhibitory effect of black tea polyphenols on hormone-resistant breast cancer cells suppressed the basal receptor tyrosine phosphorylation in HER2/neu-overexpressing MCF-7 cells. We suggest that the use of black tea polyphenols may be beneficial in the chemoprevention of hormone-dependent breast tumors and represent a possible remedy to overcome hormonal resistance of hormone-independent breast tumors (Way et al., 2004b).

6. CONCLUDING REMARKS

Intensive epidemiological studies have shown consistently that regular consumption of fruits and vegetables is associated with reduced risk of chronic diseases such as cancer and cardiovascular disease (Willett, 2002; Block et al., 1992). However, the individual antioxidants of these foods studied in clinical trials, including β-carotene, vitamin C, and vitamin E, do not appear to have consistent preventive effects comparable to the observed health benefits of diets rich in fruits and vegetables (Omenn et al., 1996).

It has been reported that fresh apples have potent antioxidant activity; and whole apple extracts inhibit the growth of colon and liver cancer cells *in vitro* in a dose-dependent manner (Eberhardt et al., 2000), suggesting that natural phytochemicals in fresh fruits could be more effective than a dietary supplement. Apples are commonly consumed and are the major contributors of phytochemicals in human diets. Apple extracts exhibit strong antioxidant and antiproliferative activities and the major part of total antioxidant activity is from the combination of phytochemicals.

Phytochemicals, including phenolics and flavonoids, are likely to be the bioactive compounds contributing to the benefits of apple. Recent studies have demonstrated that whole apple extracts prevent mammary cancer in rat models in a dose-dependent manner at doses comparable to human consumption of one, three, and six apples a day (Liu et al., 2005). This study demonstrated that whole apple extracts effectively inhibited mammary cancer growth in the rat models; thus, consumption of apples may be an effective strategy for cancer chemoprevention.

Chemopreventive studies have demonstrated that the mechanisms of action of phytochemicals and nutraceuticals in the prevention of cancer go beyond the antioxidant activity scavenging of free radicals; regulation of gene expression in cell proliferation, oncogenes, and tumor suppressor genes; induction of cell cycle arrest and apoptosis; modulation of enzyme activity in detoxification, oxidation, and reduction; stimulation of the immune system; and regulation of hormone metabolism. It is a general theme that the additive and synergistic effects of phytochemicals and nutraceuticals in fruits and vegetables are responsive for their potent antioxidant and anticancer activities and that the benefit of a diet rich in fruits and vegetables is attributed to the complex mixture of phytochemicals and nutraceuticals present in whole foods (Liu, 2004).

Recent development in the molecular mechanisms of signal transduction pathway in various cell systems has provided a strong basis for performing the synergistic effects of phytochemicals and nutraceuticals in whole foods that have been ingested by the host. Along this aspect, we have proposed that the cancer chemoprevention and antiobesity effects of tea and tea polyphenols might accomplish this through blocking the signal transduction pathways in the target cells (Lin et al., 1999; Lin, 2002).

7. ACKNOWLEDGMENTS

This study was supported by the National Science Council NSC93-2311-B-002-001; NSC93-2320-B-002-111; and NSC93-2320-B-002-127. We also thank Prof. Shoei-Yn Lin-Shiau, Institute of Pharmacology, College of Medicine, National Taiwan University for her inspiring discussion during the preparation of this article.

8. REFERENCES

Afaq, F., Adhami, V. M., Ahmad, N., and Mukhtar, H., 2003, Inhibition of ultraviolet B-mediated activation of nuclear factor κB in normal human epidermal keratinocytes by green tea constituent (-)-epigallocatechin-3-gallate, *Oncogene*, **22**:1035-1044.

Agarwal, C., Singh, R. P., Dhanalakshmi, S., Tyagi, A. K., Tecklenburg, M., Sclafani, R. A., and Agarwal, R., 2003, Silibinin upregulates the expression of cyclin-dependent kinase inhibitors and causes cell cycle arrest and apoptosis in human colon carcinoma HT-29 cells, *Oncogene*, **22**: 8271-8282.

Ahn, H. Y., Hadizadeh, K. R., Seul, C., Yun, Y. P., Vetter, H., and Sachinidis, A., 1999, Epigallocathechin-3 gallate selectively inhibits the PDGF-BB-induced intracellular signalling transduction pathway in vascular smooth muscle cells and inhibits transformation of sis-transfected NIH 3T3 fibroblasts and human glioblastoma cells (A172), *Mol Biol Cell*, **10**: 1093-1104.

Akagi, K., Hirose, M., Hoshiya, T., Mizoguchi, Y., Ito, N., and Shirai, T., 1995, Modulating effects of ellagic acid, vanillin and quercetin in a rat medium term multi-organ carcinogenesis model, *Cancer Lett*, **94**: 113-121.

Alhasan, S. A., Aranha, O., and Sarkar, F. H., 2001, Genistein elicits pleiotropic molecular effects on head and neck cancer cells, *Clin Cancer Res*, **7**: 4174-4181.

Anderson, J. J., 1999, Plant-based diets and bone health: nutritional implications, *Am J Clin Nutr*, **70**: 539S-542S.

Anderson, R. F., Fisher, L. J., Hara, Y., Harris, T., Mak, W. B., Melton, L. D., and Packer J.E., 2001, Green tea catechins partially protect DNA from (.)OH radical-induced strand breaks and base damage through fast chemical repair of DNA radicals, *Carcinogenesis*, **22**: 1189-1193.

Angel, P., and Karin, M., 1991, The role of Jun, Fos and the AP-1 complex in cell-proliferation and transformation, *Biochim Biophys Acta*, **1072**: 129-157.

Anter, E., Thomas, S. R., Schulz, E., Shapira, O. M., Vita, J. A., and Keaney, J. F., Jr., 2004, Activation of endothelial nitric-oxide synthase by the p38 MAPK in response to black tea polyphenols, *J Biol Chem*, **279**: 46637-46643.

Aoshima, H., Okita Y., Hossain, S. J., Fukue K., Mito, M., Orihara, Y., Yokoyama, T., Yamada, M., Kumagai, A., Nagaoka, Y., Uesatos, S., and Hara, Y., 2005, Effect of 3-O-octanoyl-(+)-catechin on the responses of GABAa receptors and Na+/glucose cotransporters expressed in Xenopus oocytes and on the oocyte membrane potential, *J Agric Food Chem* **53**: 1955-1959.

Arakaki, N., Toyofuku, A., Emoto, Y., Nagao, T., Kuramoto, Y., Shibata, H., and Higuti, T., 2004, Induction of G1 cell cycle arrest in human umbilical vein endothelial cells by flavone's inhibition of the extracellular signal regulated kinase cascade, *Biochem Cell Biol*, **82**: 583-588.

Arteaga, C. L., 2001, The epidermal growth factor receptor: from mutant oncogene in nonhuman cancers to therapeutic target in human neoplasia, *J Clin Oncol*, **19**: 32S-40S.

Bagli, E., Stefaniotou, M., Morbidelli, L., Ziche, M., Psillas, K., Murphy, C., and Fotsis, T., 2004, Luteolin inhibits vascular endothelial growth factor-induced angiogenesis; inhibition of endothelial cell survival and proliferation by targeting phosphatidylinositol 3'-kinase activity, *Cancer Res*, **64**: 7936-7946.

Balasubramanian, S., and Govindasamy, S., 1996, Inhibitory effect of dietary flavonol quercetin on 7,12-dimethylbenz[a]anthracene-induced hamster buccal pouch carcinogenesis, *Carcinogenesis*, **17**: 877-879.

Barthelman, M., Bair, W. B., 3rd, Stickland, K. K., Chen, W., Timmermann, B. N., Valcic, S., Dong, Z., and Bowden, G.T., 1998, (-)-Epigallocatechin-3-gallate inhibition of ultraviolet B-induced AP-1 activity, *Carcinogenesis*, **19**: 2201-2204.

Behrens, A., Sibilia, M., and Wagner, E. F., 1999, Amino-terminal phosphorylation of c-Jun regulates stress-induced apoptosis and cellular proliferation, *Nat Genet*, **21**: 326-329.

Bernstein, L. R., and Colburn, N. H., 1989, AP1/jun function is differentially induced in promotion-sensitive and resistant JB6 cells, *Science*, **244**: 566-569.

Bharti, A. C., and Agarwal, B. B., 2002, Nuclear factor-kappa B and cancer: its role in prevention and therapy, *Biochem Pharmacol*, **64**: 883-888.

Biswas, D. K., Dai, S. C., Cruz, A., Weiser, B., Graner, E., and Pardee, A. B., 2001, The nuclear factor kappa B (NF-kappa B): a potential therapeutic target for estrogen receptor negative breast cancers, *Proc Natl Acad Sci U S A*, **98**: 10386-10391.

Block, G., Patterson, B., and Subar, A., 1992, Fruit, vegetables and cancer prevention: a review of the epidemiological evidence, *Nitr Cancer*, **18**: 1-29.

Bonni, A., Brunet, A., West, A. E., Datta, S. R., Takasu, M. A., and Greenberg, M. E., 1999, Cell survival promoted by the Ras-MAPK signalling pathway by transcription-dependent and -independent mechanisms, *Science*, **286**: 1358-1362.

Booth, C., Hargreaves, D. F., Hadfield, J. A., McGown, A. T., and Potten, C. S., 1999, Isoflavones inhibit intestinal epithelial cell proliferation and induce apoptosis *in vitro*, *Br J Cancer*, **80**: 1550-1557.

Bors, W., and Saran, M., 1987, Radical scavenging by flavonoid antioxidants, *Free Radic Res Commun*, **2**: 289-294.

Boulton, T. G., Nye, S. H., Robbins, D. J., Ip, N. Y., Radziejewska, E., Morgenbesser, S. D., DePinho, R.A., Panayotatos, N., Cobb, M.H., Yancopoulos, G.D., 1991, ERKs: a family of protein-serine/threonine kinases that are activated and tyrosine phosphorylated in response to insulin and NGF, *Cell*, **65**: 663-675.

Caltagirone, S., Rossi, C., Poggi, A., Ranelletti, F. O., Natali, P. G., Brunetti, M., Aiello, F.B., and Piantelli, M., 2000, Flavonoids apigenin and quercetin inhibit melanoma growth and metastatic potential, *Int J Cancer*, **87**: 595-600.

Cantley, L. C., 2002, The phosphoinositide 3-kinase pathway, *Science*, **296**: 1655-1657.

Castelluccio, C., Paganga, G., Melikian, N., Bolwell, G. P., Pridham, J., Sampson, J., and Rice-Evans, C., 1995, Antioxidant potential of intermediates in phenylpropanoid metabolism in higher plants, *FEBS Lett*, **368**: 188-192.

Chen, W. F., Huang, M. H., Tzang, C. H., Yang, M., and Wong, M. S., 2003, Inhibitory actions of genistein in human breast cancer (MCF-7) cells, *Biochim Biophys Acta*, **1638**: 187-196.

Chen, Y. C., Liang, Y. C., Lin-Shiau, S. Y., Ho, C. T., and Lin, J. K., 1999, Inhibition of TPA-induced protein kinase C and transcription activator protein-1 binding activities by theaflavin-3,3'-digallate from black tea in NIH3T3 cells, *J Agric Food Chem*, **47**: 1416-1421.

Cohen, D. R., and Curran, T., 1988, Fra-1: a serum-inducible, cellular immediate-early gene that encodes a fos-related antigen, *Mol Cell Biol*, **8**: 2063-2069.

Cos, P., Ying, L., Calomme, M., Hu, J. P., Cimanga, K., Van Poel, B., Pieters, L., Vlietinck A.J., and Vanden Berghe, D., 1998, Structure-activity relationship and classification of flavonoids as inhibitors of xanthine oxidase and superoxide scavengers, *J Nat Prod*, **61**: 71-76.

Cowley, S., Paterson, H., Kemp, P., and Marshall, C. J., 1994, Activation of MAP kinase kinase is necessary and sufficient for PC12 differentiation and for transformation of NIH 3T3 cells, *Cell*, **77**: 841-852.

Crawford, H. C., and Matrisian, L. M., 1996, Mechanisms controlling the transcription of matrix metalloproteinase genes in normal and neoplastic cells, *Enzyme Protein*, **49**: 20-37.

Davis, J. N., Kucuk, O., and Sarkar, F. H., 1999, Genistein inhibits NF-κB activation in prostate cancer cells, *Nutr Cancer*, **35**: 167-174.

Davis, J. N., Muqim, N., Bhuiyan, M., Kucuk, O., Pienta, K. J., and Sarkar, F. H., 2000, Inhibition of prostate specific antigen expression by genistein in prostate cancer cells, *Int J Oncol*, **16**: 1091-1097.

Davis, R. J., 1994, MAPKs: new JNK expands the group, *Trends Biochem Sci*, **19**: 470-473.

Davis, R. J., 2000, Signal transduction by the JNK group of MAP kinases, *Cell*, **103**: 239-252.

DeFelice S. L., 1992, Nutraceuticals: Opportunities in an Emerging Market, Scrip Mag 9.

Deschner, E. E., Ruperto, J., Wong, G., and Newmark, H. L., 1991, Quercetin and rutin as inhibitors of azoxymethanol-induced colonic neoplasia, *Carcinogenesis* **12**: 1193-1196.

Dillard, C. J., and German, J. B., 2000, Phytochemicals: nutraceuticals and human health, *J Sci Food Agric*, **80**: 1744-1756.

Dong, Z., Birrer, M. J., Watts, R. G., Matrisian, L. M., and Colburn, N. H., 1994, Blocking of tumor promoter-induced AP-1 activity inhibits induced transformation in JB6 mouse epidermal cells, *Proc Natl Acad Sci U S A*, **91**: 609-613.

Dong, Z., Huang, C., Brown, R. E., and Ma, W. Y., 1997a, Inhibition of activator protein 1 activity and neoplastic transformation by aspirin, *J Biol Chem*, **272**: 9962-9970.

Dong, Z., Ma, W., Huang, C., and Yang, C. S., 1997b, Inhibition of tumor promoter-induced activator protein 1 activation and cell transformation by tea polyphenols, (-)-epigallocatechin gallate, and theaflavins, *Cancer Res*, **57**: 4414-4419.

Dwyer, J., 1995, Overview: dietary approaches for reducing cardiovascular disease risks, *J Nutr*, **125**:656S-665S.

Eberhardt, M. V., Lee, C. Y., and Liu, R. H., 2000, Antioxidant activity of fresh apples, *Nature*, **405**:903-904.

Elliott, A. J., Scheiber, S. A., Thomas, C., and Pardini, R. S., 1992, Inhibition of glutathione reductase by flavonoids. A structure-activity study, *Biochem Pharmacol*, **44**: 1603-1608.

Farnsworth, N. R., Bingel, A. S., Cordell, G. A., Crane, F. A., and Fong, H. S., 1975, Potential value of plants as sources of new antifertility agents II, *J Pharm Sci*, **64**: 717-754.

Favot, L., Martin, S., Keravis, T., Andriantsitohaina, R., and Lugnier, C., 2003, Involvement of cyclin-dependent pathway in the inhibitory effect of delphinidin on angiogenesis, *Cardiovasc Res*, **59**:479-487.

Ferdinandy, P., Danial, H., Ambrus, I., Rothery, R. A., and Schulz, R., 2000, Peroxynitrite is a major contributor to cytokine-induced myocardial contractile failure, *Circ Res*, **87**: 241-247.

Ferrali, M., Signorini, C., Caciotti, B., Sugherini, L., Ciccoli, L., Giachetti, D., and Comproti, M., 1997, Protection against oxidative damage of erythrocyte membrane by the flavonoid quercetin and its relation to iron chelating activity, *FEBS Lett*, **416**: 123-129.

Fotsis, T., Pepper, M. S., Aktas, E., Breit, S., Rasku, S., Adlercreutz, H., Wahala, K.,Montesano, R., and Schweigerer, L., 1997, Flavonoids, dietary-derived inhibitors of cell proliferation and in vitro angiogenesis, *Cancer Res*, **57**: 2916-2921.

Frankel, E. N., Kanner, J., German, J. B., Parks, E., and Kinsella, J. E., 1993, Inhibition of oxidation of human low-density lipoprotein by phenolic substances in red wine, *Lancet*, **341**:454-457.

Frankel, E. N., Waterhouse, A. L., and Kinsella, J. E., 1993, Inhibition of human LDL oxidation by resveratrol, *Lancet*, **341**:1103-1104.

Fresno Vara, J. A., Casado, E., de Castro, J., Cejas, P., Belda-Iniesta, C., and Gonzalez-Baron, M., 2004, PI3K/Akt signalling pathway and cancer, *Cancer Treat Rev*, **30**:193-204.

Fuhrman, B., Lavy, A., and Aviram, M., 1995, Consumption of red wine with meals reduces the susceptibility of human plasma and low-density lipoprotein to lipid peroxidation, *Am J Clin Nutr*, **61**:549-554.

Ghiselli, A., Nardini, M., Baldi, A., and Scaccini, C., 1998, Antioxidant Activity of Different Phenolic Fractions Separated from an Italian Red Wine, *J Agric Food Chem*, **46**:361-367.

Gupta, S., Afaq, F., and Mukhtar, H., 2002, Involvement of nuclear factor-κB, Bax and Bcl-2 in induction of cell cycle arrest and apoptosis by apigenin in human prostate carcinoma cells, *Oncogene*, **21**:3727-3738.

Gupta, S., Hastak, K., Afaq, F., Ahmad, N., and Mukhtar, H., 2004 Essential role of caspases in epigallocatechin-3-gallate-mediated inhibition of nuclear factor kappa B and induction of apoptosis, *Oncogene*, **23**:2507-2522.

Hagiwara, A., Yoshino, H., Ichihara, T., Kawabe, M., Tamano, S., Aoki, H., Koda, T., Nakamura, M., Imaida, K.,Ito, N., and Shirai, T., 2002, Prevention by natural food anthocyanins, purple sweet potato color and red cabbage color, of 2-amino-1-methyl-6-phenylimidazo[4,5-b]pyridine (PhIP)-associated colorectal carcinogenesis in rats initiated with 1,2-dimethylhydrazine, *J Toxicol Sci*, **27**: 57-68.

Halazonetis, T. D., Georgopoulos, K., Greenberg, M. E., and Leder, P., 1988, c-Jun dimerizes with itself and with c-Fos, forming complexes of different DNA binding affinities, *Cell*, **55**: 917-924.

Harmon, A. W., and Patel, Y. M., 2004, Naringenin inhibits glucose uptake in MCF-7 breast cancer cells: a mechanism for impaired cellular proliferation, *Breast Cancer Res Treat*, **85**: 103-110.

Hasaniya, N., Youn, K., Xu, M., Hernaez, J., and Dashwood, R., 1997, Inhibitory activity of green and black tea in a free radical-generating system using 2-amino-3-methylimidazo[4,5-f]quinoline as substrate, *Jpn J Cancer Res*, **88**: 553-558.

Hastak, K., Gupta, S., Ahmad, N., Agarwal, M. K., Agarwal, M. L., and Mukhtar, H., 2003, Role of p53 and NF-kappaB in epigallocatechin-3-gallate-induced apoptosis of LNCaP cells, *Oncogene*, **22**: 4851-4859.

Hertog, M. G., Sweetnam, P. M., Fehily, A. M., Elwood, P. C., and Kromhout, D., 1997, Antioxidant flavonols and ischemic heart disease in a Welsh population of men: the Caerphilly Study, *Am J Clin Nutr*, **65**: 1489-1494.

Heo, H.J., and Lee, C. Y., 2004, Protective effects of quercetin and vitamin C against oxidative stress-induced neurodegeneration, *J Agric Food Chem*, **52**:7514-7517.

Heo, H. J., and Lee, C. Y., 2005a, Epicatechin and catechin in cocoa inhibit amyloid beta protein induced apoptosis, *J Agric Food Chem*, **53**: 1445-1448.

Heo, H. J., and Lee, C. Y., 2005b, Strawberry and its anthocyanidins reduce oxidative stress-induced apoptosis in PC12 cells, *J Agric Food Chem*, **53**:1984-1989.

Hirai, S. I., Ryseck, R. P., Mechta, F., Bravo, R., and Yaniv, M., 1989, Characterization of junD: a new member of the jun proto-oncogene family, *Embo J*, **8**: 1433-1439.

Hirano, R., Sasamoto, W., Matsumoto, A., Itakura, H., Igarashi, O., and Kondo, K., 2001, Antioxidant ability of various flavonoids against DPPH radicals and LDL oxidation, *J Nutr Sci Vitaminol (Tokyo)*, **47**: 357-362.

Hosokawa, N., Hosokawa, Y., Sakai, T., Yoshida, M., Marui, N., Nishino, H., Kawai, K., and Aoike, A., 1990, Inhibitory effect of quercetin on the synthesis of a possibly cell-cycle-related 17-kDa protein, in human colon cancer cells, *Int J Cancer*, **45**: 1119-1124.

Hou, D. X., Kai, K., Li, J. J., Lin, S., Terahara, N., Wakamatsu, M., Fujii, M., Young, M.R., and Colburn, N., 2004, Anthocyanidins inhibit activator protein 1 activity and cell transformation: structure-activity relationship and molecular mechanisms, *Carcinogenesis*, **25**: 29-36.

Huang, C., Ma, W. Y., Dawson, M. I., Rincon, M., Flavell, R. A., and Dong, Z., 1997a, Blocking activator protein-1 activity, but not activating retinoic acid response element, is required for the antitumor promotion effect of retinoic acid, *Proc Natl Acad Sci U S A*, **94**: 5826-5830.

Huang, C., Schmid, P. C., Ma, W. Y., Schmid, H. H., and Dong, Z., 1997b, Phosphatidylinositol-3 kinase is necessary for 12-O-tetradecanoylphorbol-13-acetate-induced cell transformation and activated protein 1 activation, *J Biol Chem*, **272**: 4187-4194.

Huang, C., Ma, W. Y., Young, M. R., Colburn, N., and Dong, Z., 1998, Shortage of mitogen-activated protein kinase is responsible for resistance to AP-1 transactivation and transformation in mouse JB6 cells, *Proc Natl Acad Sci U S A*, **95**: 156-161.

Ishikawa, T., Suzukawa, M., Ito, T., Yoshida, H., Ayaori, M., Nishiwaki, M., Yonemura, A., Hara, Y., and Nakamura, H., 1997, Effect of tea flavonoid supplementation on the susceptibility of low-density lipoprotein to oxidative modification, *Am J Clin Nutr*, **66**: 261-266.

Kallunki, T., Su, B., Tsigelny, I., Sluss, H. K., Derijard, B., Moore, G., Davis, R., and Karin, M., 1994, JNK2 contains a specificity-determining region responsible for efficient c-Jun binding and phosphorylation, *Genes Dev*, **8**: 2996-3007.

Kandaswami, C., Perkins, E., Soloniuk, D. S., Drzewiecki, G., and Middleton, E., Jr., 1991, Antiproliferative effects of citrus flavonoids on a human squamous cell carcinoma in vitro, *Cancer Lett*, **56**: 147-152.

Kapiotis, S., Hermann, M., Held, I., Seelos, C., Ehringer, H., and Gmeiner, B. M., 1997, Genistein, the dietary-derived angiogenesis inhibitor, prevents LDL oxidation and protects endothelial cells from damage by atherogenic LDL, *Arterioscler Thromb Vasc Biol*, **17**: 2868-2874.

Kaplan, D. R., and Miller, F. D., 2000, Neurotrophin signal transduction in the nervous system, *Curr Opin Neurobiol*, **10**: 381-391.

Karin, M., Cao, Y., Greten, F. R., and Li, Z. W., 2002, NF-kappaB in cancer: from innocent bystander to major culprit, *Nat Rev Cancer*, **2**: 301-310.

Katso, R., Okkenhaug, K., Ahmadi, K., White, S., Timms, J., and Waterfield, M. D., 2001, Cellular function of phosphoinositide 3-kinases: implications for development, homeostasis, and cancer, *Annu Rev Cell Dev Biol*, **17**: 615-675.

Kawaii, S., Tomono, Y., Katase, E., Ogawa, K., and Yano, M., 1999, Antiproliferative activity of flavonoids on several cancer cell lines, *Biosci Biotechnol Biochem*, **63**:896-899.

Kawanishi, S., Hiraku, Y., and Oikawa, S., 2001, Mechanism of guanine-specific DNA damage by oxidative stress and its role in carcinogenesis and aging, *Mutat Res*, **488**: 65-76.

Kelm, M. A., Hammerstone, J. F., and Schmitz, H. H., 2005, Identification and quantitation of flavanols and proanthocyanidins in foods: how good are the datas? *Clin Dev Immunol*, **12**: 35-41.

Khan, M. A., and Baseer, A., 2000, Increased malondialdehyde levels in coronary heart disease, *J Pak Med Assoc*, **50**: 261-264.

Khan, W. A., Wang, Z. Y., Athar, M., Bickers, D. R., and Mukhtar, H., 1988, Inhibition of the skin tumorigenicity of (+/-)-7 beta, 8 alpha-dihydroxy-9 alpha,10 alpha-epoxy-7,8,9,10-tetrahydrobenzo[a]pyrene by tannic acid, green tea polyphenols and quercetin in Sencar mice, *Cancer Lett*, **42**: 7-12.

Khanduja, K. L., Gandhi, R. K., Pathania, V., and Syal, N., 1999, Prevention of N-nitrosodiethylamine-induced lung tumorigenesis by ellagic acid and quercetin in mice, *Food Chem Toxicol*, **37**: 313-318.

Kharbanda, S., Saxena, S., Yoshida, K., Pandey, P., Kaneki, M., Wang, Q., Cheng, K., Chen, Y. N., Campbell, A., Sudha, T., Yuan, Z. M., Narula, J., Weichselbaum, R., Nalin, C., and Kufe, D., 2000, Translocation of SAPK/JNK to mitochondria and interaction with Bcl-x(L) in response to DNA damage, *J Biol Chem*, **275**: 322-327.

Kirk, E. A., Sutherland, P., Wang, S. A., Chait, A., and LeBoeuf, R. C., 1998, Dietary isoflavones reduce plasma cholesterol and atherosclerosis in C57BL/6 mice but not LDL receptor-deficient mice, *J Nutr*, **128**: 954-959.

Ko, C. H., Shen, S. C., Lee, T. J., and Chen, Y. C., 2005, Myricetin inhibits matrix metalloproteinase-2 protein expression and enzyme activity in colorectal carcinoma cells, *Mol Cancer Ther*, **4**: 281-290.

Koh, S. H., Kim, S. H., Kwon, H., Kim, J. G., Kim, J. H., Yang, K. H., Kim, J., Kim, S. U., Yu, H. J., Do, B. R., Kim, K. S., and Jung, H. K., 2004, Phosphatidylinositol-3 kinase/Akt and GSK-3 mediated cytoprotective effect of epigallocatechin gallate on oxidative stress-injured neuronal-differentiated N18D3 cells, *Neurotoxicology*, **25**: 793-802.

Kucharczak, J., Simmons, M. J., Fan, Y., and Gelinas, C., 2003, To be, or not to be: NF-kappaB is the answer--role of Rel/NF-kappaB in the regulation of apoptosis, *Oncogene*, **22**: 8961-8982.

Kuntz, S., Wenzel, U., and Daniel, H., 1999, Comparative analysis of the effects of flavonoids on proliferation, cytotoxicity, and apoptosis in human colon cancer cell lines, *Eur J Nutr*, **38**: 133-142.

Kuo, S. M., 1996, Antiproliferative potency of structurally distinct dietary flavonoids on human colon cancer cells, *Cancer Lett*, **110**: 41-48.

Kyriakis, J. M., Banerjee, P., Nikolakaki, E., Dai, T., Rubie, E. A., Ahmad, M. F., Avruch, J., and Woodgett, J. R., 1994, The stress-activated protein kinase subfamily of c-Jun kinases, *Nature*, **369**: 156-160.

Landau, J. M., Wang, Z. Y., Yang, G. Y., Ding, W., and Yang, C. S., 1998, Inhibition of spontaneous formation of lung tumors and rhabdomyosarcomas in A/J mice by black and green tea, *Carcinogenesis*, **19**: 501-507.

Lazze, M. C., Savio, M., Pizzala, R., Cazzalini, O., Perucca, P., Scovassi, A. I., Stivala, L. A., and Bianchi, L., 2004, Anthocyanins induce cell cycle perturbations and apoptosis in different human cell lines, *Carcinogenesis*, **25**: 1427-1433.

Le Bail, J. C., Varnat, F., Nicolas, J. C., and Habrioux, G., 1998, Estrogenic and antiproliferative activities on MCF-7 human breast cancer cells by flavonoids, *Cancer Lett*, **130**: 209-216.

Lee, H. H., Ho, C. T., and Lin, J. K., 2004, Theaflavin-3,3'-digallate and penta-O-galloyl-β-D-glucose inhibit rat liver microsomal 5alpha-reductase activity and the expression of androgen receptor in LNCaP prostate cancer cells, *Carcinogenesis*, **25**: 1109-1118.

Lee, L. T., Huang, Y. T., Hwang, J. J., Lee, A. Y., Ke, F. C., Huang, C. J., Kandaswami, C., Lee, P. P., and Lee, M. T., 2004, Trans-inactivation of the epidermal growth factor receptor tyrosine kinase and focal adhesion kinase phosphorylation by dietary flavonoids: effect on invasive potential of human carcinoma cells, *Biochem Pharmacol*, **67**: 2103-2114.

Leeuwenburgh, C., Hardy, M. M., Hazen, S. L., Wagner, P., Oh-ishi, S., Steinbrecher, U. P., and Heinecke, J. W., 1997, Reactive nitrogen intermediates promote low density lipoprotein oxidation in human atherosclerotic intima, *J Biol Chem*, **272**: 1433-1436.

Lepley, D. M., Li, B., Birt, D. F., and Pelling, J. C., 1996, The chemopreventive flavonoid apigenin induces G2/M arrest in keratinocytes, *Carcinogenesis*, **17**: 2367-2375.

Lepley, D. M., and Pelling, J. C., 1997, Induction of p21/WAF1 and G1 cell-cycle arrest by the chemopreventive agent apigenin, *Mol Carcinog*, **19**: 74-82.

Li, N., Chen, X., Liao, J., Yang, G., Wang, S., Josephson, Y., Han, C., Chen, J., Huang, M. T., and Yang, C. S., 2002, Inhibition of 7,12-dimethylbenz[a]anthracene (DMBA)-induced oral carcinogenesis in hamsters by tea and curcumin, *Carcinogenesis*, **23**: 1307-1313.

Li, Y., Upadhyay, S., Bhuiyan, M., and Sarkar, F. H., 1999, Induction of apoptosis in breast cancer cells MDA-MB-231 by genistein, *Oncogene*, **18**: 3166-3172.

Lian, F., Bhuiyan, M., Li, Y. W., Wall, N., Kraut, M., and Sarkar, F. H., 1998, Genistein-induced G2-M arrest, p21WAF1 upregulation, and apoptosis in a non-small-cell lung cancer cell line, *Nutr Cancer*, **31**: 184-191.

Lian, F., Li, Y., Bhuiyan, M., and Sarkar, F. H., 1999, p53-independent apoptosis induced by genistein in lung cancer cells, *Nutr Cancer*, **33**: 125-131.

Liang, Y. C., Lin-shiau, S. Y., Chen, C. F., and Lin, J. K., 1997, Suppression of extracellular signals and cell proliferation through EGF receptor binding by (-)-epigallocatechin gallate in human A431 epidermoid carcinoma cells, *J Cell Biochem*, **67**: 55-65.

Liang, Y. C., Chen, Y. C., Lin, Y. L., Lin-Shiau, S. Y., Ho, C. T., and Lin, J. K., 1999a, Suppression of extracellular signals and cell proliferation by the black tea polyphenol, theaflavin-3,3'-digallate, *Carcinogenesis*, **20**: 733-736.

Liang, Y. C., Huang, Y. T., Tsai, S. H., Lin-Shiau, S. Y., Chen, C. F., and Lin, J. K., 1999b, Suppression of inducible cyclooxygenase and inducible nitric oxide synthase by apigenin and related flavonoids in mouse macrophages, *Carcinogenesis*, **20**: 1945-1952.

Liang, Y. C., Lin-Shiau, S. Y., Chen, C. F., and Lin, J. K., 1999c, Inhibition of cyclin-dependent kinases 2 and 4 activities as well as induction of Cdk inhibitors p21 and p27 during growth arrest of human breast carcinoma cells by (-)-epigallocatechin-3-gallate, *J Cell Biochem*, **75**: 1-12.

Lichtenstein, A. H., 1998, Soy protein, isoflavones and cardiovascular disease risk, *J Nutr*, **128**: 1589-1592.

Lin, A., and Karin, M., 2003, NF-kappaB in cancer: a marked target, *Semin Cancer Biol*, **13**: 107-114.

Lin, J. K., 2002, Cancer chemoprevention by tea polyphenols through modulating signal transduction pathways, *Arch Pharm Res* **25**: 561-571.

Lin, J. K., Chen, P. C., Ho, C. T., and Lin-Shiau, S. Y., 2000, Inhibition of xanthine oxidase and suppression of intracellular reactive oxygen species in HL-60 cells by theaflavin-3,3'-digallate, (-)-epigallocatechin-3-gallate, and propyl gallate, *J Agric Food Chem*, **48**: 2736-2743.

Lin, J.K., Liang, Y. C., and Lin-Shiau, S. Y., 1999, Cancer chemoprevention by tea polyphenols through mitotic signal transduction blockade, *Biochem Pharmacol*, **58**: 911-915.

Lin, Y. L. Cheng, C. Y. Lin, Y. P. Lau, Y. N. Juan, I. M. and Lin, J. K., 1998, Hypolipidemic effect of green tea leaves through induction of antioxidant and phase II enzymes including superoxide dismutase, catalase and glutathione S-transferase, *J. Agric. Food Chem*, **46**: 1893-1899.

Lin, Y. L., and Lin, J. K., 1997, (-)-Epigallocatechin-3-gallate blocks the induction of nitric oxide synthase by down-regulating lipopolysaccharide-induced activity of transcription factor nuclear factor-kappaB, *Mol Pharmacol*, **52**: 465-472.

Liu, J. D., Chen, S. H., Lin, C. L., Tsai, S. H., and Liang, Y. C., 2001, Inhibition of melanoma growth and metastasis by combination with (-)-epigallocatechin-3-gallate and dacarbazine in mice, *J Cell Biochem*, **83**: 631-642.

Liu, R. H., 2004, Potential synergy of phytochemicals in cancer prevention: mechanism of action, *J Nutr*, **134**: 3479s-3485s.

Liu, R. H., Liu, J., and Chen, B., 2005, Apples prevent mammary tumors in rats, *J Agric Food Chem*, **53**: 2341-2343.

Luke, M. C., and Coffey, D. S., 1994, Human androgen receptor binding to the androgen response element of prostate specific antigen, *J Androl*, **15**: 41-51.

Makita, H., Tanaka, T., Fujitsuka, H., Tatematsu, N., Satoh, K., Hara, A., Mori, H., 1996, Chemoprevention of 4-nitroquinoline 1-oxide-induced rat oral carcinogenesis by the dietary flavonoids chalcone, 2-hydroxychalcone, and quercetin, *Cancer Res*, **56**: 4904-4909.

Mallikarjuna, G., Dhanalakshmi, S., Singh, R. P., Agarwal, C., and Agarwal, R., 2004, Silibinin protects against photocarcinogenesis via modulation of cell cycle regulators, mitogen-activated protein kinases, and Akt signalling, *Cancer Res*, **64**: 6349-6356.

Marko, D., Puppel, N., Tjaden, Z., Jakobs, S., and Pahlke, G., 2004, The substitution pattern of anthocyanidins affects different cellular signalling cascades regulating cell proliferation, *Mol Nutr Food Res*, **48**: 318-325.

Martin, P. M., Horwitz, K. B., Ryan, D. S., and McGuire, W. L., 1978, Phytoestrogen interaction with estrogen receptors in human breast cancer cells, *Endocrinology*, **103**: 1860-1867.

Martin, S., Favot, L., Matz, R., Lugnier, C., and Andriantsitohaina, R., 2003, Delphinidin inhibits endothelial cell proliferation and cell cycle progression through a transient activation of ERK-1/-2, *Biochem Pharmacol*, **65**: 669-675.

Masuda, M., Suzui, M., Lim, J. T., Deguchi, A., Soh, J. W., and Weinstein, I. B., 2002, Epigallocatechin-3-gallate decreases VEGF production in head and neck and breast carcinoma cells by inhibiting EGFR-related pathways of signal transduction, *J Exp Ther Oncol*, **2**: 350-359.

McCarty, M. F., 1998, Polyphenol-mediated inhibition of AP-1 transactivating activity may slow cancer growth by impeding angiogenesis and tumor invasiveness, *Med Hypotheses*, **50**: 511-514.

Meiers, S., Kemeny, M., Weyand, U., Gastpar, R., von Angerer, E., and Marko, D., 2001, The anthocyanidins cyanidin and delphinidin are potent inhibitors of the epidermal growth-factor receptor, *J Agric Food Chem*, **49**: 958-962.

Mentor-Marcel, R., Lamartiniere, C. A., Eltoum, I. E., Greenberg, N. M., and Elgavish, A., 2001, Genistein in the diet reduces the incidence of poorly differentiated prostatic adenocarcinoma in transgenic mice (TRAMP), *Cancer Res*, **61**: 6777-6782.

Miksicek, R. J., 1993, Commonly occurring plant flavonoids have estrogenic activity, *Mol Pharmacol*, **44**: 37-43.

Miksicek, R. J., 1995, Estrogenic flavonoids: structural requirements for biological activity, *Proc Soc Exp Biol Med*, **208**: 44-50.

Minden, A., Lin, A., McMahon, M., Lange-Carter, C., Derijard, B., Davis, R. J., Johnson, G. L., and Karin, M., 1994, Differential activation of ERK and JNK mitogen-activated protein kinases by Raf-1 and MEKK, *Science*, **266**: 1719-1723.

Miura, S., Watanabe, J., Sano, M., Tomita, T., Osawa, T., Hara, Y., and Tomita, I., 1995, Effects of various natural antioxidants on the Cu(2+)-mediated oxidative modification of low density lipoprotein, *Biol Pharm Bull*, **18**: 1-4.

Miyagi, Y., Om, A. S., Chee, K. M., and Bennink, M. R., 2000, Inhibition of azoxymethane-induced colon cancer by orange juice, *Nutr Cancer*, **36**: 224-229.

Miyata, Y., Sato, T., Yano, M., and Ito, A., 2004, Activation of protein kinase C betaII/epsilon-c-Jun NH2-terminal kinase pathway and inhibition of mitogen-activated protein/extracellular signal-regulated kinase 1/2 phosphorylation in antitumor invasive activity induced by the polymethoxy flavonoid, nobiletin, *Mol Cancer Ther*, **3**: 839-847.

Moore, K. P., Darley-Usmar, V., Morrow, J., and Roberts, L. J., 2nd., 1995, Formation of F2-isoprostanes during oxidation of human low-density lipoprotein and plasma by peroxynitrite, *Circ Res*, **77**: 335-341.

Ness, A. R., and Powles, J. W., 1997, Fruit and vegetables, and cardiovascular disease: a review, *Int J Epidemiol*, **26**: 1-13.

Nigdikar, S. V., Williams, N. R., Griffin, B. A., and Howard, A. N., 1998, Consumption of red wine polyphenols reduces the susceptibility of low-density lipoproteins to oxidation in vivo, *Am J Clin Nutr*, **68**: 258-265.

Oldreive, C., Zhao, K., Paganga, G., Halliwell, B., and Rice-Evans, C., 1998, Inhibition of nitrous acid-dependent tyrosine nitration and DNA base deamination by flavonoids and other phenolic compounds, *Chem Res Toxicol*, **11**: 1574-1579.

Omenn, G. S., Goodman, G. E., Thornquist, M. D., Balmes, J., and Cullen, M. R., Glass, A., Keogh, J. P., Meyskens, F. L., Valanis, B., Williams, J. H., Barnhart, S., and Hammar, S., 1996, Effects of a combination of beta-carotene and vitamin A on lung cancer and cardiovascular disease, *N Engl J Med*, **334**: 1150-1155.

Orlowski, R. Z., and Baldwin, A. S., Jr., 2002, NF-κB as a therapeutic target in cancer, *Trends Mol Med*, **8**:385-389.

Pan, M. H., Lin-Shiau, S. Y., Ho, C. T., Lin, J. H., and Lin, J. K., 2000, Suppression of lipopolysaccharide-induced nuclear factor-kappaB activity by theaflavin-3,3'-digallate from black tea and other polyphenols through down-regulation of IkappaB kinase activity in macrophages, *Biochem Pharmacol*, **59**: 357-367.

Pannala, A. S., Rice-Evans, C. A., Halliwell, B., and Singh, S., 1997, Inhibition of peroxynitrite-mediated tyrosine nitration by catechin polyphenols, *Biochem Biophys Res Commun*, **232**: 164-168.

Ross, J. A., and Kasum, C. M., 2002, Dietary flavonoids: bioavailability, metabolic effects, and safety, *Annu Rev Nutr*, **22**: 19-34.

Salah, N., Miller, N. J., Paganga, G., Tijburg, L., Bolwell, G. P., and Rice-Evans, C., 1995, Polyphenolic flavanols as scavengers of aqueous phase radicals and as chain-breaking antioxidants, *Arch Biochem Biophys*, **322**: 339-346.

Sastre, J., Pallardo, F. V., and Vina, J., 2000, Mitochondrial oxidative stress plays a key role in aging and apoptosis, *IUBMB Life*, **49**: 427-435.

Sazuka, M., Murakami, S., Isemura, M., Satoh, K., and Nukiwa, T., 1995, Inhibitory effects of green tea infusion on in vitro invasion and in vivo metastasis of mouse lung carcinoma cells, *Cancer Lett*, **98**: 27-31.

Schreier, P., 2005, Chemopreventive compounds in the diet, *Dev Ophthalmol*, **38**: 1-58.

Seger, R., and Krebs, E. G., 1995, The MAPK signalling cascade, *Faseb J*, **9**: 726-735.

Senderowicz, A. M., 2002, Cyclin-dependent kinases as targets for cancer therapy, *Cancer Chemother Biol Response Modif*, **20**: 169-196.

Senderowicz, A. M., 2003, Cell cycle modulators for the treatment of lung malignancies, *Clin Lung Cancer*, **5**: 158-168.

Siess, M. H., Le Bon, A. M., Canivenc-Lavier, M. C., and Suschetet, M., 2000, Mechanisms involved in the chemoprevention of flavonoids, *Biofactors*, **12**: 193-199.

Silverman, N., and Maniatis, T., 2001, NF-kappaB signalling pathways in mammalian and insect innate immunity, *Genes Dev*, **15**: 2321-2342.

So, F. V., Guthrie, N., Chambers, A. F., Moussa, M., and Carroll, K. K., 1996, Inhibition of human breast cancer cell proliferation and delay of mammary tumorigenesis by flavonoids and citrus juices, *Nutr Cancer*, **26**: 167-181.

Storz, P., and Toker, A., 2003, NF-κB signalling--an alternate pathway for oxidative stress responses, *Cell Cycle*, **2**: 9-10.

Takabe, W., Niki, E., Uchida, K., Yamada, S., Satoh, K., and Noguchi, N., 2001, Oxidative stress promotes the development of transformation: involvement of a potent mutagenic lipid peroxidation product, acrolein, *Carcinogenesis*, **22**: 935-941.

Tanaka, T., Kawabata, K., Kakumoto, M., Makita, H., Ushida, J., Honjo, S., Hara, A., Tsuda, H., and Mori, H., 1999, Modifying effects of a flavonoid morin on azoxymethane-induced large bowel tumorigenesis in rats, *Carcinogenesis*, **20**: 1477-1484.

Tanaka, T., Makita, H., Kawabata, K., Mori, H., Kakumoto, M., Satoh, K., Hara, A., Murakami, A., Koshimizu, K., and Ohigashi, H., 1997, Chemoprevention of azoxymethane-induced rat colon carcinogenesis by the naturally occurring flavonoids, diosmin and hesperidin, *Carcinogenesis*, **18**: 957-965.

Tikkanen, M. J., Wahala, K., Ojala, S., Vihma, V., and Adlercreutz, H., 1998, Effect of soybean phytoestrogen intake on low density lipoprotein oxidation resistance, *Proc Natl Acad Sci U S A*, **95**: 3106-3110.

Traganos, F., Ardelt, B., Halko, N., Bruno, S., and Darzynkiewicz, Z., 1992, Effects of genistein on the growth and cell cycle progression of normal human lymphocytes and human leukemic MOLT-4 and HL-60 cells, *Cancer Res*, **52**: 6200-6208.

Tyagi, A., Agarwal, C., and Agarwal, R., 2002a, The cancer preventive flavonoid silibinin causes hypophosphorylation of Rb/p107 and Rb2/p130 via modulation of cell cycle regulators in human prostate carcinoma DU145 cells, *Cell Cycle*, **1**: 137-142.

Tyagi, A., Agarwal, C., and Agarwal, R., 2002b, Inhibition of retinoblastoma protein (Rb) phosphorylation at serine sites and an increase in Rb-E2F complex formation by silibinin in androgen-dependent human prostate carcinoma LNCaP cells: role in prostate cancer prevention, *Mol Cancer Ther*, **1**:525-532.

van het Hof, K. H., de Boer, H. S., Wiseman, S. A., Lien, N., Westrate, J. A., and Tijburg, L. B., 1997, Consumption of green or black tea does not increase resistance of low-density lipoprotein to oxidation in humans, *Am J Clin Nutr*, **66**: 1125-1132.

Verma, A. K., Johnson, J. A., Gould, M. N., and Tanner, M. A., 1988, Inhibition of 7,12-dimethylbenz(a)anthracene- and N-nitrosomethylurea-induced rat mammary cancer by dietary flavonol quercetin, *Cancer Res*, **48**: 5754-5758.

Wang, J., Eltoum, I. E., and Lamartiniere, C. A., 2002, Dietary genistein suppresses chemically induced prostate cancer in Lobund-Wistar rats, *Cancer Lett*, **186**: 11-18.

Wang, W., Heideman, L., Chung, C. S., Pelling, J. C., Koehler, K. J., and Birt, D. F., 2000, Cell-cycle arrest at G2/M and growth inhibition by apigenin in human colon carcinoma cell lines, *Mol Carcinog*, **28**: 102-110.

Wang, W., Sawicki, G., and Schulz, R., 2002, Peroxynitrite-induced myocardial injury is mediated through matrix metalloproteinase-2, *Cardiovasc Res*, **53**: 165-174.

Way, T. D., Kao, M. C., and Lin, J. K., 2004a, Apigenin induces apoptosis through proteasomal degradation of HER2/neu in HER2/neu-overexpressing breast cancer cells via the phosphatidylinositol 3-kinase/Akt-dependent pathway, *J Biol Chem*, **279**: 4479-4489.

Way, T. D., Lee, H. H., Kao, M. C., and Lin, J. K., 2004b, Black tea polyphenol theaflavins inhibit aromatase activity and attenuate tamoxifen resistance in HER2/neu-transfected human breast cancer cells through tyrosine kinase suppression, *Eur J Cancer*, **40**: 2165-2174.

Way, T. D., Kao, M. C., and Lin, J. K., 2005, Degradation of HER2/neu by apigenin induces apoptosis through cytochrome c release and caspase-3 activation in HER2/neu-overexpressing breast cancer cells, *FEBS Lett*, **579**: 145-152.

Welshons, W. V., Thayer, K. A., Judy, B. M., Taylor, J. A., Curran, E. M., and vom Saal, F. S., 2003, Large effects from small exposures. I. Mechanisms for endocrine-disrupting chemicals with estrogenic activity, *Environ Health Perspect*, **111**: 994-1006.

Wenzel, U., Kuntz, S., Brendel, M. D., and Daniel, H., 2000, Dietary flavone is a potent apoptosis inducer in human colon carcinoma cells, *Cancer Res*, **60**: 3823-3831.

White, M. F., 2003, Insulin signalling in health and disease, *Science*, **302**: 1710-1711.

Willett, W. C., 2002, Balancing life-style and genomics research for disease prevention, *Science*, **296**:695-698.

Witztum, J. L., and Steinberg, D., 1991, Role of oxidized low density lipoprotein in atherogenesis, *J Clin Invest*, **88**: 1785-1792.

Xu, Y., Ho, C. T., Amin, S. G., Han, C., and Chung, F. L., 1992, Inhibition of tobacco-specific nitrosamine-induced lung tumorigenesis in A/J mice by green tea and its major polyphenol as antioxidants, *Cancer Res*, **52**: 3875-3879.

Yang, C. S., Maliakal, P., and Meng, X., 2002, Inhibition of carcinogenesis by tea, *Annu Rev Pharmacol Toxicol*, **42**: 25-54.

Yang, C. S., and Wang, Z. Y., 1993, Tea and cancer, *J Natl Cancer Inst*, **85**: 1038-1049.

Yang, F., Oz, H. S., Barve, S., de Villiers, W. J., McClain, C. J., and Varilek, G. W., 2001, The green tea polyphenol (-)-epigallocatechin-3-gallate blocks nuclear factor-κB activation by inhibiting I κB kinase activity in the intestinal epithelial cell line IEC-6 *Mol Pharmacol*, **60**: 528-533.

Yang, G., Wang, Z. Y., Kim, S., Liao, J., Seril, D. N., Chen, X., Smith, T.J., and Yang, C.S., 1997a, Characterization of early pulmonary hyperproliferation and tumor progression and their inhibition by black tea in a 4-(methylnitrosamino)-1-(3-pyridyl)-1-butanone-induced lung tumorigenesis model with A/J mice, *Cancer Res*, **57**: 1889-1894.

Yang, G. Y., Liu, Z., Seril, D. N., Liao, J., Ding, W., Kim, S., Bondoc, F., and Yang, C.S., 1997b, Black tea constituents, theaflavins, inhibit 4-(methylnitrosamino)-1-(3-pyridyl)-1-butanone (NNK)-induced lung tumorigenesis in A/J mice, *Carcinogenesis*, **18**: 2361-2365.

Yano S., Tachibana H., and Yamada, K., 2005, Flavones suppress the expression of the high-affinity IgE receptor Fcepsilon RI in human basophilic KU812 cells, *J Agric Food Chem*, **53**: 1812-1817.

Yeh, C. W., Chen, W. J., Chiang, C. T., Lin-Shiau, S. Y., and Lin, J. K., 2003, Suppression of fatty acid synthase in MCF-7 breast cancer cells by tea and tea polyphenols: a possible mechanism for their hypolipidemic effects, *Pharmacogenomics J*, **3**: 267-276.

Yoshida, M., Yamamoto, M., and Nikaido, T., 1992, Quercetin arrests human leukemic T-cells in late G1 phase of the cell cycle, *Cancer Res*, **52**: 6676-6681.

Zhang, X. M., Chen, J., Xia, Y. G., and Xu, Q., 2005, Apoptosis of murine melanoma B16-BL6 cells induced by quercetin targeting mitochondria, inhibiting expression of PKC-alpha and translocating PKC-delta, *Cancer Chemother Pharmacol*, **55**: 251-262.

Zhou, J. R., Mukherjee, P., Gugger, E. T., Tanaka, T., Blackburn, G. L., and Clinton, S. K., 1998, Inhibition of murine bladder tumorigenesis by soy isoflavones via alterations in the cell cycle, apoptosis, and angiogenesis, *Cancer Res*, **58**: 5231-5238.

CHAPTER 9

FLAVONOIDS AS SIGNAL MOLECULES

Targets of Flavonoid Action

WENDY ANN PEER AND ANGUS S. MURPHY

Department of Horticulture, Purdue University, West Lafayette, IN 47907 USA

1. INTRODUCTION

Plant-derived foods that are rich in flavonoids are regularly touted in the popular press for their benefits in ameliorating age-related diseases. A majority of these reports focus on the antioxidant characteristics of flavonoid-rich diets and their enhancement of cardiovascular health. However, a growing number of reports in the pharmacology literature characterize flavonoid interactions with cellular components implicated in neurological pathologies and cancer. As the effective flavonoid concentrations employed in pharmacological studies utilizing cell cultures are often orders of magnitude higher than the serum concentrations seen in humans, some discrimination is required when interpreting these reports. Further, as flavonoids often mimic endogenous mammalian receptor ligands (Ciolino et al., 1999; Hunter et al., 1999; Virgili et al., 2004) or interfere with the uptake of substrates not found in plants, the applicability of such studies to plant models requires caution. However, studies of flavonoid function in animal cells often provide important insights into their functions as signal molecules in plants.

Nearly every class of flavonoid has been shown to have biological activity, with a majority related to antioxidant properties. In plants, flavonoids appear to contribute to a general reduction of reactive oxygen species and therefore impact cellular processes sensitive to REDOX effects. However, flavonoids also have been implicated in more direct interactions with transport and signal transduction pathways. One well-documented example is the role of flavonoids in fertility: while a few flavonoid-deficient plants are able to germinate, grow, and set fertile seed, most plants require flavonoids for fertility and normal pollen development. Another is flavonoid modulation of auxin transport as well as localized auxin accumulations observed during nodulation. Perhaps the best-studied example of flavonoid signaling is that of flavonoid mediation of interactions between the plants and other organisms in the environment at both competitive (allelopathy/defense) and cooperative (mycorrhizal association) levels.

In the balance of this chapter, we review potential and known molecular targets of flavonoid signaling and plant processes where flavonoids have been implicated in regulatory functions. We then describe flavonoid-dependent internal signaling processes and extraorganismal biotic interactions mediated by flavonoid signaling.

2. MOLECULAR TARGETS OF FLAVONOID ACTION

At the molecular level, potential targets of flavonoid regulation in plants range from transcription factors and kinases to ATP-binding cassette (ABC) transporters and aminopeptidases. Some of these targets are suggested primarily by similarities between plant and mammalian signaling mechanisms. Other endogenous or exogenous targets, such as receptors, ABC transporters, and hydrolases, have been directly demonstrated *in planta* or *in vivo*. Most of these interactions have been shown to be developmentally regulated. These potential and known targets are categorized below.

2.1. Transcription

Nuclear localization of flavonoids has been reported in many plant species, suggesting that flavonoids may function in transcriptional regulation of endogenous gene expression. Reports of sulfonated flavonols in nucleus in *Flaveria chloraefolia*, (Grandmaison and Ibrahim, 1996) and unidentified phenolic compounds in *Brassica napus* also were localized in the nucleus but not nucleolus (Kuras et al., 1999; Stefanowska et al., 2003). Nuclear localization of flavonols also has been shown in *Arabidopsis thaliana* (Peer et al., 2001; Buer and Muday, 2004; Saslowsky et al., 2005) and flavanols in *Tsuga canadensis, Taxus baccata, Metasequoia glyptostroboides, Coffea arabica, Prunus avium,* and *Camellia sinesis* (Feucht et al., 2004a, 2004b). Catechin binding of histone proteins has been demonstrated in plants (Polster et al., 2003; Feucht et al., 2004b), suggesting that catechins might modulate nonspecific gene transcription. Naringenin chalcone and apigenin also may influence flavonoid biosynthesis by regulating transcription of flavonoid biosynthetic enzymes (Pelletier et al., 1999) (see Chapter 4 for more details). Recent evidence that the flavonoid biosynthetic enzymes chalcone synthase (CHS) and chalcone isomerase (CHI) are localized in the nucleus, in addition to endosomal membrane surfaces, in *A. thaliana* suggests that flavonoid regulation of transcription is developmentally regulated at the subcellular level (Saslowsky et al., 2005). This observation also is consistent differential localization of phenolics in subcellular compartments observed throughout seed development and germination (Kuras et al., 1999; Stefansowska et al., 2003).

There also is evidence for flavonoid regulation of gene transcription in other organisms, like rhizobacteria. Flavonoids in root exudates from host plants are required for transcription of NodD, a bifunctional transcriptional repressor/activator of the *nod* genes in *Rhizobium leguminosarum*, in order for nodulation to occur. NodD represses its own transcription by competing with RNA polymerase (RNAP) for RNAP binding site (Hu et al., 2000). Naringenin, but not luteolin, relieves NodD binding to RNAP; thereby *nodD* can be transcribed and NodD can activate the other *nod* genes (Hu et al., 2000). The roles of flavonoids in nodulation is further developed in Section 3.2.4.

Pharmacological studies of flavonoid effects on mammalian transcription suggest other potential sites of regulation in plants. Quercetin inhibition of histone H1 and H2AX phosphorylation and genistein inhibition of H2AX phosphorylation has been documented (Notoya et al., 2004; Ye et al., 2004). Glucopyranosides of kaempferol and quercetin also specifically bind DNA polymerase α *in vitro* (Mizushina et al., 2003); however, *in vivo* binding and subsequent transcriptional induction or repression has not been shown. Flavonoids also may affect transcription by inhibiting topoisomerase (topo) activity: quercetin, myricetin, fisetin, and morin inhibited both topo I and II activity, while kaempferol, phloretin, and apigenin specifically inhibited topo II (Constantinou et al., 1995; Boege et al., 1996), and quercetin, kaempferol, and naringenin stabilized the topo II-DNA complex (Constantinou et al., 1995).

Most of the regulation of transcription by flavonoids appears to involve inhibition of phosphorylation signaling cascades or specific kinases. Quercetin specifically inhibited tumor necrosis factor α (TNFα) transcription through inhibition of phosphorylation of c-Jun N-terminal kinase/stress activated protein kinase (JNK/SAPK), thereby suppressing activator protein-1 (AP-1) from binding to TNFα promoter (Wadsworth et al., 2001; Chen et al., 2004). Apigenin inhibited inhibitor κB kinase (IκB) activity, thereby inhibiting nuclear factor κB (NF-κB)-dependent COX-2 transcription (Liang et al., 1999; O'Leary et al., 2004). The flavonols kaempferol and quercetin inhibited transcription factor AP-1 activation of nitric oxide synthase expression while the isoflavone genistein did so via NF-κB inhibition (Kim et al., 2005). Quercetin inhibited an AKT/protein kinase B (PKB) and extracellular signal-related kinase (ERK) phosphorylation and also activated transcription factor cAMP-responsive element-binding protein (Spencer et al., 2003). Quercetin activated transcription of heme oxygenase-1 via mitogen-activated protein kinase (MAPK) signaling cascade, but inhibited ERK (Lin et al., 2004). Apigenin inhibited expression of vascular endothelial growth factor and hypoxia-inducible factor-1α via PIPK/AKT and HDM2/p53 (which regulates MAPK) pathways (Fang et al., 2005). Apigenin also has been shown to inhibit Ras signaling (Klampfer et al., 2004).

2.2. Translation

As yet, there is no evidence for direct flavonoid modulation of translation in plants. However, flavonoids have been shown to affect mammalian translation by altering phosphorylation status. Genistein and quercetin were found to inhibit protein synthesis in mouse tumor cells via phosphorylative activation of eIF2-α kinases (Ito et al., 1999), and genistein decreased late viral mRNA translation via a tyrosine kinase-dependent mechanism (Xi et al., 2005).

2.3. Enzyme activity and signal transduction

There is increasing evidence that specific proteins or groups of proteins exhibit more specific interactions with flavonoids *in vivo*. Many of these interactions have been shown to depend on the B-ring substitution pattern of the interacting flavonoids (Marko et al., 2004). Flavonols are the most common of these molecules, and their high degree of activity is suggested by the tight developmental and spatial regulation of flavonol synthesis (Murphy et al., 2000; Peer et al., 2001). Several targets and potential targets have been identified in *Arabidopsis*: AtAPM1, a microsomal tyrosine aminopeptidase; the tyrosine phosphatase AtPTEN1 (phosphatase and tensin homolog); PINOID (PID), a serine/threonine kinase; ROOTS CURL IN NAPHTHYLPHTHALMIC ACID1 (RCN1); the subunit A of protein phosphatase 2A (PP2A); a mitogen-activated protein kinase (MAPK); a phosphatidyl inositol-3,4,5-triphosphate kinase, (PIPK); and auxin oxidase (IAA oxidase), a peroxidase-like enzyme. In *Arabidopsis*, flavonol accumulations have been noted in regions where AtAPM1, AtPTEN1, PID, and RCN1 are expressed (Deruere et al., 1999; Benjamins et al., 2001; Gupta et al., 2002; A. Murphy, unpublished).

AtAPM1, a dual function protein containing both enzymatic and protein trafficking domains, is a homolog of the mammalian aminopeptidase M (APM)/insulin-responsive aminopeptidase (IRAP) and aminopeptidase N (APN) proteins (Murphy et al., 2002; Muday and Murphy, 2002). AtAPM1 was isolated in a flavonoid-sensitive complex with a tyrosine aminopeptidase activity by affinity chromatography utilizing an immobilized form of the auxin efflux inhibitor naphthylphthalmic acid (NPA). AtAPM1 preferentially binds the aglycone flavonols kaempferol and quercetin, which inhibit its aminopeptidase activity (Murphy et al., 2000, 2002). The extent to which flavonoid inhibition of AtAPM1 enzymatic activity contributes to flavonoid inhibition of auxin transport activity *in vivo* (Murphy et al., 2000; Brown et al., 2001; Peer et al., 2004) is unknown (see Section 2.5 for AtAPM1 and trafficking).

PID is a protein kinase necessary for proper organ formation and has been isolated from *Arabidopsis* and pea (Christensen et al., 2000; Bai et al., 2005). PID is sensitive to NPA and therefore has a role in polar auxin transport (Benjamins et al., 2001). It also interacts with Ca^{2+} binding proteins, TOUCH3, a calmodulin-related protein, and PID-BINDING PROTEIN, an uncharacterized protein with Ca^{2+} binding motifs, suggesting a connection between auxin and Ca^{2+} signaling (Benjamins et al., 2003). This is consistent with a role for flavonoids in Ca^{2+} signaling (Lee et al., 2002; Montero et al., 2004). Analysis of the localization of PIN auxin efflux facilitator proteins and auxin transport in flavonoid-deficient mutant backgrounds suggests that PID-mediated kinase activity may be modulated by endogenous flavonols (Peer et al., 2004).

RCN1 was isolated in a screen for mutants with altered responses to NPA (Garbers et al., 1996). RCN1 is essential for PP2A activity and *rcn1* mutants have auxin-related phenotypes and increased sensitivity to okadaic acid, a phosphatase inhibitor (Garbers et al., 1996; Deruere et al., 1999; Zhou et al., 2004). Like PID, RCN1 may regulate polar auxin transport via MAPK activity (DeLong et al., 2002), and flavonoids have been shown to affect MAPK cascades in other systems (see Section 2.1).

AtPTEN1 is a pollen-specific protein that is required for normal pollen development (Gupta et al., 2002). AtPTEN1 is a homolog of the mammalian PTEN protein that regulates cell division and polarization (Perandones et al., 2004; Waite et al., 2005). *AtPTEN1* encodes a dual phosphatase acting on both phosphopeptides and phosphoinositol *in vitro*. As such, MAPK pathway components and phosphatidyl inositol-3,4,5-triphosphate (PIP3), which thereby inhibits AKT/PKB, may be endogenous AtPTEN substrates. AtPTEN1 also may regulate the MAPK pathway. Experimental evidence indicates that these signaling cascades are at least partially conserved between animals and plants with 24 putative MAPK pathways in *Arabidopsis* (Wrzaczek and Hirt, 2001; Anthony et al., 2004; Turck et al., 2004). As yet, there is no experimental evidence of flavonoid interactions with AtPTEN1. Flavonoids, however, are also required for pollen development and have been shown to affect AKT/PBK and MAPK activity in mammals, and kaempferol-treated petunia pollen has increased transcription of genes encoding regulatory or signaling proteins (Guyon et al., 2000). In a recent study in mammals, quercetin, genistein, and resveratrol increased PTEN lipid phosphatase activity but not protein phosphatase activity (Waite et al., 2005), suggesting that flavonoids may be more likely to affect AtPTEN lipophosphatase activity *in planta*.

Studies in animal cell lines have shown that flavonoids alter multiple kinase and phosphatase activities (reviewed in Williams et al., 2004). Apigenin inhibits protein kinase C (PKC) and MAPK (Kuo and Yang, 1995; Huang et al., 1996). Quercetin is routinely used as an inhibitor of mammalian PIPK, phospholipase A2, phosphodiesterases, and PKC by binding to the catalytic domain (Levy et al., 1984; Graziani and Chayoth, 1977, 1979; Graziani et al., 1982, 1983; Tammela et al., 2004). Quercetin also inhibits Ca^{2+}-dependent and phospholipid-dependent protein kinase activities (Gschwendt et al., 1983). Kaempferol inhibits mammalian monoamine oxidase/peroxidases, of which the IAA oxidase is a family member (Sloley et al., 2000). Kaempferol, quercetin, and genistein inhibit the CDC25A tyrosine phosphatase, a cell cycle-specific protein that is dephosphorylated in M phase (Aligiannis et al., 2001).

2.4. Membranes and lipids

In *Arabidopsis*, aglycone flavonols are localized at the plasma membrane (Murphy et al., 2000; Peer et al., 2001). In addition to regulating membrane proteins, flavonoids also may influence the nature of the lipid bilayer itself. Flavonoids may modify the membrane directly by changing membrane fluidity or the phosphorylation state of lipids or proteins or indirectly via signaling cascades to change the membrane composition. The orientation of hydrophobic flavonoids in the lipid bilayer can modify membrane fluidity/rigidity (Scheidt et al., 2004). Rapidly growing cells in embryonic or tumor tissues, which have less fluid membranes with lower cholesterol/phosphatidylcholine ratios and a higher degree of phosphatidyl-choline unsaturation, exhibit increased susceptibility to rigidifying flavonols like quercetin (Tsuchiya et al., 2002). The antiaggregatory and disaggregatory effects of flavonoids on human blood platelets also appear to be a function of altered membrane fluidity (Furusawa et al., 2003). In contrast to quercetin, tannins have been shown to increase membrane fluidity (Labieniec and Gabryelak, 2003).

Flavonoids also may modify the plasma membrane by altering the lipid composition. During nodulation, flavonoids induce expression of *nod* genes but also appear to alter the membrane composition of *Rizobium leguminosarum*. After nodulation induction, accumulation of the phospholipid diglycosyl diacylglycerol occurred in wild type but not in mutant bacteria lacking the *nod* genes (Orgambide et al., 1994). The root morphology of the host plants but not of nonhost plants was altered by diglycosyl diacylglycerol, which also induced a mitogenic response on the host (Orgambide et al., 1994). Flavonoids also can function in concert with ascorbic acid as antioxidants to protect the membrane from oxidative damage (Bandy and Bechara, 2001; Verstraeten et al., 2003). In addition to free radical scavenging, flavonoids induced lipid ordering at the lipid–water interface, which also reduced lipid oxidation (Erlejman et al., 2004).

2.5. Trafficking, anion channels, receptors, and cell communication

Intracellular molecular trafficking between subcellular compartments plays an important role in flavonoid biosynthesis (see Chapter 5). For example, the anion channel blocker NPPB inhibits blue light-dependent anthocyanin accumulation in *Arabidopsis* seedlings despite unaltered flavonoid biosynthetic gene transcription (Noh and Spalding, 1998), suggesting that anthocyanin accumulation requires transport of flavonoids into the vacuole. Altered pH or altered vacuolar morphology also appears to affect subcellular flavonoid accumulations, as *Arabidopsis AHA10* mutants, which harbor a defect in a plasma membrane H^+ ATPase, exhibit both altered vacuole formation and flavonoid content (Baxter et al., 2005). Similar changes in subcellular flavonol distribution also are seen in *Arabidopsis* roots when strong buffers are used to increase or decrease apoplastic pH (W. Peer and A. Murphy, unpublished). However, intercompartmental movement of flavonoids appears also to affect the abundance and distribution of some nonbiosynthetic proteins, as the transcription and subcellular localization of the flavonol-binding glutathione-*S*-transferase GSTF2 are altered in flavonoid-deficient *Arabidopsis tt4* mutants (Smith et al., 2003). Flavonoids within the membrane also may interact with membrane proteins: one study indicates that luteolin and its glucoside interacted with transmembrane domains in addition to the cytosolic loops of human MRP1 (Trompier et al., 2003).

Flavonols can interact with vesicular processes in tissues exhibiting brefeldin A (BFA)-sensitive trafficking (Peer et al., 2004); BFA is a lactone antibiotic from *Penicillium brefeldianum* that is used to inhibit protein secretion. The auxin efflux facilitator protein PIN1 has been shown to traffic to the plasma membrane through BFA-sensitive compartments via a mechanism that is also sensitive to the auxin efflux inhibitors triiodobenzoic acid (TIBA) and NPA (Geldner et al., 2001). When flavonoid-deficient *tt4 Arabidopsis* mutants were treated with flavonols, PIN1 was irreversibly retained in BFA-sensitive compartments (Peer et al., 2004). The effect was reversible in wild type, suggesting indirect flavonol modulation of trafficking in cells where they accumulate but direct interference with trafficking in cells not conditioned to their presence (Peer et al., 2004) (Figure 9.1). Flavonoids may alter the activities of proteins required for trafficking by binding them directly or altering their phosphorylation states.

Flavonoids also directly and indirectly affect vesicular cycling and protein trafficking in mammals (Almela et al., 1994; Ale-Agha et al., 2002; Lim et al., 2004). Quercetin, myricetin, and catechin-gallate were found to directly inhibit glucose uptake mediated by GLUT4; in addition, flavonoid uptake also appears to be mediated by GLUT transporters in mammalian cells (Strobel et al., 2005). GLUT4 cycling is regulated by APM/IRAP, and these data suggest that APM/IRAP is the likely mechanism regulating glucose uptake. AtAPM1 from plants has a trafficking domain and binds flavonols, suggesting that flavonol–AtAPM1 interactions also may modulate membrane trafficking in plants (Murphy et al., 2002, 2005; Muday and Murphy, 2002; Muday et al., 2003).

Flavonoid-sensitive activity of multiple drug-resistance/P-glycoprotein (MDR/PGP) ABC transporters has been demonstrated in plants and animals (Scambia et al., 1994; De Vincenzo et al., 2000; Limtrakul et al., 2005; Geisler et al., 2005; J. Blakeslee and A. Murphy, unpublished). The PGPs from plants were isolated from flavonoid-sensitive, high-affinity fractions from NPA chromatography (Murphy et al., 2002). Flavonoids alter the function of mammalian PGPs by binding to ATP binding sites or by changing the conformation of the protein (Ferte et al., 1999; Castro et al., 1999). Quercetin can bind to the first of two ATP-binding domains, resulting in inhibition of activity. In contrast, kaempferol can bind a site adjacent to the second ATP-binding domain and activate the transporter/channel. In animals cells, flavonols and isoflavones inhibit PGP activity, but only flavonols reduce PGP transcription (Limtrakul et al., 2005). Flavonol inhibition of PGP activity recently has been demonstrated in *Arabidopsis* (Geisler et al., 2005), where PGPs also appear to be required for flavonol stability on the membrane (J. Blakeslee and A. Murphy, unpublished).

The mammalian mitochondrial Ca^{2+} uniporter also can be activated by kaempferol, quercetin, and genistein (Montero et al., 2004) and quercetin can bind the Ca^{2+} release channel in sarcoplasmic reticulum, resulting in a localized Ca^{2+} flux into the cytoplasm (Lee et al., 2002). In contrast, naringenin and naringenin glycosides blocked hERG K^+ channels (Zitron et al., 2005) perhaps indirectly through kinase activity.

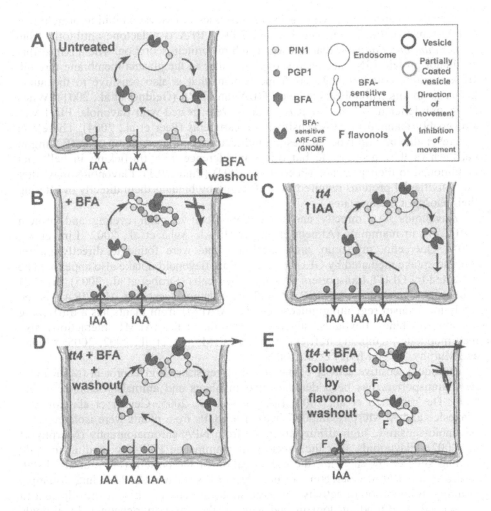

Figure 9.1 *Flavonols and the BFA-induced compartmentalization. (A) Endocytotic cycling of the PIN1 auxin efflux facilitator to and from the plasma membrane (PM) has been shown in root tips of wild-type Arabidopsis seedlings (Geldner et al., 2001). (B) When plants are treated with the fungal toxin brefeldin A (BFA), protein secretion from the Golgi complex is inhibited and the early Gogli cisternae collapse into the endoplasmic reticulum. When wild-type Arabidopsis seedlings are treated with BFA, PIN1 exocytosis is blocked, endocytosis continues, and PIN1 is sequestered in BFA-sensitive intracellular bodies (Geldner et al., 2001). When BFA is washed out of the cells with buffer or flavonols, PIN1 exocytosis resumes as in A, and PIN1 is again observed on the PM. (C) The Arabidopsis tt4 mutant has a lesion in chalcone synthase, and therefore does not synthesize flavonoids. Auxin efflux is enhanced in tt4, but PIN1 is observed in intracellular bodies and not on the PM (Peer et al., 2004). (D) When tt4 seedlings are pretreated with BFA, PIN1 is observed in the PM in addition to intracellular bodies after BFA washout (Peer et al., 2004). (E) Unlike wild-type seedlings, when tt4 seedlings are treated with BFA and then BFA is washed out with flavonols, PIN1 is*

If flavonoids bind a plasma membrane receptor, which is then internalized and trafficked to the nucleus, then the flavonoid can affect transcription. There is evidence that flavonoids are natural dietary ligands in animals. Quercetin and kaempferol have been shown to enhance and inhibit, respectively, CYP1A1 transcription mediated by the aryl hydrocarbon (AhR) receptor (Ciolino et al., 1999; Ramadass et al., 2003). Although flavonoid nuclear localization has not been investigated, quercetin-bound AhR apparently interacts with the CYP1A1 promoter to activate transcription. Flavonoids also may mimic endogenous compounds and receptor ligands for estrogen receptors α and β (Hunter et al., 1999; Virgili et al., 2004; Lee et al., 2004) and the estrogen-inducible type 11 estrogen-binding site (Garai and Aldercruetz, 2004). Flavonoids like quercetin and apigenin can activate and modulate benzodiazepene/GABA receptors, and some flavonoids act through classic GABA pathways, while other flavonoids act through a different pathway (Wasowski et al., 2002; Goutman et al., 2003; Kavvadias et al., 2004). A benzodiazepine receptor has been identified in plants (Lindemann et al., 2004), suggesting that flavonoids could be an endogenous ligand in plants.

Flavonoids also appear to enhance cell–cell communication in mammalian systems. Apigenin enhanced gap–junction communication and counteracted tumor inhibition of intercellular communication (Chaumontet et al., 1994, 1997). Kaempferol inhibited phosphorylation of signal transducer and activator of transcription 3 (STAT3) and extracellular-related kinase (ERK), thereby inducing differentiation and gap–junction communication (Nakamura et al., 2005). Cell–cell communication in plants has not been investigated, although flavonoids localized to the plasma membrane have been observed in plasmodesmata (Figure 9.2).

---

retained in intracellular bodies that are larger than those found in BFA-treated wild-type (Peer et al., 2004), suggesting that flavonols either directly bind or alter the activity of proteins necessary for cycling. Apparently, as flavonols are present in wild-type seedlings, the activities and/or phosphorylation states of proteins required for cycling are already in the functional state. When flavonols are absent, as in tt4, and then introduced after BFA treatment, the wild-type pattern of PIN1 exocytosis is not restored, reminiscent of wild-type seedlings treated with auxin efflux inhibitors (Geldner et al., 2001). Treatment with flavonols or naringenin alone restored wild type auxin transport and flavonoid patterns (Murphy et al., 2000; Brown et al., 2001). Figure is derived from Murphy et al. (2005). See Color Section for figure in colors.

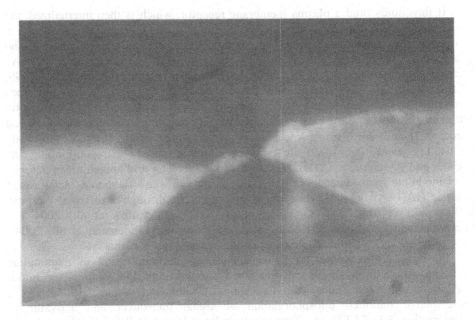

Figure 9.2 *Flavonols in the plasmodesma between two partially plasmolyzed cells in Arabidopsis roots. Roots of 5.5d wild–type seedlings were stained with DPBA and observed with an epifluorescence microscope (for methods see Peer et al., 2001). See Color Section for figure in colors.*

2.6. Cell cycle regulation, cell differentiation, and homeostasis

Flavonoids and flavonoid glycosides also may be involved in mitosis and cellular homeostasis in plants. UDP-glucuronosyltransferase1 (UGT1) is required for cell cycle regulation in pea and alfalfa, and antisense expression of *PsUGT1* under its native promoter was found to be lethal, although overexpression of antisense *PsUGT1* under the constitutive *CaMV35S* promoter was not (Woo et al., 1999). Partial inhibition of UGT1 in alfalfa resulted in an extended cell cycle and reduced growth rate, while *PsUGT1* overexpression in *Arabidopsis* resulted in a reduced life cycle (Woo et al., 2003). The likely substrate for PsUGT1 is a Flavonoid-like compound, which is reversibly converted between active and inactive aglycone and glucuronic acid forms (Woo et al., 1999, 2005), and *PsUGT1* expression colocalizes in regions of flavonoid accumulation (Woo et al., 2003; Murphy et al., 2000; Peer et al., 2001).

The activity of dihydroflavonol-4-reductase (DFR) also may contribute to metabolic homeostasis. DFR requires NADPH as a coenzyme and DFR overexpression results in increased NAD activity and transcription. This in turn alters NAD oxidation state, subsequent changes in activity, and NAD homeostasis (Hayashi et al., 2005). Although this is not a direct flavonoid target, the feedback

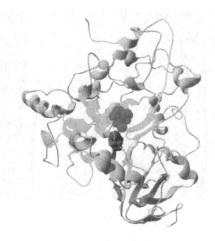

Figure 3.10

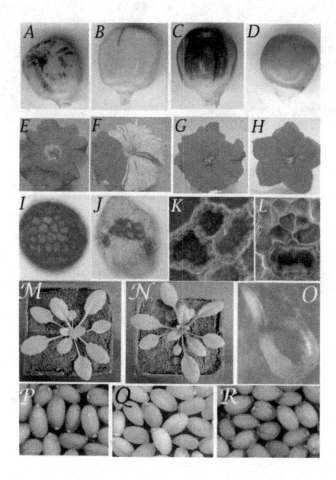

Figure 4.1

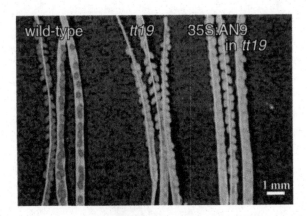

Figure 5.2

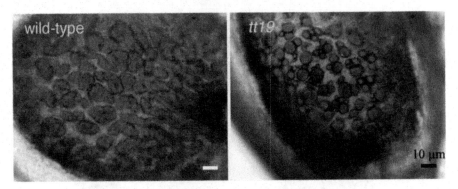

Figure 5.3

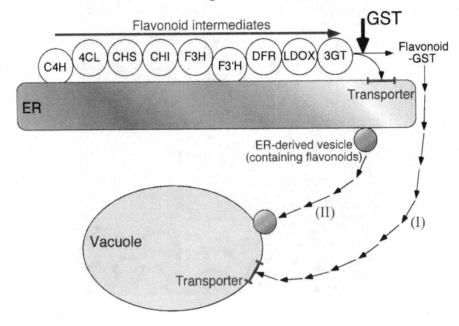

Figure 5.4

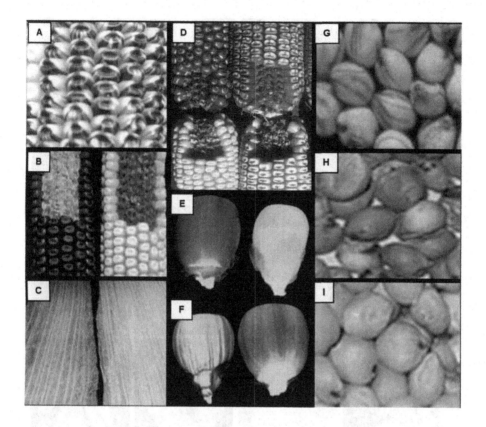

Figure 6.2

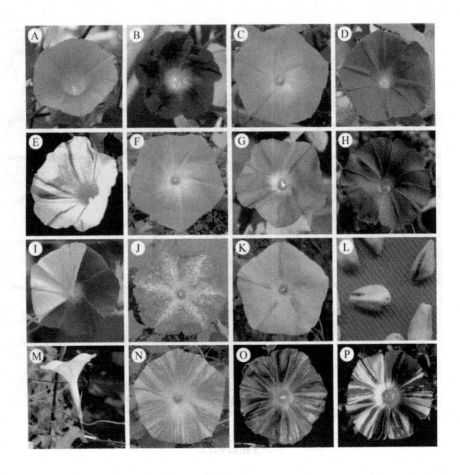

Figure 6.3

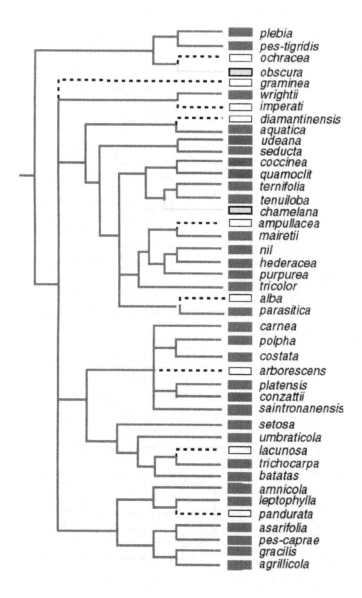

Figure 7.6

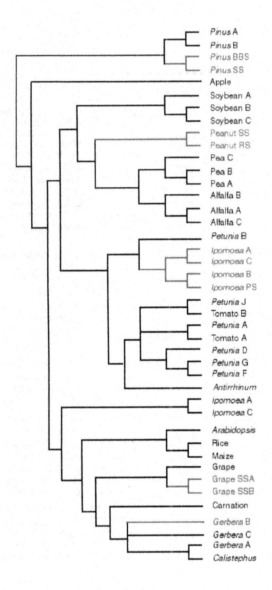

Figure 7.7

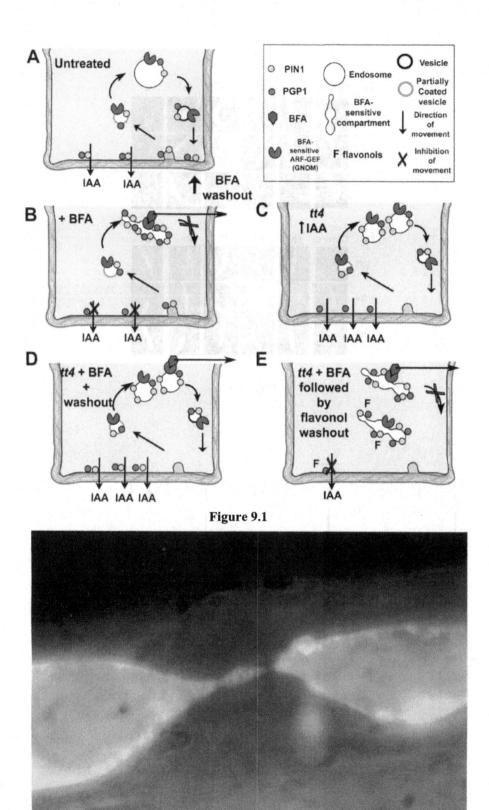

Figure 9.1

Figure 9.2

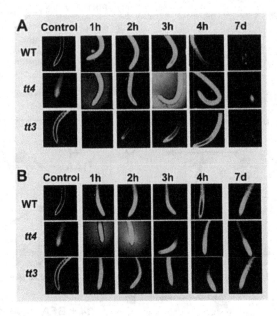

Figure 9.3

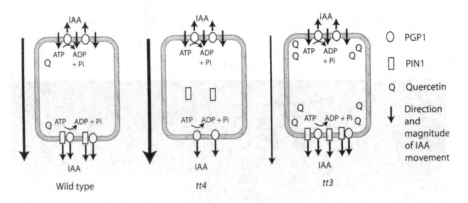

Figure 9.4

regulation of flavonoid synthesis and the activity of this flavonoid biosynthetic enzyme has global cellular effects.

Analysis of the expression of transcription factors regulating flavonoid biosynthetic genes during early embryogenesis suggests that flavonoids also function in development (Hennig et al., 2004). Flavonoids appear to be required for proper pollen and seed development (Preston et al., 2004; Debeaujon et al., 2003) but may be negative regulators at other developmental stages: soybeans that accumulate the flavonol glyscoside kaempferol-3-O-2-glycosyl-gentiobioside exhibited a significant reduction in the number of stomata compared to wild type (Liu-Gitz et al., 2000).

In mammalian cell systems, flavonoids have been shown to differentially affect the cell cycle. Tangeretin induced G1 arrest through inhibiting cyclin-dependent kinase (CDK) activity and/or elevating CDK inhibitors; however, quercetin did not arrest cells after viral infection in G1 phase (Pan et al., 2002; Beniston and Campo, 2005). Apigenin was found to induce differentiation in G2/M arrested cells, while genistein arrested cells in G2/M phase and down-regulated transcription of metalloproteinases associated with cancer metastasis (Sato et al., 1994; Kousidou et al., 2005). Quercetin suppressed differentiation and disrupted actin rings formed during cytokinesis in osteoclasts (Park et al., 2004; Woo et al., 2004).

3. FLAVONOID SIGNALING

The role of flavonoids in extraorganismal plant signaling has been described extensively, but the role of flavonoids in intra- and intercellular signaling is not as well documented. Flavonoid signaling between organisms may involve visual cues, as in pollinator attractants, as well as molecular cues. For example, flavonoid signaling via root exudates has been well studied in cases of pathogen defense, allelopathy, and nodulation where flavonoid molecules directly interact with a receptor or enzyme activity in the other organism, such as NodD in *Rhizobium spp.* Flavonoid signaling within the plant generally involves enzyme activities or cell–cell communication, but has been more difficult to characterize. However, flavonoid signaling appears to play a more direct role in auxin transport and wounding responses. Extracellular signaling of endogenous flavonoids has not been described in plants, although flavonoid-sensitive extracellular kinases have been described in mammalian systems (Spencer et al., 2003; Lin et al., 2004; Nakamura et al., 2005), suggesting that this also may be the case in discrete plant tissues.

3.1. Autocrine and paracrine effectors

Although flavonoids modulate plant signal transduction, they are not apparently endocrine hormones, as synthesis in one localization and transport and activity in another has not yet been shown. Flavonoids are best described as autocrine or paracrine effectors, as they are active within the cells in which they are synthesized and in adjacent cells. Developmental regulation of localized flavonoid accumulations and their localization in specific subcellular compartments supports this interpretation of their activity (Kuras et al., 1999; Murphy et al., 2000; Peer et al., 2001; Stefanowska et al., 2003; Saslowsky et al., 2005).

Some of the autocrine effects of flavonoids should be viewed as nonspecific. For instance, reactive oxygen species (ROS) recently have been shown to play a significant role in plant development and environmental responses (Foreman et al., 2003; Freeman et al, 2004). Heavy metal exposure induces ROS (Freeman et al., 2004), and flavonols that accumulate after heavy metal stress may act both as metal chelators and antioxidants (Skorzynska-Polit et al., 2004; Brown et al., 1998; Malinowski et al., 2004). Anthocyanins also may have a role metal chelation and metal storage in vacuole (Hale et al., 2001, 2002). However, although ROS-mediated signal transduction mechanisms may be altered in a flavonoid-deficient background, flavonoids might be better regarded as part of the cellular context in which the response takes place than as a specific signaling mechanism (Figure 9.3).

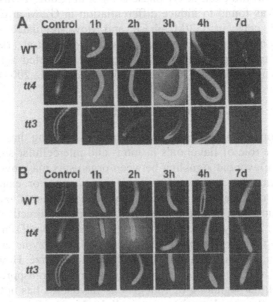

Figure 9.3 *Flavonols and oxidative stress. Wild type, tt4, and tt3 (DFR mutant that accumulates excess kaempferol and quercetin) were stained with carboxy-H2-DCFDA following treatment with (A) 1 µM IAA or (B) 10 µM CuCl₂ to induce the generation of reactive oxygen species (ROS) (for methods see Freeman et al., 2004). These data suggest that flavonols can provide short-term protection from lower levels of ROS such as those generated by IAA treatment, but not longer-term protection or protection from the greater degree of oxidative stress induced by copper treatment. These results also suggest that flavonol accumulations in regions of auxin accumulation accompanying tropic bending may be increased as localized auxin levels increase. See Color Section for figure in colors.*

3.1.1. Polar auxin transport

The phytohormone auxin, or indole-3-acetic acid (IAA), is primarily synthesized at the shoot and root apices (Ljung et al., 2002, 2005) and is transported across plasma membranes to other parts of the plant. From the acidic extracellular matrix of the apoplast, protonated IAAH may enter the cell by either diffusion or via a H^+ symporter. Once inside the neutral pH of the cell, IAA is found almost exclusively in an anionic form and may only exit the via cell polar-localized anion efflux carriers. This is often referred to as the chemiosmotic model of auxin transport (Lomax et al., 1985).

Phenolic modulation of polar auxin transport was proposed by Kenneth Thimann in the 1960s (Thimann, 1965), and studies of the effects of auxin efflux inhibitors such as NPA (Katekar and Geissler, 1979, 1980) were used as a model for the analysis of flavonoid function. Jacobs and Rubery (1988) demonstrated that quercetin, kaempferol, and genistein could displace NPA from zucchini microsomal vesicles and suggested that flavonoids could regulate auxin transport. Using a series of tyrosine kinase inhibitors, Paul Bernasconi (1996) found that genistein, but not daidzein, could displace and compete with NPA binding to membrane vesicles, thus introducing a model of flavonoid modulation of the phosphorylation state of regulatory proteins and/or lipids associated with auxin efflux. Bernasconi also showed that calmodulin antagonists, protein serine/threonine kinase inhibitors, and phosphatase inhibitors lacked the activity seen with flavonoids and tyrosine kinase inhibitors. Using wild-type and flavonoid-deficient mutants of *Arabidopsis*, Murphy et al. (2000) demonstrated that endogenous flavonoids and a tyrosine aminopeptidase regulated auxin transport *in vivo*. The flavonoid-sensitive tyrosine aminopeptidase activity was associated with low-affinity binding of NPA and the purified protein involved was identified as AtAPM1 (Murphy et al., 2000, 2002). AtAPM1 is a dual-function aminopeptidase and trafficking protein, and both functions may be involved in regulation of auxin transport (see Sections 2.3 and 2.5).

Another flavonoid-binding activity also was observed in fractions with a high affinity for NPA. These were the plasma membrane phosphoglycoproteins (PGPs) (Murphy et al., 2002) (see Section 2.5). The PGPs (also known in the mammalian literature as multiple drug resistance proteins, MDRs) are members of the ABC transporter family and have 21 functional members in *Arabidopsis*. *PGP* mutants exhibit growth defects consistent with altered auxin transport, and double mutants have more severe phenotypes, suggesting overlapping function (Noh et al., 2001; Geisler et al., 2003, 2005). Recently, IAA and IAA degradation products were shown to be substrates for AtPGP1 transport activity, and efflux of IAA was found to be inhibited by NPA and quercetin (Geisler et al., 2005). AtPGP1 is localized in shoot and root apices in light-grown seedlings where auxin is synthesized and loaded into the auxin transport stream (Sidler et al., 1998; Geisler et al., 2005). Flavonols colocalize with AtPGP1 in these regions, where the flavonols affect polar auxin transport (Peer et al., 2004) (Figure 9.4).

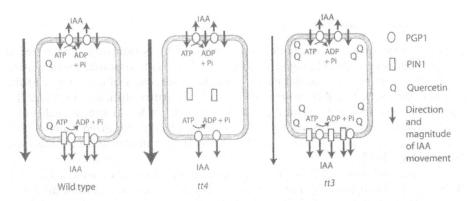

Figure 9.4 *Flavonols and auxin transport. Protonated IAA enters the cell via a H+ symporter or diffusion. Anionic IAA must exit the cell through an efflux carrier. tt4 accumulates no flavonoids and tt3 accumulates 5× more quercetin at the cotyledonary node and 10× more quercetin in the root tip than wild type (Peer et al., 2001). Auxin transport in tt4 is faster than in wild type, while transport in tt3 is slower (Peer et al., 2004). This is likely due to flavonoid regulation of auxin loading and efflux into the auxin transport stream at the apices. One mechanism through which flavonols may be acting is the plasma membrane P-glycoproteins (PGPs). The alglycone flavonol quercetin can bind to the first ATP-binding fold of PGPs, which inhibits auxin efflux (Ferte et al., 1999; Geisler et al., 2005). This is consistent with the excess quercetin accumulation in tt3 and its absence in tt4, and suggests that flavonols are involved in "fine-tuning" auxin efflux. See Color Section for figure in colors.*

The PIN family of facilitator proteins is essential for establishing the vector of auxin movement (Benkova et al., 2003; Blilou et al., 2005), but are apparently not solely responsible for auxin transport. The *pin* mutants were named after their pin-formed inflorescence phenotype (Okada et al., 1991). Transcription, subcellular localization, and tissue-specific distribution of some *PIN* family members are directly or indirectly affected by flavonoids, while other PIN proteins are unaffected by altered flavonoid levels (Peer et al., 2004). Coexpression of specific PIN-PGP combinations in heterologous systems results in enhanced auxin transport, auxin substrate specificity, and sensitivity to both NPA and flavonols (J. Blakeslee and A. Murphy, unpublished). Since specific PIN and PGP proteins differentially co-localize in the plant, PIN and PGP pairs may interact to regulate cellular auxin transport and together establish the vector and velocity of auxin efflux.

Flavonoids are present in the root cap and columella of *Arabidopsis* seedlings (Murphy et al., 2000; Peer et al., 2001, 2004). Kaempferol is present in the columella and root cap, but quercetin only accumulates after an auxin pulse or NPA treatment (Peer et al., 2004). As kaempferol inhibits mammalian monoamine oxidases that are similar to characterized IAA oxidases (Sloley et al., 2000) and flavonoid-deficient *Arabidopsis* mutants exhibit increased leakage of radiolabeled IAA or oxidized IAA from the root tip (Murphy et al., 2000; Peer et al., 2004), kaempferol may function in limiting the oxidation of auxin destined for basipetal redirection at the root tip. Quercetin accumulation in response to increased IAA levels may serve to scavenge reactive oxygen species that accumulate during IAA catabolism (Joo et al., 2001; Schopfer et al., 2002; Ljung et al., 2002).

3.1.2. Wounding and defense responses

Wounding in plants may be caused by mechanical damage, herbivory, or infection. Each of these can elicit different responses in the plant. However, two common responses among all three are the induction of auxin synthesis and *CHS* expression near the injured site (Stzein et al., 2002; Djordjevic et al., 1997; Richard et al., 2000; Cheong et al., 2002; Lo et al., 2002). Expression of other early genes of the flavonoid biosynthetic pathway is also seen after wounding, but differential expression of *CHS* genes and flavonoid accumulation between infected and uninfected plants suggest that different signaling pathways are involved (Ryder et al., 1987; Wingender et al., 1989; Lawson et al., 1996). Jasmonic acid (JA)/methyl jasmonate signaling appear to mediate some flavonoid responses to wounding, as both compounds can induce *CHS* expression (Richard et al., 2000). Both auxin-sensitive and auxin-insensitive components of JA signaling pathways have been identified, as well as species-specific responses in transcription of homologous genes involved in JA-mediated responses (Rojo et al., 1998; Seo et al., 1999; He et al., 2005). JA responses are regulated by MAPK-dependent signaling and reversible protein phosphorylation regulates JA signaling itself (Seo et al., 1999; Rojo et al., 1998), suggesting a molecular a target for flavonoid regulation.

3.1.3. Testa–embryo interactions

Flavonoids are the major color constituent in the maternally derived testa, and they have been shown to directly affect the embryo. The flavonoids in testa influence seed weight and size, dormancy, germination, and longevity among an array of *Arabidopsis* flavonoid mutants (Debeaujon et al., 2000). Although the function of flavonoids in the testa may appears to be structural by protecting the embryo from desiccation and solute leakage, damage from pathogens, and oxidative stress (Debeaujon et al., 2000), flavonoids may indirectly affect embryo development and germination. In addition to the possibility of the testa influencing the embryo, there is evidence that communication between the embryo and testa cells can alter flavonoid content, although it is unclear whether flavonoids themselves are involved in this signaling (Downie et al., 2003).

3.2. Other extraorganismal signaling

3.2.1. Plant–plant interactions

Flavonoids play a role in fertilization in many plant species. Because of alteration of generations, interactions between the sporophyte and pollen gametophyte can be regarded as extraorganismal. These interactions include pollination, self-compatibility/incompatibility chemistry and recognition (pollen allelopathy), pollen germination, and pollen tube growth (Roshchina, 2001). Flavonoids accumulate in pollen, stigmas, and petals of most flowers. Flavonoid-deficient mutants in most species are also male-sterile, as is the case with petunia, which is also self-incompatible (Napoli et al., 1999; Robbins et al., 2000). Although germination of petunia pollen can be induced with exogenous kaempferol, the basis of self-incompatibility in the Solanaceae is S-RNase activity in which the RNases within the stigma degrade pollen rRNA and inhibit pollen tube growth (Wheeler et al., 2001). Therefore, while pollen germination in petunia may be flavonoid-dependent, self-incompatibility is not. Pollination-induced or wound-induced kaempferol accumulation in petunia stigmas also enhances seed production (Vogt et al., 1994), although this is not universal in the Solanaceae (van Eldik et al., 1997).

However, flavonoid-deficient *tt4 Arabidopsis* mutants are fertile, although *tt4* exhibits reduced set seed (Burbulis et al., 1996; Ylstra et al., 1996). Like wild type, *tt4* is self-compatible. The aglycone flavonol quercetin is observed in the pollen and papillae of the stigma in wild-type *Arabidopsis* species (Peer et al., 2001; W. Peer and A. Murphy, unpublished). Application of kaempferol to the stigma in self-incompatible *Arabidopsis* species prior to hand pollination allows self-fertilization to occur (W. Peer and A. Murphy, unpublished). Since the mechanism of self-incompatibility in the Brassicaceae requires the activity of a serine/threonine protein kinase (SRK) on the stigma surface that inhibits pollen tube growth (Wheeler et al., 2001; Kusaba et al., 2001), exogenous application of kaempferol may enhance self-pollination in self-incompatible Brassicaceae by inhibiting SRK activity (see Section 2.3). It appears that quercetin, which is present in the pollen and stigma, is not involved in self-incompatibility. As such, engineered accumulation of kaempferol in the stigma of self-incompatible Brassicaceae may be of agricultural interest.

Flavonoids also are among many of the allelopathic agents that plants produce to reduce competition. Flavones from rice leaves inhibited weed growth but not rice biomass, and luteolin from chrysanthemum also inhibited weed biomass (Kong et al., 2004; Beninger and Hall, 2005). Quercetin-3-dimethylether, naringenin, and eriodictyol found in *Dittichia* root exudates induced agravitropic growth in lettuce seeds (Levizou et al., 2004). (-)-Catechin, kaempferol, and dihydroquercetin in root exudates from the invasive species *Centaurea maculosa* can trigger a wave of reactive oxygen species (ROS) and subsequent Ca^{2+} signalling, leading to root death in sensitive plant species (Bais et al., 2003a, 2003b).

3.2.2. Plant–animal interactions

In addition to pollen germination, fertilization, and seed set, flavonoids function in the attraction of animal pollinators. In flower petals, visible flavonoids such as anthocyanins, delphinidin, and cyanidin serve as attractants for pollinators like birds, small mammals, and some insects. Natural pollinators can prefer or discriminate against petal color, and therefore play an important role in the evolution of petal color; often a petal color is preferred and flowers of that color are visited more often, which enhances seed yield (Clegg and Durbin, 2000; Jones and Riethel, 2001). UV-fluorescent flavonols serve as nectar guides for bees and other insects and enhance the frequency of pollinator visits, indirectly contributing to increased seed yields (Thompson et al., 1972; Sasaki and Takahashi, 2002). Night-blooming and bird-pollinated flowers, however, often lack nectar guides (Stpiczynska et al., 2004).

Flavonoids are also one of the classes of herbivory deterrents, which may be constitutive or induced. For example, constitutively produced aglycone flavonoids, primarily 5-hydroxy-4',7-dimethoxyflavanone, in the glandular trichomes of emerging white birch leaves are mortally toxic to the larvae of the autumnal moth (Lahtinen et al., 2004). Herbivory-induced changes in flavonoid gene expression and accumulation are found in most plant species; wounded and unwounded leaves within the same aspens had increased phenylpropanoid expression (Peters and Constabel, 2002), while the herbivory-induced volatiles released from spider mite damage to lima beans leaves induce increased flavonoid biosynthetic gene expression in undamaged neighboring plants (Arimura et al., 2000).

Nematode infection induces the formation of root galls. The formation of a symbiotic nodule or a parasitic gall is similar in that auxin accumulation and flavonoid synthesis are observed at the site of infection (Hutangura et al., 1999). In contrast to symbiotic associations, a plant defense response is initiated. In oats, the accumulation of many flavonoid glycosides were induced by nematode invasion or methyl jasmonate application, and O-methylapigenin-C-deoxyhexoside-O-hexoside was identified as the active phytoalexin against two genera of parasitic nematodes (Soriano et al., 2004). In addition to inducible phytoalexins, the isoflavone medicarpin is constitutively high in roots of nematode-resistant alfalfa and is toxic to nematodes (Baldridge et al., 1998).

3.2.3. Plant–fungal interactions

Arbuscular mycorrhizae form mutualistic or symbiotic associations with plants. In addition to the novel flavonones 3,7-dihydroxy-4'-methoxyflavone and 5,6,7,8-tetrahydroxy-4'-methoxyflavone, quercetin, acacetin, and rhamnetin accumulated in roots of clover inoculated with mycorrhizae but not in noninoculated plants (Ponce et al., 2004), suggesting that flavonoids may mediate colonization. In addition, the root and shoot flavonoid composition was altered between colonized and non-colonized plants, which may be a direct or indirect effect of colonization (Ponce et al., 2004). Under low phosphate conditions, melons synthesized a C-glycosylflavone, isovotexin 2''-O-β-glusoside, which increased mycorrhizal colonization (Akiyama et al., 2002), thereby enhancing phosphate uptake. Apigenin, coumestrol, and daidzein increased mycorrhizal root colonization in soybean (Xie et al., 1995). Although the signal that initiates colonization is unknown, flavonoids modulate the development of the association.

Flavonoids also appear to provide defense against fungal infection. A flavone found in rice is allelopathic to rice fungal pathogens; quercetin, quercetin 3-methyl ether, and its glycosides inhibited conidia germination in Neurospora, and taxifolin appeared to be an antifungal agent in pine (Bonello and Blodgett, 2003; Kong et al.,

2004; Parvez et al., 2004). In addition to increased amounts of heptamethoxyflavone, nobiletin, sinensetin, and tangeretin, the amounts of glycosylated hesperetin and naringenin decreased while the amounts of the aglycone forms increased after fungal infection in citrus (Arcas et al., 2000; del Rio et al., 2004). C-Glycosylflavonoids accumulated at the plasma membrane immediately at the powdery mildew infection site and inhibition of *CHS* expression reduced resistance to fungal infection (McNally et al., 2003; Fofana et al., 2005). Theses studies suggest that different classes of flavonoids and their derivatives have differential functions.

3.2.4. Plant–microbe interactions

Recently, Ann Hirsch (2004) proposed that plant–microbe interactions occur on a continuum from commensalisms to parasitism. The multiple roles of flavonoids observed in plant-microbe interactions support this view, as flavonoid signals can attract both beneficial and parasitic bacteria. Similar to antiherbivory or anti-parasitic strategies discussed above, flavonoids also are inducible and constitutive components of the defense mechanism against infection (Dixon and Paiva, 1995; Dixon et al., 2002). Several flavones also have been shown to have antimicrobial activity (Mustafa et al., 2003; Yadava et al., 2003). Subtle changes in flavonoid speciation can determine the nature of plant microbe interactions. (+)-Catechin and its derivatives appear to be antimicrobial (Kajiya et al., 2004).

Nodulation is a special case of plant–microbe signaling. Nodulation is the formation of nitrogen-fixing nodules in roots of legumes (beans, peas, alfalfa, clover, for example) and *Parasponia*, a nonlegume, by bacteria in the Rhizobiaceae, and occurs when the host plant and the rhizobium form a symbiotic relationship. Although some rhizobia are generalists, many plant–rhizobium pairs are specialized, with species-specific flavonoids in the roots exudates that are stimulatory to compatible species of rhizobia and inhibitory to noncompatible rhizobia and other soil flora and fauna.

Flavonoids have roles both as interorganismal signaling molecules and autocrine effectors in nodulation. Flavonoids present in root exudates are perceived by the bacteria, which require flavonoids for the initiation of transcription of *nod* genes (see Section 2.1). Flavonoid levels in root exudates are sufficient to induce *nod* gene expression. For example, when flavonoid quatities were measured in root exudates from germinating bean seeds, they were found to contain 450 nmole of aglycone flavonols (myricetin, quercetin, kaempferol) and 2500 nmole aglycone anthocyanins (elphinidin, petunidin, malvidin) (Hungria et al., 1991). However, *nod* gene expression to flavonoids exhibits specificity: naringenin was found to stimulate nodulation and quercetin inhibited nodulation by *Rhizobium leguminosarum* in pea (Novak et al., 2002). Luteolin induced nodulation by *Sinorhizobium meliloti* in alfalfa (Yeh et al., 2002), while daidzein and genistein induced nodulation by *Bradyrhizobium japonicum* in soybeans (Kosslak et al., 1987) and 7,4'-dihydroxyflavone induced nodulation by *Rhizobium leguminosarum* bv. *trifolii* in white cover (Orgambide et al., 1994).

Secreted NOD factors (*N*-acetylchitooligosaccharides) also alter apoplastic pH and initiate a Ca^{2+} cascade, which is thought to either initiate or repress defense responses in the plant (Felle et al., 2000). Nodulation is initiated when the bacteria invade the root and is completed when a vascular bundle is formed in the nodule. Auxin accumulates during early nodule formation and is thought to result from flavonoid inhibition of auxin transport to the root tip (Djordjevic et al., 1997; Mathesius et al., 1998a,b). Flavonoids also likely regulate localized auxin concentrations within the nodule: 7,4'-dihydroxyflavone and its derivative inhibited IAA breakdown, while formontonin, an isoflavone, enhanced it (Mathesius, 2001). The process of rhizobial nodulation is not similar to lateral root formation, as cortical cell proliferation occurs followed by vasculature formation in the nodule.

4. CONCLUDING REMARKS

Flavonoids are bioactive molecules with specific and nonspecific effects on intra- and extraorganismal plant signaling mechanisms. However intraorganismal flavonoid signaling is probably a by-product of the evolution of plant signaling and trafficking mechanisms in an environment where flavonoids are present for purposes of extraorganismal signaling and defense and a role in initial protection from oxidative stress. Recently developed molecular biological tools and high throughput metabolic profiling technologies provide new opportunities to identify specific and nonspecific sites of flavonoid regulation.

5. ACKNOWLEDGMENTS

This work was supported by a USDA-NRI grant to A.S. Murphy.

6. REFERENCES

Akiyama, K., Matsuoka, H. and Hayashi, H., 2002, Isolation and identification of a phosphate deficiency-induced *C*-glycosylflavonoid that stimulates arbuscular mycorrhiza formation in melon roots, *MPMI* **15**: 334-340.

Ale-Agha, N., Stahl, W. and Siss, H., 2002, (-)-Epicatechin effects in rat liver epithelial cells: stimulation of gap junctional communication and counteraction of its loss due to the tumor promoter 12-O-tetradecanoylphorbol-13-acetate, *Biochem Pharmacol* **63**: 2145-2149.

Aligiannis, N., Mitaku, S., Mitrocotsa, D. and Leclerc, S., 2001, Flavonoids as cycline-dependent kinase inhibitors: Inhibition of cdc 25 phosphatase activity by flavonoids belonging to the quercetin and kaempferol series, *Planta Medica* **67**: 468-470.

Almela, M. J., Irurzun, A. and Carrasaco, L., 1994, Orobol—An inhibitor of vesicular stomatitis-virus that blocks the synthesis of viral nucleic-acids and the glycosylation of G-protein, *Antiviral Chem Chemo* **5**: 99-104.

Anthony, R. G., Henriques, R., Helfer, A., Meszaros, T., Rios, G., Testerink, C., Munnik, T., Deak, M., Koncz, C. and Bogre, L., 2004, A protein kinase target of a PDK1 signalling pathway is involved in root hair growth in Arabidopsis, *EMBO J* **23**: 572-581.

Arcas, M., Botia, J., Ortuno, A. and del Rio, J., 2000, UV irradiation alters the levels of flavonoids involved in the defence mechanism of *Citrus aurantium* fruits against *Penicillium digitatum, Euro J Plant Path* **106**: 617-622.

Arimura, G., Tashiro, K., Kuhara, S., Nishioka, T., Ozawa, R. T. and Akabayashi, J., 2000, Gene responses in bean leaves induced by herbivory and by herbivore-induced volatiles. *Biochem Biophys Res Comm* **277**: 305-310.

Bai, F., Watson, J. C., Walling, J., Weeden, N., Santner, A. A. and DeMason, D., 2005, Molecular characterization and expression of *PsPK2*, a PINOID-like gene from pea (*Pisum sativum*), *Plant Sci* **168**: 1281-1291.

Bais, H. P., Vepachedu, R., Gilroy, S., Callaway, R. M. and Vivanco, J. M., 2003a, Allelopathy and exotic plant invasion: From molecules and genes to species interactions, *Science* **301**, 1377-1380.

Bais, H. P., Walker, T. S., Kennan, A. J., Stermitz, F. R. and Vivanco, J. M., 2003b, Structure-dependent phytotoxicity of catechins and other flavonoids: Flavonoid conversions by cell-free protein extracts of *Centaurea maculosa* (spotted knapweed) roots, *J Agric Food Chem* **51**: 897-901.

Baldridge, G. D., O'Neill, N. R. and Samac, D. A., 1998, Alfalfa (*Medicago sativa* L.) resistance to the root-lesion nematode, *Pratylenchus penetrans*: defense-response gene mRNA and isoflavonoid phytoalexin levels in roots, *Plant Mol Biol* **38**: 999-1010.

Bandy, B. and Bechara, E. J. H, 2001, Bioflavonoid rescue of ascorbate at a membrane interface, *J Bioenerg Biomem* **33**: 269-277.

Baxter, I. R., Young, J. C., Armstrong, G., Foster, N., Bogenschutz, N., Cordova, T., Peer, W. A., Hazen, S.P., Murphy, A.S. and Harper, J., 2005, A plasma membrane H+-ATPase is required for the formation of proanthocyanidins in the seed coat endothelium of *Arabidopsis thaliana, PNAS* **102**: 2649-2654.

Beninger, C. W. and Hall, J., 2005, Allelopathic activity of luteolin 7-O-beta-glucuronide isolated from *Chrysanthemum morifolium* L, *Biochem Sys Ecol*, **33**: 103-111.

Beniston, R. G. and Campo, M., 2005, HPV-18 transformed cells fail to arrest in G1 in response to quercetin treatment, *Virus Res* **109**: 203-209.

Benjamins, R., Ampudia, C. S. G., Hooykaas, P. J. J. and Offringa, R., 2003, PINOID-mediated signalling involves calcium-binding proteins, *Plant Physiol* **132**: 1623-1630.

Benjamins, R., Quint, A., Weijers, D., Hooykaas, P. and Offringa, R., 2001, The PINOID protein kinase regulates organ development in Arabidopsis by enhancing polar auxin transport, *Develop* **128**: 4057-4067.

Benkova, E., Michniewicz, M., Sauer, M., Teichmann, T., Seifertova, D., Jurgens, G. and Friml, J., 2003, Local efflux-dependent auxin gradients as a common module for plant organ formation, *Cell* **115**: 591-602.

Bernasconi, P., 1996, Effect of synthetic and natural protein tyrosine kinase inhibitors on auxin efflux in zucchini (*Cucurbita pepo*) hypocotyls, *Physiol Plant* **96**: 205-210.

Blilou, I., Xu, J., Wildwater, M., Willemsen, V., Paponov I., Friml J., Heidstra R., Aida M., Palme K. and Scheres B, 2005, The PIN auxin efflux facilitator network controls growth and patterning in *Arabidopsis* roots. *Nature* **433**: 39-44.

Boege, F., Straub, T., Kehr, A., Boesenberg, C., Christiansen, K., Andersen, A., Jakob, F. and Kohrle, J., 1996, Selected novel flavones inhibit the DNA binding or the DNA religation step of eukaryotic topoisomerase I, *J Biol Chem* **271**: 2262-2270.

Bonello, P. and Blodgett, J., 2003, *Pinus nigra-Sphaeropsis sapinea* as a model pathosystem to investigate local and systemic effects of fungal infection of pines, *Physiol Mol Plant Path* **63**: 249-261.

Brown, D. E., Rashotte, A. M., Murphy, A. S., Normanly, J., Tague, B. W., Peer, W. A., Taiz, L. and Muday, G. K., 2001, Flavonoids act as negative regulators of auxin transport in vivo in Arabidopsis, *Plant Physiol* **126**: 524-535.

Brown, J. E., Khodr, H., Hider, R. C. and Rice-Evans, C.A., 1998, Structural dependence of flavonoid interactions with Cu2+ ions: implications for their antioxidant properties, *Biochem J* **330**: 1173-1178.

Buer, C. S. and Muday, G. K., 2004, The *transparent testa4* mutation prevents flavonoid synthesis and alters auxin transport and the response of Arabidopsis roots to gravity and light, *Plant Cell* **16**: 1191-1205.

Burbulis, I. E., Iacobucci, M. and Shirley, B. W., 1996, A null mutation in the first enzyme of flavonoid biosynthesis does not affect male fertility in Arabidopsis, *Plant Cell* **8**: 1013-1025.

Castro, A., Horton, J., Vanoye, C. and Altenberg, G., 1999, Mechanism of inhibition of P-glycoprotein-mediated drug transport by protein kinase C blockers, *Biochem Pharm* **58**: 1723-1733.

Chaumontet, C., Bex, V., Gaillardsanchez, I., Seillanheberden, C., Suschetet, M. and Martel, P., 1994, Apigenin and tangeretin enhance gap junctional intercellular communication in rat-liver epithelial-cells, *Carcinogen* **15**: 2325-2330.

Chaumontet, C., Droumaguet, C., Bex, V., Heberden, C., Gaillard-Sanchez, I. and Martel, P., 1997, Flavonoids (apigenin, tangeretin) counteract tumor promoter-induced inhibition of intercellular communication of rat liver epithelial cells, *Cancer Lett* **114**: 207-210.

Chen, C. C., Chow, M. P., Huang, W. C., Lin, Y. C. and Chang, Y. J., 2004, Flavonoids inhibit tumor necrosis factor-alpha-induced up-regulation of intercellular adhesion molecule-1 (ICAM-1) in respiratory epithelial cells through activator protein-1 and nuclear factor-kappa B: Structure-activity relationships, *Mol Pharm* **66**: 683-693.

Cheong, Y. H., Chang, H. S., Gupta, R., Wang, X., Zhu, T. and Luan, S., 2002, Transcriptional profiling reveals novel interactions between wounding, pathogen, abiotic stress, and hormonal responses in Arabidopsis, *Plant Physiol* **129**: 661-677.

Christensen, S, K., Dagenais, N., Chory, J. and Weigel, D., 2000, Regulation of auxin response by the protein kinase PINOID, *Cell* **100**: 469-478.

Ciolino, H. P., Daschner. P. J. and Yeh, G. C., 1999, Dietary flavonols quercetin and kaempferol are ligands of the aryl hydrocarbon receptor that affect *CYP1A1* transcription differentially, *Biochem J* **340**: 715-722.

Clegg, M. T. and Durbin, M. L., 2000, Flower color variation: A model for the experimental study of evolution, *PNAS* **97**: 7016-7023.

Constantinou, A., Mehta, R., Runyan, C., Rao, K., Vaughan, A. and Moon, R., 1995, Flavonoids as DNA topoisomerase antagonists and poisons-structure-activity-relationships, *J Nat Prod-L* **58**: 217-225.

Debeaujon, I., Leon-Kloosterzeil, K.M., and Koornneef, M., 2000, Influence of the testa on seed dormancy, germination, and longevity in Arabidopsis. *Plant Physiol* **122**: 403-413.

Debeaujon, I., Leon-Kloosterziel, K.M., Koornneef, M., 2003a, Influence of the testa on seed dormancy, germination and longevity in Arabidopsis, *Plant Physiol* **122**: 403-413.

Debeaujon, I., Nesi, N., Perez, P., Devic, M., Grandjean, O., Caboche, M., Lepiniec, L., 2003b, Proanthocyanidin-accumulating cells in Arabidopsis testa: Regulation of differentiation and role in seed development, *Plant Cell* **15**: 2514-2531.

del Rio, J., Gomez, P., Baidez, A., Arcas, M., Botia, J. and Ortuno, A., 2004, Changes in the levels of polymethoxyflavones and flavanones as part of the defense mechanism of *Citrus sinensis* (cv. Valencia Late) fruits against *Phytophthora citrophthora*, *J Agric Food Chem* **52**: 1913-1917.

DeLong, A., Mockaitis, K. and Christensen, S., 2002, Protein phosphorylation in the delivery of and response to auxin signals, *Plant Mol Biol* **49**: 285-303.

Deruere, J., Jackson, K., Garbers, C., Soll, D. and DeLong, A., 1999, The RCN1-encoded A subunit of protein phosphatase 2A increases phosphatase activity *in vivo*, *Plant J* **20**: 389-399.

De Vincenzo, R., Ferlini, C., Distefano, M., Gaggini, C., Riva, A., Bombardelli, E., Morazzoni, P., Valenti, P., Belluti, F., Ranelletti, F. O., Mancuso, S. and Scambia, G., 2000, In vitro evaluation of newly developed chalcone analogues in human cancer cells, *Cancer Chemother Pharm* **46**: 305-312.

Dixon, R. and Paiva, N., 1995, Stress-Induced Phenylpropanoid Metabolism, *Plant Cell* **7**: 1085-1097.

Dixon, R.A., Achnine, L., Kota, P., Liu, C. J., Reddy, M. S. S. and Wang, L. J., 2002, The phenylpropanoid pathway and plant defence - a genomics perspective, *Mol Plant Path* **3**: 371-390.

Djordjevic, M. A., Mathesius, U., Arioli, T., Weinman, J. J. and Gartner, E., 1997, Chalcone synthase gene expression in transgenic subterranean clover correlates with localised accumulation of flavonoids, *Aus J Plant Physiol* **24**: 119-132.

Downie, A. B., Zhang, D., Dirk, L. M. A., Thacker, R. R., Pfeiffer, J. A., Drake, J. L., Levy, A. A., Butterfield, D. A., Buxton, J. W. and Snyder, J. C., 2003, Communication between the maternal testa and the embryo and/or endosperm affect testa attributes in tomato, *Plant Physiol* **133**: 145-160.

Erlejman, A. G., Verstraeten, S. V., Fraga, C. G. and Oteiza, P. I., 2004, The interaction of flavonoids with membranes: Potential determinant of flavonoid antioxidant effects, *Free Radical Res* **38**: 1311-1320.

Fang, J., Xia, C., Cao, Z. X., Zheng, J. Z., Reed, E. and Jiang, B. H., 2005, Apigenin inhibits *VEGF* and *HIF-1* expression via PI3K/AKT/p70S6K1 and HDM2/p53 pathways, *FASEB J* **19**: 342-353.

Felle, H. H., Kondorosi, E., Kondorosi, A. and Schultze, M., 2000, How alfalfa root hairs discriminate between Nod factors and oligochitin elicitors, *Plant Physiol* **124**: 1373-1380.

Ferte, J., Kuhnel, J., Chapuis, G., Rolland, Y., Lewin, G. and Schwaller, M., 1999, Flavonoid-related modulators of multidrug resistance: Synthesis, pharmacological activity, and structure-activity relationships, *J Med Chem* **42**: 478-489.

Feucht, W., Dithmar, H. and Polster, J., 2004a, Nuclei of tea flowers as targets for flavanols, *Plant Biol* **6**: 696-701.

Feucht, W., Treutter, D. and Polster, J., 2004b, Flavanol binding of nuclei from tree species, *Plant Cell Rep* **22**: 430-436.

Fofana, B., Benhamou, N., McNally, D. J., Labbé, C., Séguin, A. and Bélanger, R. R., 2005, Suppression of induced resistance in cucumber through disruption of the flavonoid pathway, *Phytopath* **95**: 114-123.

Foreman, J., Demidchik, V., Bothwell, J. H. F., Mylona, P., Miedema, H., Torres, M. A., Linstead, P., Costa, S., Brownlee, C., Jones, J. D. G., Davies, J. M. and Dolan, L., 2003, Reactive oxygen species produced by NADPH oxidase regulate plant cell growth, *Nature* **422**: 442-446.

Freeman, J. L., Persans, M. W., Nieman, K., Albrecht, C., Peer, W., Pickering, I. J. and Salt D. E., 2004, Increased glutathione biosynthesis plays a role in nickel tolerance in thlaspi nickel hyperaccumulators, *Plant Cell* **16**: 2176-2191.

Furusawa, M., Tsuchiya, H., Nagayama, M., Tanaka, T., Nakaya, K. and Iinuma, M., 2003, Anti-platelet and membrane-rigidifying flavonoids in brownish scale of onion, *J Health Sci* **49**: 475-480.

Garai, J. and Adlercreutz, H., 2004, Estrogen-inducible uterine flavonoid binding sites: is it time to reconsider? *J Steroid Biochem Mol Biol* **88**: 377-381.

Garbers, C., DeLong, A., Deruere, J., Bernasconi, P. and Soll, D., 1996, A mutation in protein phosphatase 2A regulatory subunit A affects auxin transport in Arabidopsis, *EMBO J* **15**: 2115-2124.

Geisler, M., Blakeslee, J. J., Bouchard, R., Lee, O. R., Vincenzetti, V., Bandyopadhyay, A., Peer, W. A., Bailly, A., Richards, E. L., Edjendal, K. F., Smith, A. P., Baroux, C., Grossniklaus, U., Muller, A., Hrycyna, C. A., Dudler, R., Murphy, A. S. and Martinoia, E, 2005, Cellular export of auxin by MDR-type ATP-binding cassette transporters of *Arabidopsis thaliana*, *Plant J* in press.

Geisler, M., Kolukisaoglu, H. U., Bouchard, R., Billion, K., Berger, J., Saal, B., Frangne, N., Koncz-Kalman, Z., Koncz, C., Dudler, R., Blakeslee, J. J., Murphy, A. S., Martinoia, E. and Schulz, B., 2003,

TWISTED DWARF1, a unique plasma membrane-anchored immunophilin-like protein, interacts with Arabidopsis multidrug resistance-like transporters AtPGP1 and AtPGP19, *Mol Biol Cell* **14**: 4238-4249.

Geldner, N., Friml, J., Stierhof, Y. D., Jurgens, G. and Palme, K., 2001, Auxin transport inhibitors block PIN1 cycling and vesicle trafficking, *Nature* **413**: 425-428.

Goutman, J. D., Waxemberg, M. D., Donate-Oliver, F., Pomata, P. E. and Calvo, D.J., 2003, Flavonoid modulation of ionic currents mediated by GABA(A) and GABA(C) receptors. *Eur J Pharm* **461**: 79-87.

Grandmaison, J. and Ibrahim, R., 1996, Evidence for nuclear protein binding of flavonol sulfate esters in *Flaveria chloraefolia*, *J Plant Physiol* **147**: 653-660.

Graziani, Y. and Chayoth, R., 1977, Quercetin—inhibitor of phosphodiesterase activity in Ehrlich ascites tumor-cells, *Israel J Med Sc* **13**: 969.

Graziani, Y. and Chayoth, R., 1979, Regulation of cyclic-amp level and synthesis of DNA, RNA and protein by quercetin in Ehrlich Ascites tumor-cells, *Biochem Pharm* **28**: 397-403.

Graziani, Y., Chayoth, R., Karny, N., Feldman, B. and Levy, J., 1982, Regulation of protein-kinases activity by quercetin in Ehrlich Ascites tumor-cells, *Bioch et Biophys Acta* **714**: 415-421.

Graziani, Y., Erikson, E. and Erickson, R. I., 1983, The effect of quercetin on the phosphorylation activity of the rous-sarcoma virus transforming gene-product *in vitro* and *in vivo*, *Eur J Biochem* **135**: 583-589.

Gschwendt, M., Horn, F., Kittstein, W. and Marks, F, 1983, Inhibition of the calcium-dependent and phospholipid-dependent protein-kinase activity from mouse-brain cytosol by quercetin, *Biochem Biophys Res Comm* **117**: 444-447.

Gupta, R., Ting, J. T. L., Sokolov, L. N., Johnson, S. A. and Luan, S., 2002, A tumor suppressor homolog, *AtPTEN1*, is essential for pollen development in Arabidopsis. *Plant Cell* **14**: 2495-2507.

Guyon, V. N., Astwood, J. D., Garner, E. C., Dunker, A. K. and Taylor, L. P., 2000, Isolation and characterization of cDNAs expressed in the early stages of flavonol-induced pollen germination in Petunia, *Plant Physiol* **123**: 699-710.

Hale, K. L., McGrath, S. P., Lombi, E., Stack, S. M., Terry, N., Pickering, I. J., George, G. N. and Pilon-Smits, E. A., 2001, Molybdenum sequestration in Brassica species. A role for anthocyanins? *Plant Physiol* **126**: 1391-1402.

Hale, K. L., Tufan, H. A., Pickering, I. J., George, G. N., Terry, N., Pilon, M. and and Pilon-Smits, E. A., 2002, Anthocyanins facilitate tungsten accumulation in Brassica, *Physiol Plant* **116**: 351-358.

Hayashi, M., Takahashi, H., Tamura, K., Huang, J., Yu, L. H., Kawai-Yamada, M., Tezuka, T. and Hirofumi, U., 2005, Enhanced dihydroflavonol-4-reductase activity and NAD homeostasis leading to cell death tolerance in transgenic rice, *PNAS* **102**: 7020-7025.

He, G., Tarui, Y. and Iino, M., 2005, A novel receptor kinase involved in jasmonate-mediated wound and phytochrome signalling in maize coleoptiles, *Plant Cell Physiol* doi: 10.1093/pcp/pci092.

Hennig, L., Gruissem, W., Grossniklaus, U. and Kohler, C., 2004, Transcriptional programs of early reproductive stages in Arabidopsis, *Plant Physiol* **135**: 1765-1775.

Hirsch, A., 2004, Plant-microbe symbioses: A continuum from commensalism to parasitism. *Symbiosis* **37**: 345-363.

Hu, H. L., Liu, S. T., Yang, Y., Chang, W. Z. and Hong, G. F., 2000, In *Rhizobium leguminosarum*, NodD represses its own transcription by competing with RNA polymerase for binding sites, *Nuc Acids Res* **28**: 2784-2793.

Huang, Y. T., Kuo, M. L., Liu, J. Y., Huang, S. Y. and Lin, J. K., 1996, Inhibitions of protein kinase C and proto-oncogene expressions in NIH 3T3 cells by apigenin, *Eur J Cancer* **32A**: 146-151.

Hungria, M., Jospeh, C. and Phillips, D., 1991, Anthocyanidins and flavonols, major nod gene inducers from seeds of a black-seeded common bean (*Phaseolus vulgaris* L), *Plant Physiol* **97**: 751-758.

Hunter, D. S. Hodges, L. C., Vonier, P. M., Fuchs-Young, R., Gottardis, M. M. and Walker, C. L., 1999, Estrogen receptor activation via activation function 2 predicts agonism of xenoestrogens in normal and neoplastic cells of the uterine myometrium. *Cancer Res* **59**: 3090-3099.

Hutangura, P., Mathesius, U., Jones, M. G. K. and Rolfe, B. G., 1999, Auxin induction is a trigger for root gall formation caused by root-knot nematodes in white clover and is associated with the activation of the flavonoid pathway, *Aus J Plant Physiol* **26**: 221-231.

Ito, T., Warnken, S. P. and May, W. S., 1999, Protein synthesis inhibition by flavonoids: Roles of eukaryotic initiation factor 2 alpha kinases, *Biochem Biophys Res Comm* **265**: 589-594.

Jacobs, M. and Rubery, P. H, 1988, Naturally-occurring auxin transport regulators, *Science* **241**: 346-349.

Jones, K. N. and Reithel, J. S., 2001, Pollinator-mediated selection on a flower color polymorphism in experimental populations of Antirrhinum (Scrophulariaceae), *Am J Bot* **88**: 447-454.

Joo, J. H., Bae, Y. S. and Lee, J. S, 2001, Role of auxin-induced reactive oxygen species in root gravitropism, *Plant Physiol* **126**: 1055-1060.

Kajiya, K., Hojo, H., Suzuki, M., Nanjo, F., Kumazawa, S. and Nakayama, T., 2004, Relationship between antibacterial activity of (+)-catechin derivatives and their interaction with a model membrane, *J Agric Food Chem* **52**: 1514-1519.

Katekar, G. F. and Geissler, A. E., 1979, Evidence of a common-mode of action for a class of auxin transport inhibitors, *Plant Physiol* **63**: 22.

Katekar, G. F. and Geissler A. E., 1980, Auxin transport inhibitors. 4. Evidence of a common-mode of action for a proposed class of auxin transport inhibitors - the phytotropins, *Plant Physiol* **66**: 1190-1195.

Kavvadias, D., Sand, P., Youdim, K. A., Qaiser, M. Z., Rice-Evans, C., Baur, R., Sigel, E., Rausch, W. D., Riederer, P. and Schrerer, P., 2004, The flavone hispidulin, a benzodiazepine receptor ligand with positive allosteric properties, traverses the blood-brain barrier and exhibits anticonvulsive effects, *Brit J Pharm* **142**: 811-820.

Kim, A. R., Cho, J. Y., Zou, Y., Choi, J. S. and Ching, H. Y., 2005, Flavonoids differentially modulate nitric oxide production pathways in lipopolysaccharide-activated RAW264.7 cells, *Arch Pharm Res* **28**: 297-304.

Klampfer, L. Huang, J., Sasazuki, T., Shirasawa, S. and Augenlicht, L, 2004, Oncogenic Ras promotes butyrate-induced apoptosis through inhibition of gelsolin expression, *J Biol Chem* **279**: 36680-36688.

Kong, C. H., Xu, X. H., Zhou, B., Hu, F., Zhang, C. X. and Zhang, M. X., 2004, Two compounds from allelopathic rice accession and their inhibitory activity on weeds and fungal pathogens, *Phytochem* **65**: 1123-1128.

Kosslak, R., Bookland, R., Barkei, J., Paaren, H. and Appelbaum, E., 1987, Induction of *Bradyrhizobium japonicaum* common nod genes by isoflavones isolated from *Glycine max*, *PNAS* **84**: 7428-7432.

Kousidou, O. C., Mitropoulou, T. N., Roussidis, A. E., Kletsas, D., Theocharis, A. D. and Karamanos, N. K., 2005, Genistein suppresses the invasive potential of human breast cancer cells through transcriptional regulation of metalloproteinases and their tissue inhibitors, *Int J Oncol* **26**: 1101.

Kuo, M. L. and Yang, N. C., 1995, Reversion of v-H-*ras*-transformed NIH 3T3 cells by apigenin through inhibiting mitogen-activated protein-kinase and its downstream oncogenes, *Biochem Biophys Res Com* **212**: 767-775.

Kuras, M., Stefanowska-Wronka, M., Lynch, J. M. and Zobel, A. M., 1999, Cytochemical localization of phenolic compounds in columella cells of the root cap in seeds of *Brassica napus* - Changes in the localization of phenolic compounds during germination. *Annal Bot* **84**: 135-143.

Kusaba, M., Dwyer, K., Hendershot, J., Vrebalov, J., Nasrallah, J. and Nasrallah, M., 2001, Self-incompatibility in the genus *Arabidopsis*: characterization of the S locus in the outcrossing *A. lyrata* and its autogamous relative *A. thaliana*, *Plant Cell* **13**: 627-643.

Labieniec, M. and Gabryelak, T., 2003, Effects of tannins on Chinese hamster cell line B14, *Mut Res-Fund Molr Mech Mutagen* **539**: 127-135.

Lahtinen, M., Salminen, J., Kapari, L., Lempa, K., Ossipov, V., Sinkkonen, J., Valkama, E., Haukioja, E. and Pihlaja, K., 2004, Defensive effect of surface flavonoid aglycones of *Betula pubescens* leaves against first instar *Epirrita autumnata* larvae, *J of Chem Ecol* **30**: 2257-2268.

Lawson, C. G. R., Rolfe, B. G. and Djordjevic, M. A., 1996, *Rhizobium* inoculation induces condition-dependent changes in the flavonoid composition of root exudates from *Trifolium subterraneum*, *Aus J Plant Physiol* **23**: 93-101.

Lee, E. H., Meissner, G. and Kim, D. H., 2002, Effects of quercetin on single Ca^{2+} release channel behavior of skeletal muscle, *Biophys J* **82**: 1266-1277.

Lee, G. S., Choi, K. C., Kim, H. J. and Jeung, E. B., 2004, Effect of genistein as a selective estrogen receptor beta agonist on the expression of Calbindin-D-9k in the uterus of immature rats, *Toxicol Sci* **82**: 451-457.

Levizou, E., Karageorgou, P., Petropoulou, Y., Grammatikopoulos, G. and Manetas, Y., 2004, Induction of ageotropic response in lettuce radicle growth by epicuticular flavonoid aglycons of *Dittrichia viscose*, *Biol Plant* **48**: 305-307.

Levy, J., Teuerstein, I., Marbach, M., Radian, S. and Sharoni, Y., 1984, Tyrosine protein-kinase activity in the DMBA-induced rat mammary-tumor - inhibition by quercetin. *Biochem Biophys Res Com* **123**: 1227-1233.

Liang, Y. C., Huang, Y. T., Tsai, S. H., Lin-Shiau, S. Y., Chen, C. F. and Lin, J. K., 1999, Suppression of inducible cyclooxygenase and inducible nitric oxide synthase by apigenin and related flavonoids in mouse macrophages *Carcinogen* **20**: 1945-1952.

Lim, M., McKenzie, K., Floyd, A. D., Kwon, E. and Zeitlin, P. L., 2004, Modulation of Delta F508 cystic fibrosis transmembrane regulator trafficking and function with 4-phenylbutyrate and flavonoids, *Am J Resp Cell Mol Biol* **31**: 351-357.

Limtrakul, P., Khantamat, O. and Pintha, K., 2005, Inhibition of P-glycoprotein function and expression by kaempferol and quercetin, *J Chemother* **17**: 86-95.

Lin, H. C., Cheng, T. H., Chen, Y. C. and Juan, S.H., 2004, Mechanism of heme oxygenase-1 gene induction by quercetin in rat aortic smooth muscle cells, *Pharmacology* **71**: 107-112.

Lindemann, P., Koch, A., Degenhardt, B., Hause, G., Grimm, B. and Papadopoulos, V., 2004, A novel *Arabidopsis thaliana* protein is a functional peripheral-type benzodiazepine receptor, *Plant Cell Physiol* **45**: 723-733.

Liu-Gitz, L., Britz, S. J. and Wergin, W. P., 2000, Blue light inhibits stomatal development in soybean isolines containing kaempferol-3-O-2(G)-glycosyl-gentiobioside (K9), a unique flavonoid glycoside, *Plant Cell Environ* **23**: 883-891.

Ljung, K., Hull, A., Kowalczyk, M., Marchant, A., Celenza, J., Cohen, J. D. and Sandberg, G., 2002, Biosynthesis, conjugation, catabolism, and homeostasis of indole-3-acetic acid in *Arabidopsis thaliana*, *Plant Mol Biol* **49**: 249-272.

Ljung, K., Hull, A. K., Celenza, J., Yamada, M., Estelle, M., Normanly, J. and Sandberg, G., 2005, Sites and regulation of auxin biosynthesis in Arabidopsis roots, *Plant Cell* **17**: 1090-1104.

Lo, C., Coolbaugh, R. and Nicholson, R. L., 2002, Molecular characterization and *in silico* expression analysis of a chalcone synthase gene family in *Sorghum bicolor*, *Physiol Mol Plant Path* **61**: 179-188.

Lomax, T. L., Mehlhorn, R. J. and Briggs, W. R, 1985, Active auxin uptake by zucchini membrane-vesicles - Quantitation using electron-spin-resonance volume and delta-pH determinations, *PNAS* **82**: 6541-6545.

Malinowski, D. P., Zuo, H., Belesky, D. P. and Alloush, G. A., 2004, Evidence for copper binding by extracellular root exudates of tall fescue but not perennial ryegrass infected with *Neotyphodium spp.* endophytes, *Plant Soil* **267**: 1-12.

Marko, D., Puppel, N., Tjaden, Z., Jakobs, S. and Pahlke, G., 2004, The substitution pattern of anthocyanidins affects different cellular signalling cascades regulating cell proliferation, *Mol Nutr Food Res* **48**: 318-325.

Mathesius, U., 2001, Flavonoids induced in cells undergoing nodule organogenesis in white clover are regulators of auxin breakdown by peroxidase, *J Exper Bot* **52**: 419-426.

Mathesius, U., Bayliss, C., Weinman, J. J., Schlaman, H. R. M., Spaink, H. P., Rolfe, B. G., McCully, M. E. and Djordjevic, M. A., 1998a, Flavonoids synthesized in cortical cells during nodule initiation are early developmental markers in white clover, *Mol Plant-Microbe Interact* **11**: 1223-1232.

Mathesius, U., Schlaman, H. R. M., Spaink, H. P., Sautter, C., Rolfe, B. G. and Djordjevic, M.A., 1998b, Auxin transport inhibition precedes root nodule formation in white clover roots and is regulated by flavonoids and derivatives of chitin oligosaccharides, *Plant J* **14**: 23-34.

McNally, D. J., Wurms, K. V., Labbe, C. and Belanger, R. R., 2003, Synthesis of C-glycosyl flavonoid phytoalexins as a site-specific response to fungal penetration in cucumber, *Physiol Mol Plant Path* **63**: 293-303.

Mizushina, Y., Ishidoh, T., Kamisuki, S., Nakazawa, S., Takemura, M., Sugawara, F., Yoshida, H. S. and Sakaguchi, K., 2003, Flavonoid glycoside: a new inhibitor of eukaryotic DNA polymerase alpha and a new carrier for inhibitor-affinity chromatography, *Biochem Biophys Res Com* **301**: 480-487.

Montero, M., Lobaton, C.D., Hernandez-Sanmiguel, E., Santodomingo, J., Vay, L., Moreno, A. and Alverez, J., 2004, Direct activation of the mitochondrial calcium uniporter by natural plant flavonoids, *Biochem J* **384**: 19-24.

Muday, G. K. and Murphy, A. S, 2002, An emerging model of auxin transport regulation, *Plant Cell* **14**: 293-299.

Muday, G. K., Peer, W. A. and Murphy, A. S., 2003, Vesicular cycling mechanisms that control auxin transport polarity, *TIPS* **8**: 301-304.

Murphy, A. S., Peer, W. A. and Taiz, L., 2000, Regulation of auxin transport by aminopeptidases and endogenous flavonoids, *Planta* **211**: 315-324.

Murphy, A. S., Bandyopadhyay, A., Holstein, S. E. and Peer, W. A, 2005, Endocytotic cycling of PM proteins, *Ann Rev Plant Biol* **56**: 221-251.

Murphy, A. S., Hoogner, K. R., Peer, W. A. and Taiz, L., 2002, Identification, purification, and molecular cloning of N-1-naphthylphthalmic acid-binding plasma membrane-associated aminopeptidases from Arabidopsis, *Plant Physiol* **128**: 935-950.

Mustafa, K. A., Perry, N. B. and Weavers, R. T., 2003, 2-Hydroxyflavanones from *Leptospermum polygalifolium* subsp *polygalifolium* - Equilibrating sets of hemiacetal isomers, *Phytochem* **64**: 1285-1293.

Nakamura, Y., Chang, C. C., Mori, T., Sato, K., Ohtsuki, B., Upham, L. and Trosko, J.E., 2005, Augmentation of differentiation and gap junction function by kaempferol in partially differentiated colon cancer cells, *Carcinogen* **26**: 665-671.

Napoli, C. A., Fahy, D., Wang, H. Y. and Taylor, L.P., 1999, *White anther*: A petunia mutant that abolishes pollen flavonol accumulation, induces male sterility, and is complemented by a chalcone synthase transgene, *Plant Physiol* **120**: 615-622.

Noh, B. and Spalding, E., 1998, Anion channels and the stimulation of anthocyanin accumulation by blue light in Arabidopsis seedlings, *Plant Physiol* **116**: 503-509.

Noh, B., Murphy, A. S. and Spalding, E. P., 2001, Multidrug resistance-like genes of Arabidopsis required for auxin transport and auxin-mediated development, *Plant Cell* **13**: 2441-2454.

Notoya, M., Tsukamoto, Y., Nishimura, H., Woo, J. T. Nagai, K., Lee, I. S. and Hagiwara, H., 2004, Quercetin, a flavonoid, inhibits the proliferation, differentiation, and mineralization of osteoblasts *in vitro*, *Eur J Pharm* **485**: 89-96.

Novak, K., Chovanec, P., Skrdleta, V., Kropacova, M., Lisa, L. and Nemcova, M., 2002, Effect of exogenous flavonoids on nodulation of pea (*Pisum sativum* L.), *J Exp Bot* **53**: 1735-1745.

O'Leary, K. A., de Pascual-Tereasa, S., Needs, P. W., Bao, Y. P., O'Brien, N. M. and Wiliiamson, G., 2004, Effect of flavonoids and Vitamin E on cyclooxygenase-2 (COX-2, transcription, *Mut Res-Fund Mol Mech Mutagen* **551**: 245-254.

Okada, K., Ueda, J., Komaki, M., Bell, C. and Shimura, Y., 1991, Requirement of the auxin polar transport-system in early stages of Arabidopsis floral bud formation, *Plant Cell* **3**: 677-684.

Orgambide, G. G., Phillip-Hollingsworth, S., Hollingsworth, R. I. and Dazzo, F. B., 1994, Flavone-enhanced accumulation and symbiosis-related biological-activity of a diglycosyl diacylglycerol membrane glycolipid from *Rhizobium leguminosarum* biovar. *trifolii*, *J Bacteriol* **176**: 4338-4347.

Pan, M.H., Chen, W.-J., Lin-Shiau, S.-Y., Ho, C.-T. and Lin, J.-K., 2002, Tangeretin induces cell-cycle G1 arrest through inhibiting cyclin-dependent kinases 2 and 4 activities as well as elevating Cdk inhibitors p21 and p27 in human colorectal carcinoma cells, *Carcinogen* **23**: 1677-1684.

Park, E. K., Kim, M. S., Lee, S. H., Kim, K. H., Park, J. Y., Kim, T. H., Lee, I. S., Woo, J. T., Jung, J. C., Shin, H. I., Choi, J. Y. and Kim, S. Y., 2004, Furosin, an ellagitannin, suppresses RANKL-induced osteoclast differentiation and function through inhibition of MAP kinase activation and actin ring formation, *Biochem Biophys Res Com* **325**: 1475-1480.

Parvez, M. M., Tomita-Yokotani, K., Fujii, Y., Konishi, T. and Iwahina, T., 2004, Effects of quercetin and its seven derivatives on the growth of *Arabidopsis thaliana* and *Neurospora crassa*, *Biol System Ecol* **32**: 631-635.

Peer, W. A., Bandyopadhyay, A., Blakeslee, J. J., Makam, S. N., Chen, R., Mason, P. and Murphy, A., 2004, Variation in expression and protein localization of the PIN family of auxin efflux facilitator proteins in flavonoid mutants with altered auxin transport in *Arabidopsis thaliana*, *Plant Cell* **16**: 1898-1911.

Peer, W. A., Brown, D. E., Tague, B. W., Muday, G. K., Taiz, L. and Murphy, A. S., 2001, Flavonoid accumulation patterns of transparent testa mutants of Arabidopsis, *Plant Physiol* **126**: 536-548.

Pelletier, M. K., Burbulis, I. E. and Winkel-Shirley, B., 1999, Disruption of specific flavonoid genes enhances the accumulation of flavonoid enzymes and end-products in Arabidopsis seedlings, *Plant Mol Biol* **40**: 45-54.

Perandones, C., Costanzo, R. V., Kowaljow, V., Pivetta, O. H., Carminatti, H. and Radrizzani, M., 2004, Correlation between synaptogenesis and the PTEN phosphatase expression in dendrites during postnatal brain development, *Brain Res Mol Brain Res* **128**: 8-19.

Peters, D. and Constabel, C., 2002, Molecular analysis of herbivore-induced condensed tannin synthesis: cloning and expression of dihydroflavonol reductase from trembling aspen (*Populus tremuloides*), *Plant J* **32**: 701-712.

Polster, J., Dithmar, H. and Feucht, W., 2003, Are histones the targets for flavan-3-ols (catechins) in nuclei? *Biol Chem* **384**: 997-1006.

Ponce, M. A., Scervino, J. M., Erra-Balsells, R., Ocampo, J. A. and Godeas, A. M., 2004, Flavonoids from shoots and roots of *Trifolium repens* (white clover) grown in presence or absence of the arbuscular mycorrhizal fungus *Glomus intraradices*, *Phytochem* **65**: 1925-1930.

Preston, J., Wheeler, J., Heazlewood, J., Li, S. F. and Parish, R. W., 2004, AtMYB32 is required for normal pollen development in *Arabidopsis thaliana*, *Plant J* **40**: 979-995.

Ramadass, P., Meerarani, P., Toborek, M., Robertson, L. W. and Hennig, B., 2003, Dietary flavonoids modulate pcb-induced oxidative stress, CYP1A1 induction, and AhR-DNA binding activity in vascular endothelial cells, *Toxicol Sci* **76**: 212-219.

Richard, S., Lapointe, G., Rutledge, R. G. and Seguin, A., 2000, Induction of chalcone synthase expression in white spruce by wounding and jasmonate, *Plant Cell Physiol* **41**: 982-987.

Robbins, T. P., Harbord, R. M., Sonneveld, T. and Clarke, K., 2000, The molecular genetics of self-incompatibility in *Petunia* hybrida, *Ann Bot* **85**: 105-112.

Rojo, E., Titarenko, E., Leon, J., Berger, S., Vancanneyt, G. and Sanchez-Serrano, J. J., 1998, Reversible protein phosphorylation regulates jasmonic acid-dependent and -independent wound signal transduction pathways in *Arabidopsis thaliana*, *Plant J* **13**: 153-165.

Roshchina, V., 2001, Molecular-cellular mechanisms in pollen allelopathy, *Allelopathy J*, **8**: 11-28.

Ryder, T. B., Hedrick, S. A., Bell, J. N., Liang, X., Clouse, S. D., and Lamb, C. J., 1987, Organization and differential activation of a gene family encoding the plant defense enzyme chalcone synthase in *Phaseolus vulgaris*, *Mol Gen Genet* **210**: 219-233.

Sasaki, K. and Takahashi, T., 2002, A flavonoid from Brassica rapa flower as the UV-absorbing nectar guide, *Phytochem* **61**: 339-343.

Saslowsky, D. E., Warek, U. and Winkel, B. S. J., 2005, Nuclear localization of flavonoid enzymes in Arabidopsis, *J Biol Chem*, in press.

Sato, F., Matsukawa, Y., Matsumoto, K., Nishino, H. and Sakai, T., 1994, Apigenin induces morphological-differentiation and G2-M arrest in rat neuronal cells, *Biochem Biophys Res Com* **204**: 578-584.

Scambia, G., Ranelletti, F. O., Panici, P. B., de Vincenzo, R., Bonanno, G., Ferrandina, G., Piantelli, M., Bussa, S., Rumi, C., Cianfriglia, M. and Mancuso, S., 1994, Quercetin potentiates the effect of Adriamycin in a multidrug-resistant mcf-7 human breast-cancer cell-line - P-glycoprotein as a possible target, *Cancer Chemother Pharm* **34**: 459-464.

Scheidt, H. A., Pampel, A., Nissler, L., Gebhardt, R. and Huster, D., 2004, Investigation of the membrane localization and distribution of flavonoids by high-resolution magic angle spinning NMR spectroscopy, *Bioch Biophys Acta - Biomembranes* **1663**: 97-107.

Schopfer, P., Liszkay, A., Bechtold, M., Frahry, G. and Wagner, A., 2002, Evidence that hydroxyl radicals mediate auxin-induced extension growth, *Planta* **214**: 821-828.

Seo, S., Sano, H. and Ohashi, Y., 1999, Jasmonate-based wound signal transduction requires activation of WIPK, a tobacco mitogen-activated protein kinase, *Plant Cell* **11**: 289-298.

Sidler, M., Hassa, P., Hasan, S., Ringli, C. and Dudler, R., 1998, Involvement of an ABC transporter in a developmental pathway regulating hypocotyl cell elongation in the light, *Plant Cell* **10**: 1623-1636.

Skorzynska-Polit, E., Drazkiewicz, M., Wianowska, D., Maksymiec, W., Dawidowicz, A. L. and Tukiendorf, A., 2004, The influence of heavy metal stress on the level of some flavonols in the primary leaves of *Phaseolus coccineus*, *Acta Physiol Plant* **26**: 247-254.

Sloley, B. D., Urichuk, L. J., Morley, P., Durkin, J., Shan, J. J., Pang, P. K. T. and Coutts, R. T., 2000, Identification of kaempferol as a monoamine oxidase inhibitor and potential neuroprotectant in extracts of *Ginkgo biloba* leaves, *J Pharm Pharmacol* **52**: 451-459.

Smith, A. P., Nourizadeh, S. D., Peer, W. A., Xu, J. H., Bandyopadhyay, A., Murphy, A. S. and Goldsbrough, P. B., 2003, Arabidopsis *AtGSTF2* is regulated by ethylene and auxin, and encodes a glutathione S-transferase that interacts with flavonoids, *Plant J* **36**: 433-442.

Soriano, I. R., Asenstorfer, R. E., Schmidt, O. and Riley, I. T., 2004, Inducible flavone in oats (*Avena sativa*) is a novel defense against plant-parasitic nematodes, *Phytopath* **94**: 1207-1214.

Spencer, J. P. E., Rice-Evans, C. and Williams, R. J., 2003, Modulation of pro-survival Akt/protein kinase B and ERK1/2 signalling cascades by quercetin and its in vivo metabolites underlie their action on neuronal viability, *J Biol Chem* **278**: 34783-34793.

Stefanowska, M., Zobel, A. M. and Kuras, M., 2003, Cytochemical localization of phenolic compounds in columella cells of the root cap during maturation of seeds of *Brassica napus* L, *Plant Biol* **5**: 378-382.

Stpiczynska, M., Davies, K. L. and Gregg, A., 2004, Nectary structure and nectar secretion in *Maxillaria coccinea* (Jacq.) L.O. Williams ex Hodge (Orchidaceae), *Ann Bot* **93**: 87-95.

Strobel, P., Allard, C., Perez-Acle, T., Calderon, R., Aldunate, R. and Leighton, F., 2005, Myricetin, quercetin and catechin-gallate inhibit glucose uptake in isolated rat adipocytes, *Biochem J* **386**: 471-478.

Sztein, A. E., Ilic, N., Cohen, J. D. and Cooke, T. J., 2002, Indole-3-acetic acid biosynthesis in isolated axes from germinating bean seeds: The effect of wounding on the biosynthetic pathway, *Plant Growth Regulat* **36**: 201-207.

Tammela, P., Ekokoski, E., Garcia-Horsman, A., Talman, V., Finel, M., Tuominen, R. and Vuorela, P., 2004, Screening of natural compounds and their derivatives as potential protein kinase C inhibitors, *Drug Dev Res* **63**: 76-87.

Thimann, K, 1965, Toward an endocrinology of higher plants, *Recent Prog Horm Res* **21**: 579-596.

Thompson, W. R., Meinwald, J., Aneshansley, D. and Eisner, T., 1972, Flavonols: pigments responsible for ultraviolet absorption in nectar guide of flower, *Science* **177**: 528-530.

Trompier, D., Baubichon-Cortay, H., Chang, X. B., Maitrejean, M., Barron, D., Riordon, J. R. and Di Pietro, A., 2003, Multiple flavonoid-binding sites within multidrug resistance protein MRP1, *Cell Mol Life Sci* **60**: 2164-2177.

Tsuchiya, H., Nagayama, M., Tanaka, T, Furusawa, M., Kashimata, M. and Takeuchi, H., 2002, Membrane-rigidifying effects of anti-cancer dietary factors, *Biofactors* **16**: 45-56.

Turck, F., Zilbermann, F., Kozma, S. C., Thomas, G. and Nagy, F., 2004, Phytohormones participate in an S6 kinase signal transduction pathway in Arabidopsis, *Plant Physiol* **134**: 1527-1155.

van Eldik, G J., Reijnen, W. H., Ruiter, R. K., van Herpen, M. M., Schrauwen, J. A. and Wullems, G. J., 1997, Regulation of flavonol biosynthesis during anther and pistil development, and during pollen tube growth in *Solanum tuberosum*, *Plant J* **11**: 105-113.

Verstraeten, S. V., Keen, C. L., Schmitz, H. H., Fraga, C. G. and Oteiza, P. I., 2003, Flavan-3-ols and procyanidins protect liposomes against lipid oxidation and disruption of the bilayer structure, *Free Radical Biol Med* **34**: 84-92.

Virgili, F., Acconcia, F., Ambra, R., Rinna, A., Totta, P. and Marino, M., 2004, Nutritional flavonoids modulate estrogen receptor alpha signaling, *IUBMB Life* **56**: 145-151.

Vogt, T., Pollak, P., Tarlyn, N. and Taylor, L. P, 1994, Pollination-Induced or wound-induced kaempferol accumulation in petunia stigmas enhances seed production, *Plant Cell* **6**: 11-23.

Wadsworth, T. L., McDonald, T. L. and Koop, D. R., 2001, Effects of *Ginkgo biloba* extract (EGb 761) and quercetin on lipopolysaccharide-induced signalling pathways involved in the release of tumor necrosis factor-alpha *Biocheml Pharmacol* **62**: 963-974.

Waite, K. A., Sinden, M. R. and Charis, E., 2005, Phytoestrogen exposure elevates PTEN levels, *Human Mol Genet* **14**: 1457-1463.

Wasowski, C., Marder, M., Viola, H., Medina, J. H. and Paladini, A. C., 2002, Isolation and identification of 6-methylapigenin, a competitive ligand for the brain GABA(A) receptors, from *Valeriana wallichii*, *Planta Med* **68**: 934-936.

Wheeler, M., Frankin-Tong, V. and Fanklin, F., 2001, The molecular and genetic basis of pollen-pistal interactions, *New Phyt* **151**: 565-584.

Williams, R. J., Spencer, J. P. E. and Rice-Evans, C., 2004, Flavonoids: Antioxidants or signalling molecules? *Free Radical Biol Med* **36**: 838-849.

Wingender, R., Rohrig, H., Horicke, C., Wing, D. and Schell, J., 1989, Differential regulation of soybean chalcone synthase genes in plant defence, symbiosis and upon environmental stimuli, *Mol Gen Genet* **218**: 315-322.

Woo, H., Jeong, B. and Hawes, M., 2005, Flavonoids: from cell cycle regulation to biotechnology *Biotechnol Lett* **27**: 365-374.

Woo, H. H., Faull, K. F., Hirsch, A. M. and Hawes, M C., 2003, Altered life cycle in arabidopsis plants expressing *PsUGT1*, a UDP-glucuronosyltransferase-encoding gene from pea, *Plant Physiol* **133**: 538-548.

Woo, H. H., Orbach, M. J., Hirsch, A. M. and Hawes, M. C., 1999, Meristem-localized inducible expression of a UDP-glycosyltransferase gene is essential for growth and development in pea and alfalfa, *Plant Cell* **11**: 2303-2315.

Woo, J. T., Nakagawa, H., Notoya, M., Yonezawa, T., Udagawa, N., Lee, I. S., Ohnishi, M., Hagiwara, H. and Nagai, K., 2004, Quercetin suppresses bone resorption by inhibiting the differentiation and activation of osteoclasts, *Biol Pharmacol Bull* **27**: 504-509.

Wrzaczek, M. and Hirt, H., 2001, Plant MAP kinase pathways: how many and what for? *Biol Cell* **93**: 81-87.

Xi, Q. R., Cuesta, R. and Schneider, R. J., 2005, Regulation of translation by ribosome shunting through phosphotyrosine-dependent coupling of adenovirus protein 100k to viral mRNAs, *J Virol* **79**: 5676-5683.

Xie, Z. P., Staehelin, C., Vierheilig, H., Wiemken, A., Jabbouri, S., Broughton, W. J., Vogeli-Lange, R. and Boller, T., 1995, Rhizobial nodulation factors stimulate mycorrhizal colonization of nodulating and nonnodulating soybeans, *Plant Physiol* **108**: 1519-1525.

Yadava, R. N. and Verma, V., 2003, A new biologically active flavone glycoside from the seeds of *Cassia fistula* (Linn.), *Asian Nat Prod Res* **5**: 57-61.

Ye, R. Q., Goodarzi, A. A., Kurz, E. U., Saito, S., Higashimoto, Y., Lavin, M. F., Appella, E., Anderson, C. W. and Lees-Miller, S. P., 2004, The isoflavonoids genistein and quercetin activate different stress signalling pathways as shown by analysis of site-specific phosphorylation of ATM, p53 and histone H2AX, *DNA Repair* **3**: 235-244.

Yeh, K., Peck, M. and Long, S., 2002, Luteolin and GroESL modulate in vitro activity of NodD, *J Bacteriol* **184**: 525-530.

Ylstra, B., Muskens, M. and van Tunen, A. J., 1996, Flavonols are not essential for fertilization in *Arabidopsis thaliana*, *Plant Mol Biol* **32**: 1155-1158.

Zhao, J. and Sakai, K., 2002, Multiple signalling pathways mediate fungal elicitor-induced ß-thujaplicin biosynthesis in *Cupressus lusitanica* cell cultures. *J Exper Bot* **54**: 647-656.

Zhou H.W., Nussbaumer C., Chao Y., DeLong A., 2004, Disparate roles for the regulatory A subunit isoforms in *Arabidopsis protein phosphatase 2A*. *Plant Cell* **16**:709-22.

Zitron E., Scholz E., Owen R.W., Luck S., Kiesecker C., Thomas D., Kathofer S., Niroomand F., Kiehn J., Kreye V.A., Katus H.A., Schoels W., Karle C.A., 2005, QTc prolongation by grapefruit juice and its potential pharmacological basis: HERG channel blockade by flavonoids. *Circulation* **111**:835-838.

INDEX

2-oxoglutarate-dependent dioxygenase
(2-ODD), 71, 77, 78, 79, 82, 88, 179
3-deoxy flavonoid, 99, 101, 134
4-coumaroyl
CoA-ligase (4CL), 72, 150
7,2'-dihydroxy, 4'-methoxyisoflavanol
dehydratase (DMID), 72
Ac (Activator), 149, 151, 152, 161, 162,
163
acetyl transferases (ACT), 72, 86
acylation, 22, 24, 133
adaptation, 176, 188, 193, 204, 205
affinity chromatography, 125, 242
aglycone, 49, 52, 54, 55, 56, 58, 60, 61,
242, 243, 248, 254, 255, 256
aha10, 244
AKT, 226, 241, 243
aleurone, 74, 98, 106, 157, 158, 201
algae, 177
allelopathy, 240, 249, 254
aminopeptidase, 242, 251
AN1, 98, 100, 101, 102, 106, 107, 108,
111, 112, 113, 114, 115, 158, 162
AN2, 101, 102, 104, 106, 107, 108, 115,
158, 189
AN9, 106, 130, 131, 133, 158
ANT1, 129, 135
anthocyanic vacuolar inclusions (AVI),
133
anthocyanidin reductase (ANR), 32, 33,
34, 36, 72, 76, 77, 88, 103, 150, 159,
160
anthocyanidin synthase (ANS), 32, 33,
34, 72, 75, 76, 77, 78, 88, 105, 112,
150, 158, 159, 178, 181, 182, 183,
185, 194, 203
anthocyanin, 73, 76, 80, 84, 85, 86, 97,
98, 100, 101, 102, 105, 107, 108, 109,
112, 114, 115, 116, 117, 127, 128,
129, 130, 131, 133, 134, 135, 136,
139, 148, 155, 156, 157, 159, 160,
162, 163, 164, 175, 176, 182, 187,
188, 189, 190, 191, 193, 194, 195,
196, 197, 200, 201, 202, 204, 244
anthocyanoplasts, 133, 134, 139
Antirrhinum, 73, 148, 182, 183, 187, 194

apiforol, 99
apigenin, 56, 214, 217, 219, 223, 224,
228, 240, 241, 243, 247, 249, 255
Arabidopsis, 73, 75, 76, 78, 79, 84,
86, 87, 88, 89, 90, 97, 98, 100, 101,
102, 103, 104, 105, 111, 112, 113,
114, 115, 124, 125, 128, 129, 131,
132, 133, 134, 135, 136, 138, 139,
140, 148, 155, 156, 157, 159, 160,
161, 183, 187, 190, 197, 201, 202,
203, 204, 240, 242, 243, 244, 245,
246, 248, 251, 252, 253, 254
AtAPM1, 242, 245, 251
AtMYB12, 105
AtMYB4, 112
ATP-binding cassette (ABC), 126, 127,
128, 131, 136, 139, 240, 245, 251
AtPTEN, 243
aurones, 4, 12, 72, 150, 178, 182
auronols, 4
autocrine, 223, 249, 250, 256
auxin, 87, 123, 124, 178, 240, 242, 243,
245, 246, 247, 249, 250, 251, 252,
253, 255, 257
BAHD, 86
BANYLUS (BAN), 103, 104, 109, 114,
159, 160
berries, 63
Black Mexican Sweet (BMS), 134, 140
Brassica napus, 87, 240
Brassicaceae, 85, 196, 203, 254
brefeldin A (BFA), 245, 246, 247
Bronze2 (BZ2), 106, 127, 130, 131, 132,
133, 136, 137, 138, 158
Buckwheat, 125
C1, 98, 99, 101, 103, 105, 106, 109, 111,
112, 115, 134, 152, 157, 158, 159,
163, 197, 201
Caerphilly Study, 217
calmodulin, 242, 251
Camellia sinesis, 240
cancer, 213, 215, 218, 219, 221, 222, 223,
224, 226, 227, 228, 229, 239, 249
carnation, 131, 133, 155
carrot, 126, 133
Caryophyllales, 77